FIELD GUIDE TO THE
WATER LIFE
OF BRITAIN

WATER LIFE OF BRITAIN

was edited and designed by
The Reader's Digest Association Limited, London

First edition Copyright © 1984
The Reader's Digest Association Limited
25 Berkeley Square, London W1X 6AB

Printed in Great Britain

The typeface used for text in this book
is 8 point Bembo roman.

The picture of the perch on the cover was painted by Mick Loates

READER'S DIGEST
NATURE LOVER'S LIBRARY

FIELD GUIDE TO THE
WATER LIFE
OF BRITAIN

PUBLISHED BY THE READER'S DIGEST ASSOCIATION LIMITED

LONDON · NEW YORK · MONTREAL · SYDNEY · CAPE TOWN

Contributors

The publishers wish to express their gratitude to the following people for their major contributions to this Field Guide to the Water Life of Britain

PRINCIPAL CONSULTANTS AND AUTHORS

Dr Frances Dipper
Dr Anne Powell

ADDITIONAL CONSULTANTS
AND WRITERS
Dr Trevor Beebee
Dr Pat Morris

PHOTOGRAPHERS

For a full list of acknowledgments to the photographers whose work appears in this book, see page 336

ARTISTS

Robin Armstrong	Mick Loates
Bob Bampton	Tricia Newell
Peter Barrett	Colin Newman
Dick Bonson	D. W. Ovendon
Wendy Bramall	Andrew Robinson
Jim Channell	Jim Russell
Kevin Dean	Sue Stitt
Norman Lacey	Phil Weare
Stuart Lafford	Sue Wickison
Ann Winterbotham	

A full list of the paintings contributed by each artist appears on page 336

Contents

Understanding water life

Water is essential to life, accounting for between 60 and 99 per cent of the body weight of all plants and animals. Because of this, water is the ideal environment for living organisms. It was in the water of the seas that life first appeared, some 3,500 million years ago. Water still supports a much greater diversity of life than the land, and living organisms are to be found in virtually all the flowing and standing waters on Earth, from the deepest seas and largest lakes and rivers, to the smallest pools and highest mountain streams. Even the ice of glaciers and the water between soil particles supports single-celled algae.

Unlike the land surface, aquatic environments must be thought of as three-dimensional living space. Many animals and plants live permanently in the open water, or pelagic zone. Some, like the jellyfish, are floaters (plankton); and some, like fish and squids, are swimmers (nekton). Others live on or in the sea-bed or on each other (benthos). This means that life exists throughout all water depths in the same area of ocean.

Deep waters such as the seas and big lakes are stable places to live. The temperature, for example, changes very slowly compared to that of small pools, which may freeze in winter and dry out in summer. The chemical balance is also very constant compared with that of small bodies of water, and organisms are much less likely to suffer from sudden changes in oxygen or salt content.

Saltwater oceans and seas cover about two-thirds of the Earth's surface. Marine creatures live even in the deepest parts, but it is around the coasts, in shallow seas down to about 600 ft (180 m) deep – the edge of the continental shelf – that the greatest number of species is to be found. Here the water is rich in plant nutrients which drain off the land with the river water, encouraging abundant growth of seaweeds and microscopic floating plants (plankton). These in turn support a great variety of animals. Seaweeds will grow only in shallow water down to about 100 ft (30 m) where sufficient light can penetrate, although this depth limit varies with the clarity of the water. Most of the plant production in the open sea is from the minute plant plankton floating near the surface where the sunlight reaches them.

Life on the seashore varies with the type of shoreline, which is determined by the geology of the area and its exposure to wave action. Stable rocky shores of bedrock and large boulders support the greatest variety of marine life since they provide a hard surface for attachment and plenty of cracks and crevices for shelter. In contrast, few plants and animals can survive on shifting pebble shores where the stones are constantly ground together by the waves.

Sediment shores vary from coarse sand to very fine mud, depending on how exposed they are to the stirring action of the waves. Few seaweeds or encrusting animals can live on the sediments, but hidden beneath the surface there is often a wealth of burrowing animal life. Rich muddy sands support the greatest variety of species.

The zones of the seashore

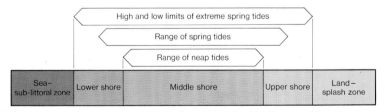

Animals and plants on the seashore are not randomly distributed. Different species occupy different levels, depending mainly on their ability to withstand exposure when the tide goes out. Above the average tide range, the **upper shore** is flooded for only a few hours each month by high spring tides. The same tides, when they recede, expose parts of the **lower shore** which are normally under water. The **middle shore** is flooded and exposed twice daily, even by the smallest tides of the month – the neap tides – when the high water level is at its lowest.

All fresh waters originate as rain, which is water evaporated from the sea and carried inland as clouds. The rain falls as almost pure water, the salt having been left behind in the sea. If it falls on hard, insoluble rocks it drains off again almost as pure as it fell. But if the rain lands on soluble rocks or on soils, the water which drains off on the surface or from springs contains many dissolved minerals, agricultural fertilisers and floating particles. Fresh waters therefore vary considerably in character, and this affects the life they support. Shell-building animals, for example, are most abundant in calcium-rich hard waters which drain off chalk and limestone. In contrast, upland streams of very pure soft water support relatively little life. Lowland waters which drain off fertilised soils are often choked with vegetation. When it dies the rotting process can de-oxygenate the water and kill many animals. Such 'organic pollution' has become more common with increased use of artificial fertilisers.

The water life is also affected by the size and depth of the water body and its rate of flow, or pattern of water movement. Large lakes are usually poorly vegetated compared with small shallow ponds, for the light level on the bed is too low to support rooted plants, and the wave action on exposed shores prevents plant growth. A slow-flowing lowland river supports animals and plants which will not tolerate turbulent conditions, while some freshwater animals can survive only in fast moving streams.

The naming of water life

The water life of Britain is very diverse, and this book covers creatures ranging from whales to water fleas. Many have common names, such as alderfly or sea urchin, but common names can sometimes be misleading. In British waters there are two different alderflies and four different sea urchins. Sometimes the same common name may be used for different animals, or one animal may have several common names. Although colourful and descriptive, many of these names are misleading – the sea mouse, for example, is a worm. To overcome such problems, every animal has a scientific name which is in two parts, the genus name (like a surname) and the species name. The salmon, for example, is called *Salmo salar*, and the trout is *Salmo trutta*. We know that they are closely related because they both have the same genus or generic name, *Salmo*.

Scientific names are internationally understood and often describe particular attributes of the animal or plant. The species or specific name *edulis* or *edule* is often used for edible animals such as the edible cockle *Cerastoderma edule* and the mussel *Mytilus edulis*. Unfortunately, scientific names are sometimes changed by specialists who make new discoveries about the relationships between different species. In this book the old scientific name of some common species has been given as well as the new name. In other cases the new names have not been used, for they are yet to be fully accepted and are not in regular use.

How to use this book

The water life in this book is divided into species normally found in fresh inland waters, and species normally found in sea waters. To identify a species, decide on the water type and turn either to the key to freshwater life on pp. 16–19, or the key to shallow sea life on pp. 20–29. For species found in both water types, both keys may have to be consulted. In these keys the animals and plants described in the book are grouped according to obvious identification features such as overall shape; whether or not it has limbs or a shell; or the number of fins. In many cases these groupings correspond to the scientific classification, but this is not always so. For example, animals which are shaped like worms are grouped together in the book for easy comparison.

Having identified from the keys the group to which the plant or animal belongs, turn next to the pages indicated to identify the species. For some difficult groups such as sea mats and some freshwater insects it has only been possible to include a few easily identified species. Each illustrated page includes a painting of the whole creature or plant, and smaller paintings of particular aspects which are useful in identification. In many groups the commonest species has been illustrated on one page, and a chart of similar species is given on the facing page. It is important to read the descriptions which accompany the illustrations, for many creatures vary their colours and even their shapes according to their surroundings.

How animals and plants live in water

The variety of life in seas, rivers, lakes and ponds ranges through the whole animal and plant kingdoms – from microscopic, single-celled protozoa to whales that may weigh over 100 tons.

The size, shape and colour are often the first and most obvious clues to the identity of a water creature. But how it moves, where it burrows or clings, how it feeds and reproduces, and how it reacts to external stimuli can be as distinctive as the colour of its fins or the markings on its body. The time of year, and the place where it is seen are also valuable clues to its identity.

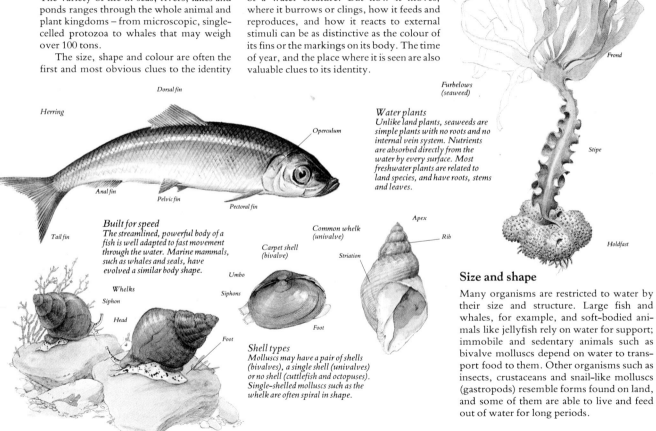

Herring

Dorsal fin

Operculum

Anal fin

Pelvic fin

Pectoral fin

Tail fin

Built for speed
The streamlined, powerful body of a fish is well adapted to fast movement through the water. Marine mammals, such as whales and seals, have evolved a similar body shape.

Whelks

Siphon

Head

Foot

Water plants
Unlike land plants, seaweeds are simple plants with no roots and no internal vein system. Nutrients are absorbed directly from the water by every surface. Most freshwater plants are related to land species, and have roots, stems and leaves.

Furbelows
(seaweed)

Frond

Stipe

Holdfast

Carpet shell
(bivalve)

Umbo

Siphons

Foot

Common whelk
(univalve)

Striation

Apex

Rib

Shell types
Molluscs may have a pair of shells (bivalves), a single shell (univalves) or no shell (cuttlefish and octopuses). Single-shelled molluscs such as the whelk are often spiral in shape.

Size and shape

Many organisms are restricted to water by their size and structure. Large fish and whales, for example, and soft-bodied animals like jellyfish rely on water for support; immobile and sedentary animals such as bivalve molluscs depend on water to transport food to them. Other organisms such as insects, crustaceans and snail-like molluscs (gastropods) resemble forms found on land, and some of them are able to live and feed out of water for long periods.

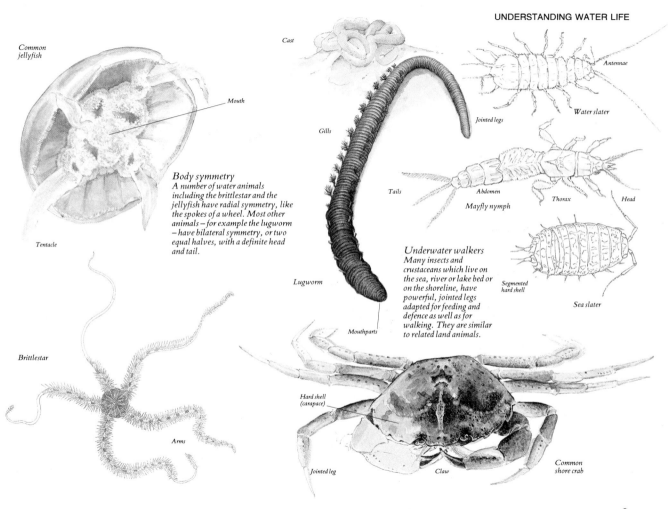

Common jellyfish

Mouth

Tentacle

Cast

Gills

Tails

Lugworm

Mouthparts

Jointed legs

Antennae

Water slater

Abdomen

Mayfly nymph

Thorax

Head

Segmented hard shell

Sea slater

Body symmetry
A number of water animals including the brittlestar and the jellyfish have radial symmetry, like the spokes of a wheel. Most other animals – for example the lugworm – have bilateral symmetry, or two equal halves, with a definite head and tail.

Underwater walkers
Many insects and crustaceans which live on the sea, river or lake bed or on the shoreline, have powerful, jointed legs adapted for feeding and defence as well as for walking. They are similar to related land animals.

Brittlestar

Arms

Hard shell (carapace)

Jointed leg

Claw

Common shore crab

9

Shells for protection
Molluscs such as the winkle build their
shells from calcium obtained from the
water. Soft animals like the sea cucumber
have no skeleton or shell.

Sea cucumber

Common winkle

Great diving beetle

Common lobster

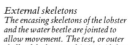

External skeletons
The encasing skeletons of the lobster
and the water beetle are jointed to
allow movement. The test, or outer
shell, of the heart urchin is rigid, but
has holes for soft, flexible tube feet.

Test of sea urchin

Supporting skeletons

Many water animals have soft bodies which
are supported by the water; some, such as
worms, obtain protection by burrowing or
living in tubes made of sand grains. Molluscs also have soft bodies, but most can
retreat into a hard shell. Insects and crustaceans are completely enclosed in hard,
jointed external skeletons which both support and protect them, while vertebrates
such as fish, reptiles and mammals are supported by skeletons within their bodies.

Skeleton of lemon sole

Internal skeleton
The bones of a fish provide
support for large, powerful
swimming muscles. Thin
bony scales provide
protection.

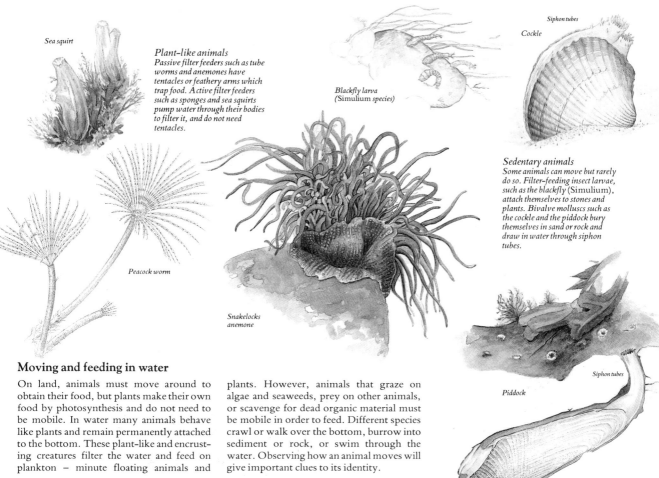

Sea squirt

Plant-like animals
Passive filter feeders such as tube worms and anemones have tentacles or feathery arms which trap food. Active filter feeders such as sponges and sea squirts pump water through their bodies to filter it, and do not need tentacles.

Blackfly larva
(Simulium *species*)

Siphon tubes

Cockle

Peacock worm

Sedentary animals
Some animals can move but rarely do so. Filter-feeding insect larvae, such as the blackfly (Simulium), attach themselves to stones and plants. Bivalve molluscs such as the cockle and the piddock bury themselves in sand or rock and draw in water through siphon tubes.

Snakelocks
anemone

Siphon tubes

Piddock

Moving and feeding in water

On land, animals must move around to obtain their food, but plants make their own food by photosynthesis and do not need to be mobile. In water many animals behave like plants and remain permanently attached to the bottom. These plant-like and encrusting creatures filter the water and feed on plankton – minute floating animals and plants. However, animals that graze on algae and seaweeds, prey on other animals, or scavenge for dead organic material must be mobile in order to feed. Different species crawl or walk over the bottom, burrow into sediment or rock, or swim through the water. Observing how an animal moves will give important clues to its identity.

11

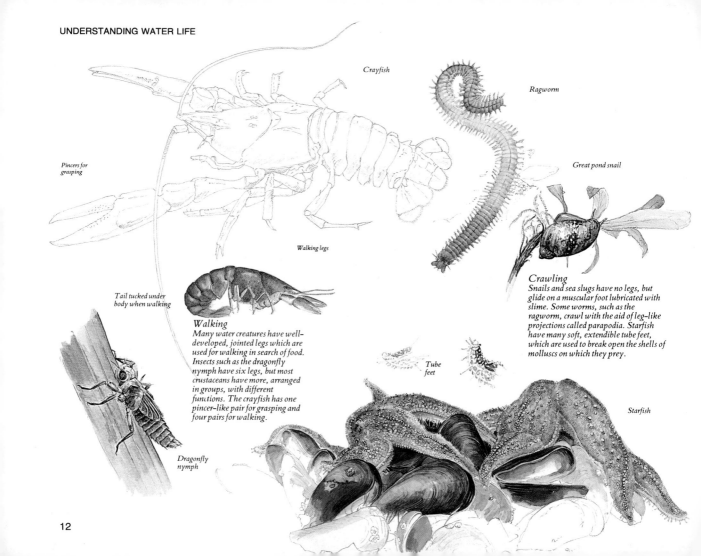

Crayfish

Ragworm

Pincers for grasping

Great pond snail

Walking legs

Crawling
Snails and sea slugs have no legs, but glide on a muscular foot lubricated with slime. Some worms, such as the ragworm, crawl with the aid of leg-like projections called parapodia. Starfish have many soft, extendible tube feet, which are used to break open the shells of molluscs on which they prey.

Tail tucked under body when walking

Walking
Many water creatures have well-developed, jointed legs which are used for walking in search of food. Insects such as the dragonfly nymph have six legs, but most crustaceans have more, arranged in groups, with different functions. The crayfish has one pincer-like pair for grasping and four pairs for walking.

Tube feet

Starfish

Dragonfly nymph

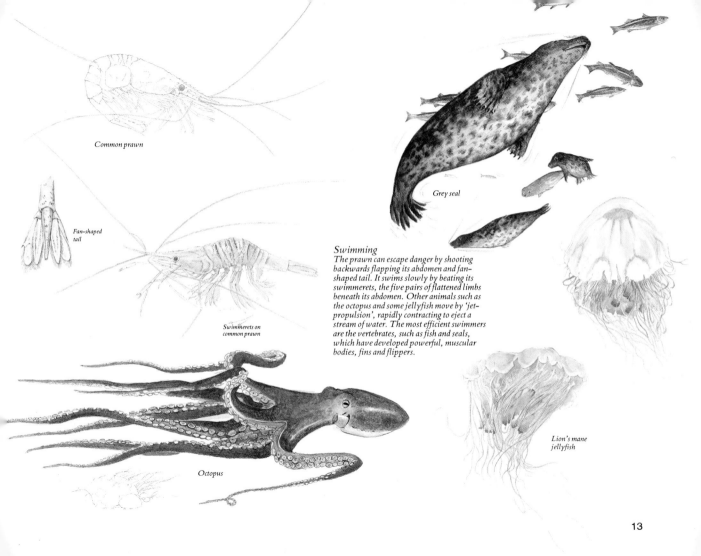

Common prawn

Fan-shaped tail

Swimmerets on common prawn

Grey seal

Swimming

The prawn can escape danger by shooting backwards flapping its abdomen and fan-shaped tail. It swims slowly by beating its swimmerets, the five pairs of flattened limbs beneath its abdomen. Other animals such as the octopus and some jellyfish move by 'jet-propulsion', rapidly contracting to eject a stream of water. The most efficient swimmers are the vertebrates, such as fish and seals, which have developed powerful, muscular bodies, fins and flippers.

Octopus

Lion's mane jellyfish

Water flea bearing eggs

Asexual and hermaphrodite
In summer the water flea may reproduce asexually from eggs which do not need fertilisation. Many worms such as the sludge worm are hermaphrodite, and must mate to fertilise each other's eggs.

Mating sludge worms

Spawning char

Hydra

Spawning and mating
Most female fish, such as the char, lay eggs which are then fertilised externally by the male. Some male sharks, however, such as the dogfish, implant sperm in the female during mating, and fertilisation occurs internally.

Budding hydra
Young hydra may develop as buds on the body of the parent, eventually detaching to begin independent life.

Mating dogfish

Regeneration
Some animals, such as starfish, have such powers of regeneration that a severed limb may grow into a new individual, providing part of the central disc is present.

Starfish

Reproduction and growth

Most animals reproduce sexually at some stage in their life-cycle, but many small species such as water fleas are also able to reproduce by producing unfertilised eggs – a process known as parthenogenesis. Some of the less-complex animals can reproduce by splitting into two, or growing buds which develop into new individuals.

Sexual reproduction involves the fertilisation of eggs with sperm. Hermaphrodite animals such as worms and many molluscs produce both, although self-fertilisation is uncommon. Separate sexes are more usual, and many water-dwelling animals release eggs and sperm into the water where they mingle, enabling fertilisation to take place. The young may hatch as miniatures of the parent, with the same habits, or as larvae which eat different foods and have a different way of life.

Barnacles

Floating larvae
Barnacles, like many immobile animals, have floating larvae which spread the species by drifting in the plankton to other areas.

Barnacle larva

Winged adult

Emerging adult

Eggs

Larva

Mating dragonflies

Winged adults
Many insects, such as dragonflies, spend much of their lives as aquatic larvae, finally changing into flying adults to mate and lay eggs.

Elver

Strange offspring
The larval stages of many water animals are quite unlike the adults. Some, such as eel larvae, were once thought to be different species.

Eel larva

Adult eel

Growth rings
Insects and crustaceans with enveloping external skeletons can grow only by shedding the old skeleton and producing a larger one. Other animals simply add to existing hard material, and since they grow fast in summer and slowly in winter, the annual additions are visible as growth rings on shells, scales and bones.

Fish scales

Painter's mussel

15

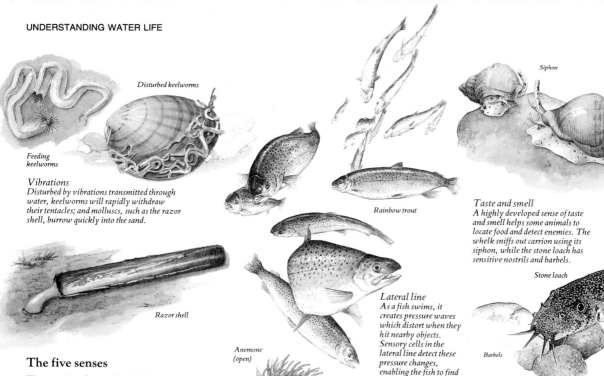

Disturbed keelworms

Feeding keelworms

Siphon

Whelks

Rainbow trout

Vibrations

Disturbed by vibrations transmitted through water, keelworms will rapidly withdraw their tentacles; and molluscs, such as the razor shell, burrow quickly into the sand.

Razor shell

Taste and smell

A highly developed sense of taste and smell helps some animals to locate food and detect enemies. The whelk sniffs out carrion using its siphon, while the stone loach has sensitive nostrils and barbels.

Stone loach

Barbels

Lateral line

As a fish swims, it creates pressure waves which distort when they hit nearby objects. Sensory cells in the lateral line detect these pressure changes, enabling the fish to find its way in the dark and swim in unison with others.

Anemone (open)

Anemone (closed)

The five senses

Water animals need well-developed senses to find their way, locate food, detect predators and recognise potential mates. Creatures which live on tidal shores must also avoid drying out when exposed to the air by the receding tide. Like land animals, water animals use sight, hearing, smell, taste and touch; fish are also sensitive to pressure waves. The importance of sight is evident from the camouflage adopted by many species to confuse their enemies.

Exposure

When the tide is out, many shoreline animals, such as anemones, barnacles and winkles, contract and close up to conserve moisture.

Goose barnacles

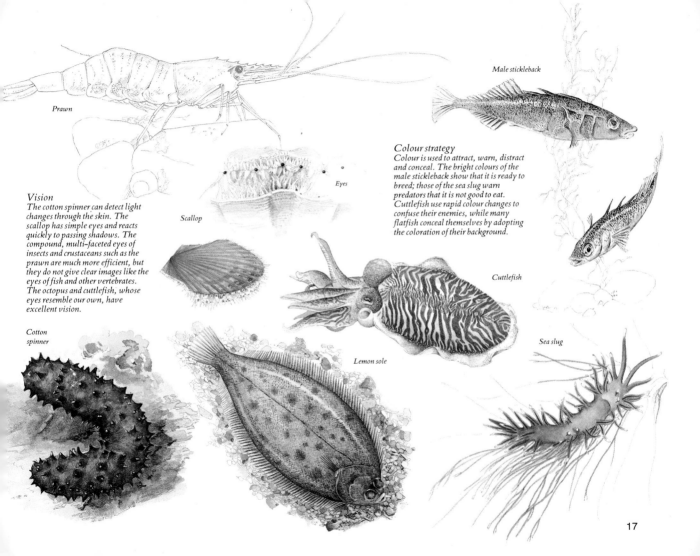

Prawn

Male stickleback

Scallop

Eyes

Colour strategy
Colour is used to attract, warn, distract and conceal. The bright colours of the male stickleback show that it is ready to breed; those of the sea slug warn predators that it is not good to eat. Cuttlefish use rapid colour changes to confuse their enemies, while many flatfish conceal themselves by adopting the coloration of their background.

Vision
The cotton spinner can detect light changes through the skin. The scallop has simple eyes and reacts quickly to passing shadows. The compound, multi-faceted eyes of insects and crustaceans such as the prawn are much more efficient, but they do not give clear images like the eyes of fish and other vertebrates. The octopus and cuttlefish, whose eyes resemble our own, have excellent vision.

Cuttlefish

Cotton spinner

Sea slug

Lemon sole

Key to freshwater life

In the identification key on the following pages, the freshwater animals and plants described in this book are divided into 12 main groups, based on distinctive characteristics such as body shape or number of legs. Each group is sub-divided according to more detailed features such as fin structure. To identify a specimen, examine it closely and proceed through the key to find which group it belongs to. Then turn to the page numbers given, and study the illustrations until the species is found. The groups are:

1 FISH WITH ELONGATE BODIES
2 FISH WITH TWO DORSAL FINS
3 FISH WITH ONE DORSAL FIN
4 AMPHIBIANS
5 INSECTS – THREE PAIRS OF JOINTED LIMBS
6 FOUR OR MORE PAIRS OF JOINTED LIMBS
7 ANIMALS WITH SHELLS
8 WORMS AND WORM-LIKE ANIMALS
9 VERY SMALL FLOATING ORGANISMS
10 VERY SMALL ATTACHED AND CRAWLING ORGANISMS
11 SMALL PLANT-LIKE ORGANISMS
12 SIMPLE PLANTS

Perch

Dorsal fins more or less equal in size.
Perch p. 48
Ruffe p. 49
Zander p. 50
Bullhead p. 51
Rare freshwater fish pp. 74–75

1 Fish with elongate bodies
Bodies long, thin and snakelike; scaleless or with minute scales embedded in the skin. No pelvic fins.

River lamprey

Eel

Bodies lack paired fins.
Lampreys pp. 30–31

Bodies with paired fins.
Eel pp. 32–33

2 Fish with two dorsal fins
First dorsal fin may be soft with branched rays, or strong and spined.

Trout

Grayling

Second dorsal fin is small and lacks rays (adipose fin).
Salmon pp. 34–35   Powan p. 45
Trout pp. 36–37   Grayling p. 46
Rainbow trout pp. 40–41 Smelt p. 47
Char pp. 42–43   Rare freshwater fish pp. 74–75
Vendace p. 44

Bullhead

Three-spined stickleback

First dorsal fin is reduced to a series of isolated spines.
Sticklebacks pp. 52–53

3 Fish with one dorsal fin

Fins consist usually of branched or jointed soft fin rays.

Gudgeon

Chub

4 Amphibians

Eggs laid in water; larvae are known as tadpoles.

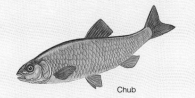
Common frog

Barbels round the mouth.
Common carp and varieties pp. 54–55
Barbel p. 58
Gudgeon p. 59
Tench pp. 60–61
Rare freshwater fish pp. 74–75

No barbels round the mouth.
Crucian carp p. 56   Rudd p. 67
Goldfish p. 57   Chub p. 68
Bream pp. 62–63   Dace and orfe p. 69
Minnow p. 64   Loaches pp. 70–71
Bleak p. 65   Pike pp. 72–73
Roach p. 66   Rare freshwater fish pp. 74–75

Four limbs, leathery skin.
Globular, jelly-like eggs.
Frogs pp. 76–77
Toads pp. 78–79
Newts pp. 80–81

5 Insects – three pairs of jointed limbs

Body divided into three sections – head, thorax and abdomen. Adults usually have two pairs of wings.

Stonefly nymph

Mayfly nymph

One tail appendage.
Creature does not live in a case.
Alderfly larvae p. 84
Pond insects with one tail p. 85

Alderfly larva

Two long tail appendages.
Does not live in a case.
Stonefly nymphs pp. 86–87

Three tail appendages.
Does not live in a case.
Mayfly nymphs pp. 90–91
Damselfly nymphs pp. 92–93
Screech beetle larva p. 101

Caddis larva

Pond skater

Water boatman

Two short tail appendages.
Creature lives in case.
Caddis larvae pp. 96–97

Long antennae. Lives on water surface.
Pond skater p. 99
Water cricket p. 99
Water measurer p. 99

Antennae concealed, no tail;
lives below water surface.
Dragonfly nymphs pp. 94–95
Backswimmer p. 98
Water boatman p. 98

5 Insects – three pairs of jointed limbs (continued)

Great diving beetle

Two pairs of wings; front pair made of hard, horny material.
Water beetles pp. 100–1

6 Four or more pairs of jointed limbs

No wings present. May be spider-like in shape or may have articulated, plated bodies.

Water mite

Four pairs of legs; no antennae. May be very small.
Water spider p. 105
Water mites p. 105

Native crayfish

More than four pairs of legs; hard-plated body.
Native crayfish p. 104
Freshwater shrimp p. 106
Water slater p. 107

7 Animals with shells

Molluscs including gastropods (snails and limpets) and bivalves (mussels and cockles).

One-shelled, snail-like with spiral-coiled or cone-shaped shell.
Water snails pp. 110–11
Limpets p. 112
Shell chart p. 130

Great pond snail

Swan mussel

Two hinged shells; only one shell may be found.
Orb shell cockle p. 113
Freshwater mussels pp. 114–15
Shell chart pp. 131–2

Caddis larva cases

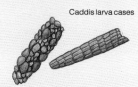

Other shell-like objects.
Caddis larvae cases pp. 96–97

8 Worms and worm-like animals

Soft, elongated bodies with no skeletal structure and lacking legs.

Polycelis species

Tubifex species

Bodies segmented or unsegmented; flat, round, or thin and hair-like in shape.
Freshwater worms pp. 116–17

Piscicola species

Segmented bodies with suckers at both ends.
Leeches pp. 118–19

Bloodworm (non-biting midge larva)

Segmented larvae with small heads, some with false legs (prolegs); otherwise worm-like or maggot-like.
Bloodworms and other legless larvae pp. 120–1

9 Very small floating organisms

Range from single-celled, jelly-like amoeba to very small crustaceans. Can usually only be seen with a hand lens.

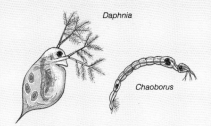

Daphnia

Chaoborus

Asplanchna

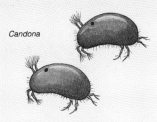

Candona

Limbless; many have a wheel-like crown of hair-like organs called cilia.
Floating animals and plants pp. 122–3

Transparent bodies; may be flea-like with jointed limbs or may have segmented bodies; sometimes carry black air sacs or egg sacs.
Legless larvae p. 121
Floating animals and plants pp. 122–3

Resemble minute two-shelled animals. Jointed limbs protrude from between the shells encasing the flea-shaped body.
Floating animals and plants pp. 122–3

10 Very small attached and crawling organisms

Found under water on plants or on mud and silt. Can usually only be seen with a hand lens.

Forms include: sac-like bodies with fringe of tentacles at free end; mat-like overgrowths on surfaces; sponge-like masses; slow-moving flea-like bodies; and minute bear-shaped creatures.
Small submerged surface-dwellers p. 125

Vorticella

Pond sponge

11 Small plant-like organisms

Usually found attached to stems and leaves of plants under water.

Green hydra

Small, columnar bodies, single or colonial, each with tentacles.
Hydras p. 124

12 Simple plants

Non-flowering primitive plants found in, on and near ponds, streams, rivers, lakes and marshy areas.

Fontanalis antipyretica

Conocephalum conicum

Azolla filiculoides

Mossy plants lacking differentiated roots, stems and leaves.
Water ferns, mosses and liverworts pp. 126–7

Usually of more or less uniform tissue undivided into stems and leaves.
Water ferns, mosses and liverworts pp. 126–7

Fern-like in structure.
Water ferns, mosses and liverworts pp. 126–7

21

Key to marine life

In the identification key on the following pages, the animals and plants described in the marine section of this book are divided into 12 groups based on distinctive characteristics, such as body shape, whether they have shells, do they float or swim or do they resemble plants. Each group is sub-divided according to more detailed features – for example, the number of shells, soft jelly-like body or fern-like appearance.

To identify a specimen, examine it closely, then proceed through the key to discover to which group it belongs. After this preliminary identification, turn to the pages referred to in that section and study the illustrations on those pages until the particular species is found. The 12 groups are:

1 ACTIVE SWIMMING ANIMALS WITH FLIPPERS
2 ACTIVE SWIMMING ANIMALS WITH FINS
3 ANIMALS WITH JOINTED LIMBS
4 ANIMALS WITH SHELLS
5 SLUG-LIKE ANIMALS
6 ACTIVE SWIMMING ANIMALS WITH TENTACLES
7 SPINY-SKINNED ANIMALS
8 WORMS AND WORM-LIKE ANIMALS
9 FLOATING ANIMALS
10 PLANT-LIKE ANIMALS ATTACHED TO ROCKS
   OR SEAWEEDS
11 OTHER ATTACHED AND ENCRUSTING ANIMALS
   AND PLANTS
12 PLANTS

1 Active swimming animals with flippers

Large warm-blooded animals that suckle their young (mammals).

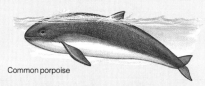

Common porpoise

Grey seal

Naked skin; forelimbs (flippers) but no hindlimbs; horizontally flattened tail; nostrils (blowhole) on top of head.
Whales, dolphins and porpoises pp. 144–9

Dog-like head, hairy coat, streamlined body, no dorsal fin.
Seals pp. 150–3

2 Active swimming animals with fins

Fish with skeletons consisting of gristle or cartilage (sharks and rays) or of bones.

Plaice

Skate

Lesser dogfish

Shark-like fish with the mouth under the head; rough skin like sandpaper, no scales.
Sharks and dogfish pp. 200–9

Flattened fish, usually found on the sea-bed.
Flatfish pp. 154–9   Dragonet p. 188
Skates and rays pp. 160–5   Angler fish p. 199

Fish with elongated bodies.
Eels pp. 166–7
Pipefish pp. 168–9
Sand eels p. 170   Butterfish
Butterfish p. 171
Blennies and similar fish p. 173
Shore-pool fish pp. 180–1
Visiting marine fish pp. 198–9

Ballan wrasse

Fish with one dorsal fin.
Blennies and similar fish pp. 172–3
Herring p. 174
Lumpsucker p. 175
Wrasses pp. 176–7
Visiting marine fish pp. 198–9

Cod

Common goby

Thick-lipped
grey mullet

Fish with two dorsal fins.

Grey mullets p. 182	Tub gurnard p. 188
Bass p. 183	Weevers p. 189
Ling and hake p. 184	Visiting marine fish pp. 198–9
Rockling p. 185	Sharks and dogfish pp. 200–9
Gobies pp. 186–7	

Fish with more than two dorsal fins.

Cod pp. 190–1	Haddock p. 195
Whiting p. 192	Mackerel pp. 196–7
Bib and poor cod p. 193	
Pollack and saithe p. 194	

3 Animals with jointed limbs

Walking and swimming animals with
many pairs of jointed limbs; bodies
segmented; hard outer shell which is
periodically moulted.

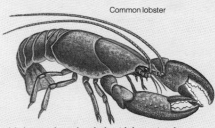
Common lobster

Spider crab

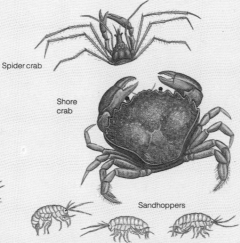
Shore
crab

**Ovoid or rounded body with four pairs of
walking legs and a large pair of pincers.
The small segmented tail is tucked under
the body.**
Crabs pp. 210–17

**Moderate to large, long body with four pairs of
walking legs and a pair of large pincers. In hermit
crabs the body is soft and hidden inside an empty
mollusc shell.**
Lobsters pp. 218–21
Hermit crabs p. 213

Common prawn

Sandhoppers

Sea spider

**Small, long-bodied animals with five pairs of walking
legs which may have small pincers, and five pairs of
swimming legs under the tail. Dart quickly
backwards through the water if frightened.**
Prawns and shrimps pp. 222–3

**Other small crawling animals with three or more
pairs of walking legs with no pincers or only very
small ones. Body is flattened vertically or
horizontally. Insect-like or spider-like.**
Small creatures of the seashore pp. 224–5

4 Animals with shells

Soft-bodied animals with a protective outer case. A broad, flattened foot is used for crawling or digging.

Tellin shells

Great scallop

Whelk

One-shelled, snail-like animals with spiral-coiled or cone-shaped shells.

Shell chart pp. 132–5   European cowrie p. 234
Limpets pp. 226–7   Topshells p. 235
Winkles pp. 230–1   Tall-shelled animals pp. 236–7
Whelks pp. 232–3

Two-hinged shells, or valves. When the animal is dead, only one shell may be found.

Shell chart pp. 135–43   Carpet, Venus and tellin
Mussels pp. 238–9   shells pp. 244–5
Oysters pp. 240–1   Sand gaper and razor shells
Cockles pp. 242–3   pp. 246–7

Scallops pp. 248–9
Common piddock and boring molluscs pp. 250–1

Chitons

Common acorn barnacles

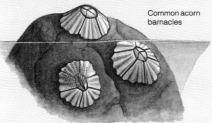

Shipworm tubes

Many-shelled animals. Either small flat shells attached to rocks, with eight shell-plates along the back, or small, volcano-like shells or stalked shells made up of many plates.

Chitons p. 252
Barnacles p. 253

Other shell-like objects.

Tusk shell p. 237
Shipworm tubes p. 251
Tests of sea urchins pp. 272–5
Tubes of marine worms pp. 281–3

5 Slug-like animals

Slow-moving and soft-bodied. Body may be wide and flattened (sea slugs) or cylindrical (sea cucumbers).

Common grey sea slug

Often brightly coloured creatures with gills and small lumps, called processes, on the back.

Sea slugs pp. 256–7

Sea gherkin

Cucumber-shaped animals with branched tentacles at one end, which can be retracted. Adhere to rocks with tube feet.

Sea cucumbers pp. 258–9

6 Active swimming animals with tentacles

Animals move by jet propulsion.

Common cuttlefish

Soft, streamlined body with eight or ten long tentacles, covered in suckers, surrounding the mouth.

Octopuses and cuttlefish pp. 260–1, 264–5

7 Spiny-skinned animals

Move by means of hydraulic
tube feet which carry
suction discs at their ends.

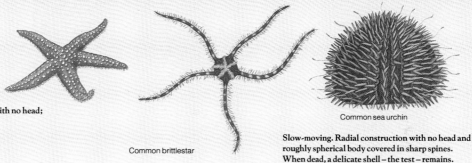

Common
starfish

Common brittlestar

Common sea urchin

**Slow-moving. Radial construction with no head;
spiny skin. Star-shaped body.**
Starfish and brittlestars pp. 266–71

**Slow-moving. Radial construction with no head and
roughly spherical body covered in sharp spines.
When dead, a delicate shell – the test – remains.**
Sea urchins pp. 272–5

8 Worms and worm-like animals

Group contains both active and
sedentary creatures. Sedentary
forms live in tubes or buried in
mud and sand.

Lugworm

Red ribbon worm

Common
ragworm

Worms with the body clearly divided into segments.
Ragworms and other predatory worms pp. 278–9
Lugworms and other burrowing worms pp. 280–1

**Worms and worm-like animals with
smooth, unsegmented bodies.**
Sea cucumbers pp. 258–9
Ribbon worms p. 277

Flatworm
(*Oligocladus
sanguinolentus*)

Peacock worms

Keel worms

Worms with small, flat, leaf-like bodies.
Flatworms p. 276

**Animals hidden in tubes made of sand grains, mud,
parchment-like material or a hard, limy substance.
Tubes bent, straight or coiled. Worm tubes may have
tentacles appearing from the top if undisturbed in
water.**

Tusk shell p. 237
Teredo (boring mollusc) p. 251
Tube worms (including peacock and keel worms) pp. 281–3

9 Floating animals

Found on or near the surface. Many are soft–bodied and transparent.

Common jellyfish

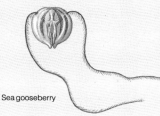

Sea gooseberry

Arrow worms

Body jelly-like and almost transparent with trailing tentacles.
Jellyfish pp. 286–7
Portuguese man-of-war p. 288
Wind and current-borne drifters p. 289
Sea gooseberry p. 290

Other small floating animals on or near the surface.
Small animals in the plankton p. 291

10 Plant-like animals attached to rocks or seaweeds

Animals, often brightly coloured, that at first sight resemble plants.

Beadlet anemone

Dead man's fingers colony

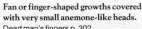

Detail of polyp

Soft body with many tentacles surrounding the mouth; resembles blob of jelly when tentacles are retracted.
Sea anemones pp. 296–9

Fan or finger-shaped growths covered with very small anemone-like heads.
Dead man's fingers p. 302
Sea fans p. 303

Obelia polyps

Obelia species

Bugula species

Common featherstar

Small, delicate growths resembling ferns, miniature trees or thin twigs. Covered with minute anemone-like heads only clearly visible with a hand lens. Heads may be withdrawn if disturbed.
Hydroids (sea firs) p. 300

Small, brownish, branching, bushy or clumped growths. Seen through a hand lens the surface is made up of many individual oblong chambers, each with a head of tentacles that withdraws if disturbed.
Moss animals p. 295

Animals with five pairs of feather-like arms. They can swim but often hold onto rocks for long periods.
Featherstars p. 271

11 Other attached and encrusting animals and plants

Often brightly coloured, the growths usually consist of colonies of many individuals. Plants are barely recognisable as such.

Sea orange sponge

Star sea squirts

Hairy sea mat on red seaweed

Marine lichens pp. 316–17

Variously shaped growths including lumps, branches and crusts that look and feel spongy.
Sponges pp. 292–3
Sea mat and moss animals pp. 294–5
Dead man's fingers p. 302

Ross coral

Variously coloured growths closely encrusting rocks or seaweeds.
Sponges pp. 292–3
Sea mat and moss animals pp. 294–5

Colonial sea squirts p. 301
Lithothamnion/Lithophylum group p. 315

Common sea squirts

Sea fan

Hard growths resembling coral.
Sabellaria worm p. 283   Cup coral p. 297
Hornwrack p. 295   Sea fan p. 303
Ross coral p. 295   *Lithothamnion/Lithophylum* group p. 315

Firm or soft sac-like body with two openings called siphons, out of which water squirts if the body is pressed.
Sea squirts pp. 304–5

---

12 Plants

Range in colour from blue-green and green to reds, violets and browns.

Sea lettuce

Dulse

May be anchored by roots like land plants or may be attached to the sea-bed, to rocks or to other plants by a holdfast.
Green seaweeds p. 306
Eel grass p. 307
Kelps, wracks and other brown seaweeds pp. 308–11
Red seaweeds pp. 312–15

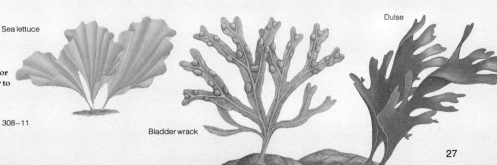

Bladder wrack

27

WATER LIFE
OF BRITAIN

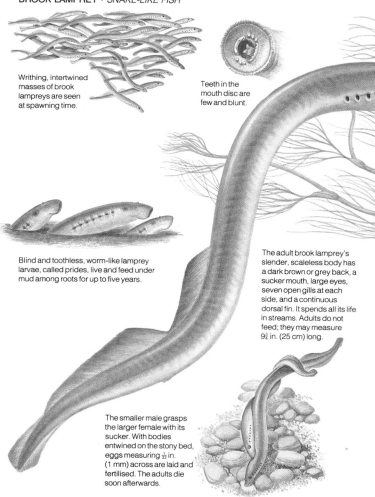

Writhing, intertwined masses of brook lampreys are seen at spawning time.

Teeth in the mouth disc are few and blunt.

Blind and toothless, worm-like lamprey larvae, called prides, live and feed under mud among roots for up to five years.

The adult brook lamprey's slender, scaleless body has a dark brown or grey back, a sucker mouth, large eyes, seven open gills at each side, and a continuous dorsal fin. It spends all its life in streams. Adults do not feed; they may measure 9¾ in. (25 cm) long.

The smaller male grasps the larger female with its sucker. With bodies entwined on the stony bed, eggs measuring ½ in. (1 mm) across are laid and fertilised. The adults die soon afterwards.

Brook lampreys, as their name implies, spend all their lives in brooks or streams, usually in the upper reaches.

Brook lamprey *Lampetra planeri*

Lampreys are primitive, jawless vertebrates which possess only a simple skeleton composed of cartilage. They have no paired fins and no scales or protective plates. They have an eel-like form, but can easily be distinguished from eels by their sucker and seven pairs of gill slits. The brook lamprey, or Planer's lamprey, is the smallest and the commonest of the three lampreys found in the British Isles. Apart from its use as bait in angling, it is of no commercial value.

The fish spawn between March and June, the exact time depending on the water temperature. Each female lays up to 1,200 eggs in a shallow nest excavated in the sandy or gravelly bed. After spawning the adults die. The eggs take about three weeks to hatch into larvae, or prides. It was once thought that a pride was a separate species from the brook lamprey and in 19th-century fish books it is often referred to by the Latin name of *Ammocoetes branchialis*. The prides strain organic particles from the mud. They grow slowly for four or five years, then the transition to the adult stage takes place during the autumn and winter prior to spawning. This transition includes the atrophy of the gut, so the adults do not feed.

Few but sharp teeth and two larger tooth plates stud the river lamprey's mouth disc.

River and sea lampreys are blood suckers. They scrape a hole in the prey with their teeth, then cling on with the sucker and draw out blood. Haddock, saithe, cod, shad and herrings are common prey.

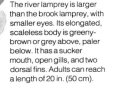

The river lamprey is larger than the brook lamprey, with smaller eyes. Its elongated, scaleless body is greeny-brown or grey above, paler below. It has a sucker mouth, open gills, and two dorsal fins. Adults can reach a length of 20 in. (50 cm).

The fish come from coastal waters to breed in clean, fast rivers with stony or sandy beds.

Sea lamprey
Petromyzon marinus

Mottled dark and light brown, with two separate dorsal fins, and growing to 36 in. (90 cm) long, the sea lamprey feeds in coastal waters, entering rivers to spawn. It has radiating rows of sharp teeth and one tooth plate.

River lamprey *Lampetra fluviatilis*

After about five years in fresh water the river lamprey or lampern spends one or two years in the sea. It is a parasite on other fish during its sea life, but the lamprey – or the scar it makes – is found chiefly on migrating or estuarine fish so it probably stays near the shore. Attached to a migrating trout or salmon, the lamprey gets not only food but an easy ride to fresh water. Once there it stops feeding and reaches sexual maturity. It spawns the next spring. Male and female together scoop the hollow into which the female lays up to 20,000 eggs. The larvae feed on organic debris and bacteria on the river bed. More than a million river lampreys a year were caught in the Thames for food in the 18th century, but since the Industrial Revolution pollution has severely reduced their numbers.

The larger sea lamprey has a life history similar to the river lamprey's but it ventures further out to sea and attacks a wider range of fish, often causing debility and death. On return to fresh water it spawns in sandy or gravelly shallows. Females lay up to 200,000 eggs. Sea lampreys are not common enough in British waters to ruin fisheries as happened in the North American Great Lakes after access to the sea was provided.

The leptocephalus, or larva of the common eel, is flat, leaf-shaped, and ½ in. (5 mm) long when hatched. It grows to 3 in. (75 mm) before changing into an elver.

Heart, spine and gills are visible in newly metamorphosed elvers, which are transparent and often called glass eels.

During the summer the elvers acquire the yellow-brown pigmentation of river eels as they move upstream.

Freshwater eels are mainly nocturnal and hunt by smell. They use the pair of tube-like nostrils on the upper lip for scenting out prey.

The slimy coat of mucus prevents drying and water loss, and helps eels to survive out of water for long periods. They can wriggle over land from one stretch of water to another – and often occur in land-locked lakes.

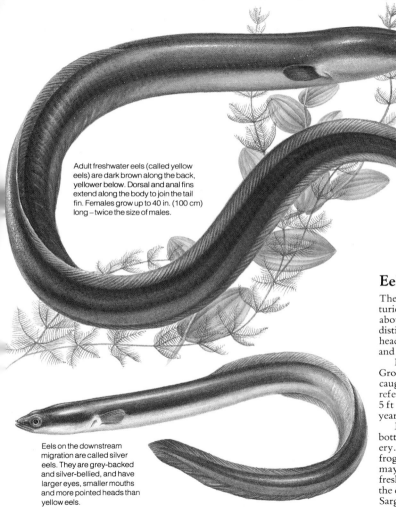

Adult freshwater eels (called yellow eels) are dark brown along the back, yellower below. Dorsal and anal fins extend along the body to join the tail fin. Females grow up to 40 in. (100 cm) long – twice the size of males.

Eels on the downstream migration are called silver eels. They are grey-backed and silver-bellied, and have larger eyes, smaller mouths and more pointed heads than yellow eels.

The River Severn is famous for its eel run, and the elvers returning from the sea are netted in great numbers to supply eel farms.

Eel *Anguilla anguilla*

The eel has been fished, farmed and eaten in Europe for centuries – yet scientifically there is still much that is mysterious about its life. For example, the yellow eel is found as two distinct types, by far the most common with a broad, blunt head, the other with a slim, sharp snout. Why two types exist and how they are related is not known.

Lowland rivers, ditches and fens are where eels thrive best. Growth there may be prodigious – an eel of over 20 lb (9 kg) was caught near Norwich in 1839, while other East Anglian records refer to eels of 23 lb (10·5 kg) and 27 lb (12 kg); both were over 5 ft (152 cm) long. Ely is said to have been named from the yearly rent of 100,000 eels paid to the lord of the manor.

Eels are far commoner today than is generally realised; their bottom-dwelling, nocturnal habits enable them to avoid discovery. While in fresh water their food includes snails, crayfish, frogs, tadpoles and fish eggs. The freshwater stage of their lives may last up to 30 years. Some young eels never move right into fresh water, but remain in estuaries. However, they all return to the coast and most authorities now agree that they go back to the Sargasso Sea, in mid-Atlantic, to reproduce.

The female uses its tail to scoop a long redd, or hollow, 4–12 in. (10–30 cm) deep, in the gravel bed, usually in November or December.

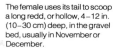

The fry leave the gravel and feed on insect larvae, worms, and other small animals.

Fry become parr at about 4 in. (10 cm) long, when 8–11 dark 'fingerprints' mark their flanks.

Nudging and swimming side by side, male and female shed batches of sperm and eggs into the redd. A second male may be in attendance. The female uses its tail to push a covering of gravel over the eggs, which are $\frac{1}{4}$ in. (6 mm) in diameter.

Between one and four years old, the fish are ready to migrate to sea as smolts. They are about 7 in. (18 cm) long and silver with a few black spots.

Near the centre the scale displays the life-span of the parr. The next rings were formed during the sea life. This scale is from a fish that spawned and returned to sea.

Dark kelts, salmon after spawning, are exhausted and many die.

In spring alevins emerge from the eggs but stay in the gravel for up to six weeks, living on the yolk.

Scotland's Dee is typical of salmon rivers all over Britain – clean, cool, well-oxygenated, and with clear passage to spawning grounds in its headwaters.

Salmon *Salmo salar*

Thousands of miles and up to four years may separate a salmon from its river of birth, but when the time comes for spawning the fish will find its way from the Atlantic to the waters where it hatched. The earth's magnetic field or even the stars may control the salmon's direction finding in the ocean. At the coast, a chemical memory enables the fish to 'smell' its own river.

A salmon arriving in fresh water at the end of winter is silver and sleekly plump from its diet of small herrings, sand eels and crustaceans. It does not eat again until the autumn spawning is finished, but it will snap at small items in the river – including anglers' bait. The journey to the headwaters is strenuous, often through rapids and up waterfalls. Large salmon can leap heights up to 10 ft (3 m), jumping best from deep water. A salmon can lose almost half its weight from these exertions.

During the journey the salmon's skin becomes thicker and rough, and develops a pink tinge. The flush fades after spawning and males gradually lose the kype, or hook, on the lower jaw. Salmon that survive spawning to return to the sea become silvery again, with a sprinkling of black spots. They grow rapidly and are ready for spawning again after a year or two.

Its larger size and slightly forked, narrower-based tail differentiate the salmon from the closely related trout. As it swims upstream to spawn, the salmon becomes pink-flushed and red-blotched on the flanks, and the male's lower jaw develops a kype, or hook. Large salmon may reach 4 ft (120 cm) in length and weigh 64 lb (29 kg).

35

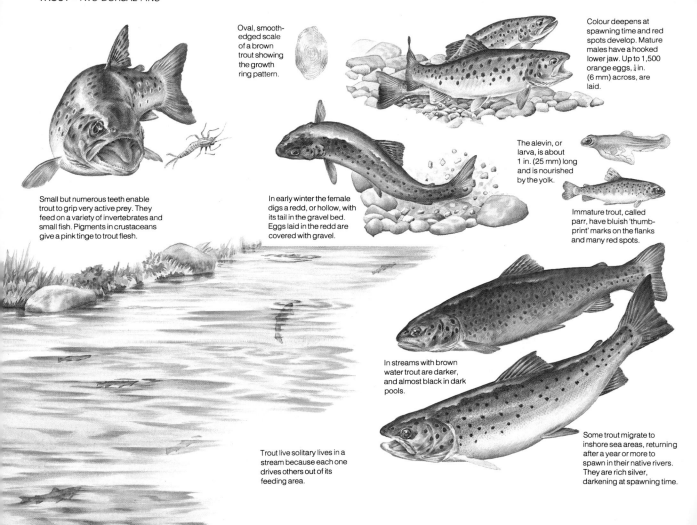

Oval, smooth-edged scale of a brown trout showing the growth ring pattern.

Colour deepens at spawning time and red spots develop. Mature males have a hooked lower jaw. Up to 1,500 orange eggs, ¼ in. (6 mm) across, are laid.

Small but numerous teeth enable trout to grip very active prey. They feed on a variety of invertebrates and small fish. Pigments in crustaceans give a pink tinge to trout flesh.

In early winter the female digs a redd, or hollow, with its tail in the gravel bed. Eggs laid in the redd are covered with gravel.

The alevin, or larva, is about 1 in. (25 mm) long and is nourished by the yolk.

Immature trout, called parr, have bluish 'thumbprint' marks on the flanks and many red spots.

In streams with brown water trout are darker, and almost black in dark pools.

Trout live solitary lives in a stream because each one drives others out of its feeding area.

Some trout migrate to inshore sea areas, returning after a year or more to spawn in their native rivers. They are rich silver, darkening at spawning time.

Trout prefer cold, well-oxygenated water running over a clean gravel bottom. They occur throughout Britain, especially in chalk and mountain streams.

Trout vary in colour, depending on their habitat, from silver through shades of brown with reddish-orange spots on their flanks. The tail is scarcely forked at all. Trout in fresh waters grow up to 39 in. (99 cm). Trout which migrate to sea may grow up to 55 in. (140 cm).

Trout *Salmo trutta*

Wide variations in colour and growth rate, depending on the local environment, have resulted in many forms of the trout – for example, the silvery sea trout, the dark-spotted Loch Leven trout, Orkney trout and Irish trout. However, they are all one species which has a migratory habit over parts of its range.

The large eggs of the trout contain copious reserves of yolk. They may take over six weeks to hatch into alevins, or sac-fry (so-called because the yolk sac remains attached). For two or three weeks – or longer if the water is cold – the alevins obtain nourishment from the yolk. Gradually they start searching for food, which at first consists of very small stream animals.

Growth depends on food availability and whether the young fish migrate to estuaries and other inshore areas. Sea trout, as they are then called, feed on small fish such as sprats. They grow much faster than river-dwelling trout, whose diet is chiefly invertebrates such as freshwater shrimps and insect larvae – though some freshwater trout turn to feeding on small fish and grow much faster. After two or three years, both sea and freshwater trout move upriver to spawn. Trout live for about five or six years, but 20-year-old specimens have been caught.

37

The rich chalk streams of southern England

The tree-lined, lime-rich rivers of southern England, flowing down from the chalk hills and downs, are the trout streams beloved by anglers. Here the native brown trout grow fat and fast, living on an abundance of aquatic insect larvae and fresh-water shrimps. They are lowland rivers with a gentle slope, usually crystal clear and seldom more than 3–5 ft (1–1·5 m) deep and not more than 30–40 ft (9–12 m) wide. The river bed is stone and gravel with occasional patches of sand and mud which support dense swathes of plant growth such as water crowfoot, providing cover for trout and other animals.

The shallow, clear water of chalk streams may become choked with water-weeds, particularly the narrow-leaved species of water-crowfoot. In trout streams the weed is regularly cut to ensure a steady flow of water and to prevent flooding.

The shallows near the bank of the river are often overgrown with water plants such as watercress, water forget-me-not and unbranched bur-reed. On the bank are rushes and marsh plants such as water-celery, water-parsnip, brooklime and meadowsweet. The thick vegetation provides plenty of nesting sites for water fowl.

A characteristic fish of these waters is the grayling. Like the trout, it is often seen feeding on winged insects which have fallen into the water after mating.

In spring and summer, hatching insects attract insect-eating birds such as the swift and swallow, which hawk low over the water to take mayflies, midges and damselflies. Moorhens, ducks and swans feed on the abundant waterlife.

Chalk streams are often carefully managed for angling. The vegetation is cut back to give access to the bank, which may be reinforced with planking to reduce erosion by the river and by the feet of fishermen. In spring and early summer the planks may become dotted with the discarded skins of mayfly nymphs, while the adults swarm over the water prior to mating. Once the insects have mated and laid their eggs they fall into the water to be devoured by fish.

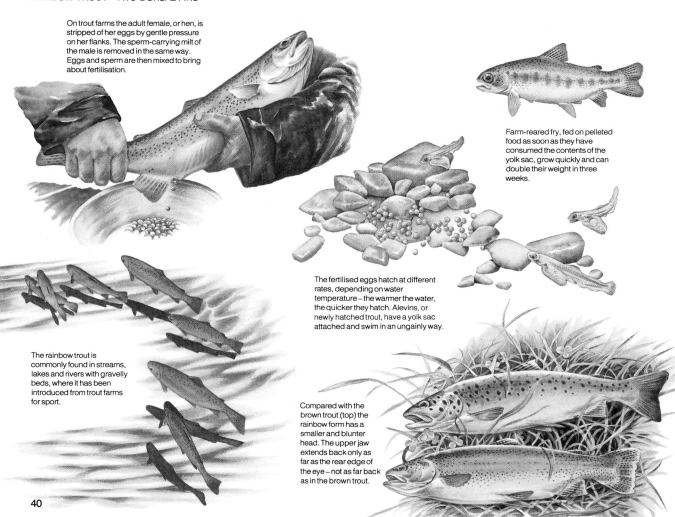

On trout farms the adult female, or hen, is stripped of her eggs by gentle pressure on her flanks. The sperm-carrying milt of the male is removed in the same way. Eggs and sperm are then mixed to bring about fertilisation.

Farm-reared fry, fed on pelleted food as soon as they have consumed the contents of the yolk sac, grow quickly and can double their weight in three weeks.

The fertilised eggs hatch at different rates, depending on water temperature – the warmer the water, the quicker they hatch. Alevins, or newly hatched trout, have a yolk sac attached and swim in an ungainly way.

The rainbow trout is commonly found in streams, lakes and rivers with gravelly beds, where it has been introduced from trout farms for sport.

Compared with the brown trout (top) the rainbow form has a smaller and blunter head. The upper jaw extends back only as far as the rear edge of the eye – not as far back as in the brown trout.

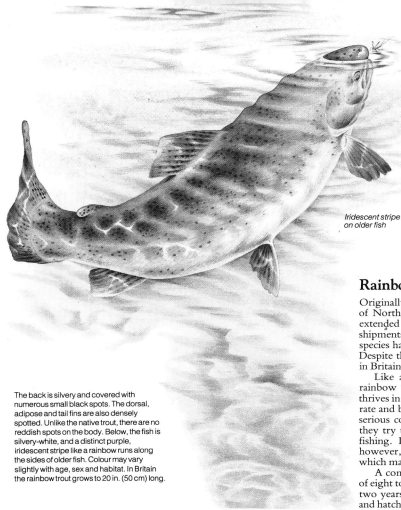

Iridescent stripe
on older fish

Rainbow trout have become naturalised in very few British rivers and lakes – the River Misbourne in Buckinghamshire is one place where they breed.

The back is silvery and covered with numerous small black spots. The dorsal, adipose and tail fins are also densely spotted. Unlike the native trout, there are no reddish spots on the body. Below, the fish is silvery-white, and a distinct purple, iridescent stripe like a rainbow runs along the sides of older fish. Colour may vary slightly with age, sex and habitat. In Britain the rainbow trout grows to 20 in. (50 cm) long.

Rainbow trout *Salmo gairdneri*

Originally restricted to the rivers draining the Rocky Mountains of North America, the range of the rainbow trout has been extended by man throughout much of the world. The first shipments of eggs arrived in Britain in 1884, and since then the species has been introduced into more than 800 British waters. Despite this there are only 40 naturalised breeding populations in Britain, of which seven are completely self-sustaining.

Like all members of the salmon family (salmonids), the rainbow trout requires high levels of dissolved oxygen, and thrives in traditional trout streams, but owing to its fast growth rate and big appetite it is regarded by many fishery keepers as a serious competitor to the native brown trout. Because of this they try to keep it out, and so maintain the quality of the fly fishing. Its tolerance of relatively high water temperatures, however, makes it a useful fish for stocking shallower waters which may warm up in summer.

A comparatively short-lived fish, reaching a maximum age of eight to nine years, the rainbow trout begins to breed at about two years old. On fish farms the eggs are incubated artificially and hatch after about 35 days at 50°F (10°C).

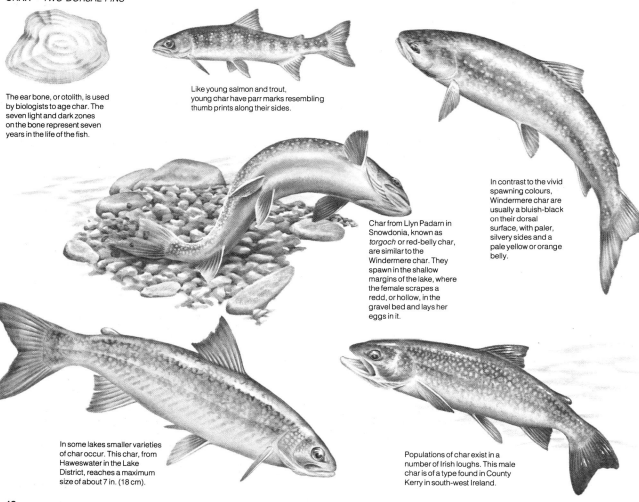

The ear bone, or otolith, is used by biologists to age char. The seven light and dark zones on the bone represent seven years in the life of the fish.

Like young salmon and trout, young char have parr marks resembling thumb prints along their sides.

Char from Llyn Padarn in Snowdonia, known as *torgoch* or red-belly char, are similar to the Windermere char. They spawn in the shallow margins of the lake, where the female scrapes a redd, or hollow, in the gravel bed and lays her eggs in it.

In contrast to the vivid spawning colours, Windermere char are usually a bluish-black on their dorsal surface, with paler, silvery sides and a pale yellow or orange belly.

In some lakes smaller varieties of char occur. This char, from Haweswater in the Lake District, reaches a maximum size of about 7 in. (18 cm).

Populations of char exist in a number of Irish loughs. This male char is of a type found in County Kerry in south-west Ireland.

Char prefer cold, clear waters like those of Hawes-water in the Lake District, where they feed on plank-tonic crustaceans, insects and molluscs.

Char *Salvelinus alpinus*

Char closely resemble their relatives the trout and salmon, but occur less widely. Several of the lakes in the Lake District, including Buttermere, Windermere, Haweswater and Conis-ton, have char populations. The fish are also found in a number of Scottish lochs, Irish loughs, and in Llyn Peris and Llyn Padarn in Wales. Most populations show minor differences in size and colouring, and in some cases they have been given different names. The Welsh char is called *torgoch* (red-belly) and those of Loch Killin in Scotland are known as haddy.

In different lakes, spawning habits, feeding and growth rates differ. In Britain there appear to be two groups; those that spawn in deep water in late winter or spring and those that spawn in shallow water in autumn. Both groups may occur in one lake, as in Windermere. Yellow eggs measuring $\frac{1}{8}$ in. (3 mm) are shed on gravel in still or flowing waters.

All types of char are carnivorous, living mainly on small planktonic crustaceans. They grow slowly, reaching maturity in four to five years, though autumn spawners usually mature more quickly. Maximum sizes vary between lakes: 1 lb (0·5 kg) is a large fish in Wales while Windermere char are slightly larger.

In autumn prior to spawning, the colouring of Windermere char is heightened, especially in the male, whose ventral surface and fins become a rich orange-red. Females are duller and have a smaller head. In both sexes, the back and sides are covered in red-and-white spots. Size is also variable, with adult fish usually 10–12 in. (25–30 cm) long.

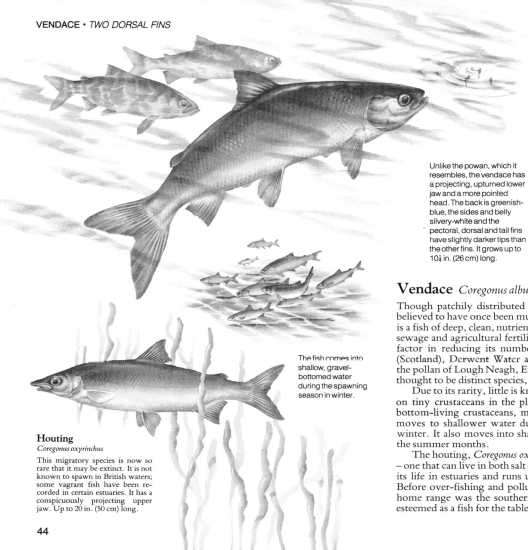

Unlike the powan, which it resembles, the vendace has a projecting, upturned lower jaw and a more pointed head. The back is greenish-blue, the sides and belly silvery-white and the pectoral, dorsal and tail fins have slightly darker tips than the other fins. It grows up to 10¼ in. (26 cm) long.

The vendace is not common and is now found in only a few lakes, such as Bassenthwaite in the Lake District.

The fish comes into shallow, gravel-bottomed water during the spawning season in winter.

Houting
Coregonus oxyrinchus

This migratory species is now so rare that it may be extinct. It is not known to spawn in British waters; some vagrant fish have been recorded in certain estuaries. It has a conspicuously projecting upper jaw. Up to 20 in. (50 cm) long.

Vendace *Coregonus albula*

Though patchily distributed today, the vendace, or pollan, is believed to have once been much more widespread in Britain. It is a fish of deep, clean, nutrient-poor waters, and pollution from sewage and agricultural fertilisers may have been an important factor in reducing its numbers. The vendace of Lochmaben (Scotland), Derwent Water and Bassenthwaite (England) and the pollan of Lough Neagh, Erne, Derg and Ree in Ireland, were thought to be distinct species, but are now considered one.

Due to its rarity, little is known of the fish's biology. It feeds on tiny crustaceans in the plankton for part of the year, and bottom-living crustaceans, molluscs and insect larvae when it moves to shallower water during the spawning migration in winter. It also moves into shallower water for a period during the summer months.

The houting, *Coregonus oxyrinchus*, is a euryhaline white fish – one that can live in both salt and fresh water. It spends much of its life in estuaries and runs up-river to spawn during winter. Before over-fishing and pollution made this fish very rare, its home range was the southern North Sea. It was once highly esteemed as a fish for the table.

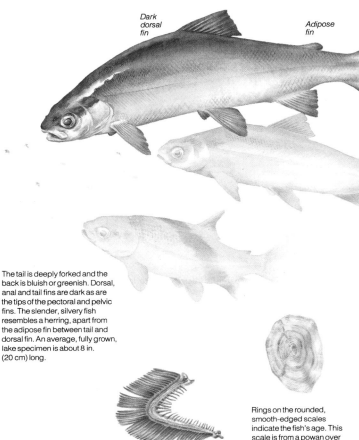

Dark dorsal fin

Adipose fin

Deeply forked tail

The tail is deeply forked and the back is bluish or greenish. Dorsal, anal and tail fins are dark as are the tips of the pectoral and pelvic fins. The slender, silvery fish resembles a herring, apart from the adipose fin between tail and dorsal fin. An average, fully grown, lake specimen is about 8 in. (20 cm) long.

The fish is a plankton feeder and has sieve-like gill rakers on each gill arch to prevent small food particles escaping over the gills when water is expelled during respiration.

Rings on the rounded, smooth-edged scales indicate the fish's age. This scale is from a powan over three years old.

Coregonus lavaretus is found in some Scottish lochs and Cumbrian lakes, as well as Llyn Tegid (Bala Lake) in Wales.

Powan *Coregonus lavaretus*

In the 17th and 18th centuries there were important local fisheries for the powan, which was regarded as the poor man's herring. It is, however, very susceptible to pollution, and many European lakes that contained powan before the Industrial Revolution now no longer do so.

Coregonus lavaretus is the most widespread of the five fresh-water species of white fish. Because it is found isolated in widely separated areas, it is called by different names in different places – gwyniad in Wales, skelly in the Lake District and powan in Loch Lomond. There is some variation in colour between the geographic races; for example, the Lake District type has distinct spots on its flanks.

Spawning occurs from December to March but there is little detailed information about the life history. One problem is its inability to survive capture, so that captive rearing and observation is not possible. Because they are essentially plankton feeders, they are not normally caught by anglers. *C. lavaretus* can reach a greater size than the vendace, but growth depends very much on local conditions. For example, gwyniad up to 18 in. (45 cm) long and 2 lb (900 g) in weight have been caught.

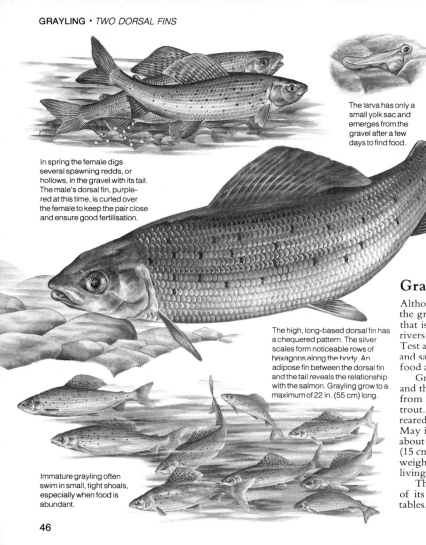

In spring the female digs several spawning redds, or hollows, in the gravel with its tail. The male's dorsal fin, purple-red at this time, is curled over the female to keep the pair close and ensure good fertilisation.

The larva has only a small yolk sac and emerges from the gravel after a few days to find food.

The high, long-based dorsal fin has a chequered pattern. The silver scales form noticeable rows of hexagons along the body. An adipose fin between the dorsal fin and the tail reveals the relationship with the salmon. Grayling grow to a maximum of 22 in. (55 cm) long.

Immature grayling often swim in small, tight shoals, especially when food is abundant.

Clean, cool brooks and streams running swiftly over stones suit the grayling. It is patchily distributed throughout Britain.

Grayling *Thymallus thymallus*

Although it is now found as far apart as Somerset and the Tay, the grayling has only a patchy, local distribution and much of that is due to introduction by man into lakes and rivers. Even rivers where the grayling is well known, such as Hampshire's Test and Itchen, are not part of its original distribution. In trout and salmon waters, it is thought to compete with small fish for food and because of this is often removed as a pest.

Grayling are extremely sensitive to pollution, both chemical and thermal. In areas of intensive agriculture and near effluents from electricity generating stations they disappear even before trout. They are difficult to farm and so are not replaced by fish reared artificially. Spawning takes place between March and May in gravelly shallows and the deep yellow eggs hatch after about three weeks. The young grow rapidly, reaching 6 in. (15 cm) in about a year. In southern rivers adults may reach a weight of 5 lb (2·3 kg). The fish feed mainly on small creatures living on the river bed, and may also eat trout and salmon eggs.

The thyme-like smell a fresh grayling has may be the origin of its Latin name, *Thymallus*. It is a delicacy on continental tables, but in Britain is regarded chiefly as a good sporting fish.

46

Adipose
fin

Small yellow eggs
are shed between
March and May.
Sticky stalks
allow them to
cling to weeds
and stones.

The large mouth
has many needle-
like teeth, used to
capture crustaceans,
sprats and fish fry.

Smelts live in coastal waters and estuaries, but spawn in rivers (mainly south-
eastern) over sand or gravel.

The slim, silvery smelt, greenish
above and with the dorsal fin well
back, is related to the salmon family
– as the small, fleshy, rayless
adipose fin on the back suggests.
When fresh, it gives off a cucumber-
like odour. Adults may reach 12 in.
(30 cm) or more in length.

Smelt *Osmerus eperlanus*

This inshore migratory fish occurs on the east coast from the
Tay to the Thames, on the west from the Clyde to the Conwy,
and in south-west Ireland. Land-locked populations occur in
fresh water in other parts of Europe, but the last inland smelts in
the British Isles, at Rostherne Mere in Cheshire, are thought to
have died out in the 1920s. The smelt fishery of the Thames was
once one of the major tidal fisheries, one of several on the east
coast. The fish was prized for its unusual cucumber flavour.
Pollution in the early 19th century destroyed many of the
fisheries, but since the early 1970s smelts have returned to rivers
like the Thames and breed in freshwater stretches.

The smelt is relatively short-lived – seven years is about
average – and its spawning success is variable, so that numbers
fluctuate widely over two or three years. Success may depend
on water temperature and river flow between March and July
when first the eggs, then the fry, abound. The young eat
copepods (small crustaceans) and various young fish. After a
summer in fresh water, they move to brackish water, returning
to fresh water only to spawn. In the sea they feed voraciously on
shrimps, the young of the cod family, sprats and herrings.

47

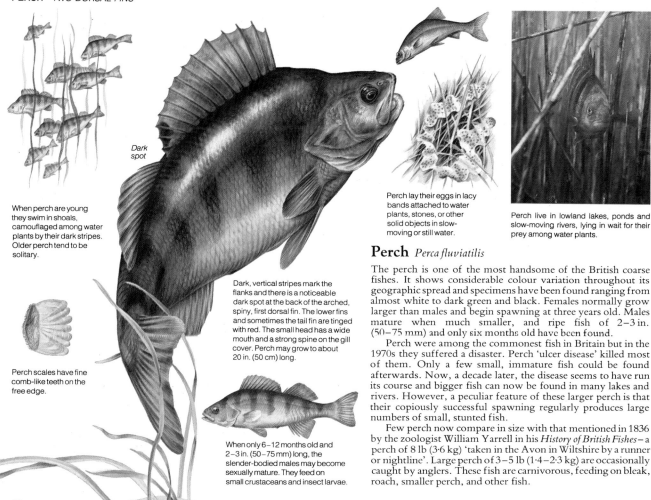

Dark spot

When perch are young they swim in shoals, camouflaged among water plants by their dark stripes. Older perch tend to be solitary.

Perch scales have fine comb-like teeth on the free edge.

Dark, vertical stripes mark the flanks and there is a noticeable dark spot at the back of the arched, spiny, first dorsal fin. The lower fins and sometimes the tail fin are tinged with red. The small head has a wide mouth and a strong spine on the gill cover. Perch may grow to about 20 in. (50 cm) long.

When only 6–12 months old and 2–3 in. (50–75 mm) long, the slender-bodied males may become sexually mature. They feed on small crustaceans and insect larvae.

Perch lay their eggs in lacy bands attached to water plants, stones, or other solid objects in slow-moving or still water.

Perch live in lowland lakes, ponds and slow-moving rivers, lying in wait for their prey among water plants.

Perch *Perca fluviatilis*

The perch is one of the most handsome of the British coarse fishes. It shows considerable colour variation throughout its geographic spread and specimens have been found ranging from almost white to dark green and black. Females normally grow larger than males and begin spawning at three years old. Males mature when much smaller, and ripe fish of 2–3 in. (50–75 mm) and only six months old have been found.

Perch were among the commonest fish in Britain but in the 1970s they suffered a disaster. Perch 'ulcer disease' killed most of them. Only a few small, immature fish could be found afterwards. Now, a decade later, the disease seems to have run its course and bigger fish can now be found in many lakes and rivers. However, a peculiar feature of these larger perch is that their copiously successful spawning regularly produces large numbers of small, stunted fish.

Few perch now compare in size with that mentioned in 1836 by the zoologist William Yarrell in his *History of British Fishes* – a perch of 8 lb (3·6 kg) 'taken in the Avon in Wiltshire by a runner or nightline'. Large perch of 3–5 lb (1·4–2·3 kg) are occasionally caught by anglers. These fish are carnivorous, feeding on bleak, roach, smaller perch, and other fish.

Numerous brown speckles are scattered on the light olive back and form rows on the tail and two spiny dorsal fins, which are joined. The belly is white. The body is high and narrow and the head blunt with noticeable sensory canals below the large eyes. Ruffe may reach 7 in. (18 cm) in length.

The shield-shaped bone which protects the gills bears concentric rings that indicate the fish's age, like the rings on a tree.

Canals, lakes and slow, lower reaches of rivers, mainly in East Anglia and the Midlands, are where ruffe live.

Spawning shoals move in spring to shallower water to lay clusters of pale yellow eggs, less than $\frac{1}{32}$ in. (1 mm) across, which stick to weeds or stones in slow water.

Ruffe swim in shoals in bottom waters with a sand or gravel bed. They feed by day and lie at the bottom during the night.

Ruffe *Gymnocephalus cernua*

Ruffe soup and fried ruffe are reported to be delicious, but this is not a popular eating fish. This may be because it is fairly small and the rough scales – hence the name – make it difficult to prepare. Also known as the pope, it is quite rare and is found only in a few areas of England, excluding the south-west. Anglers consider the ruffe a nuisance because it will attempt to take almost any bait that is presented when they are angling for other fish. Crustaceans and insect larvae form much of its diet and small fish are included when it grows larger – about 5 in. (12·5 cm) or more.

During spawning in April and May, the sticky eggs may stretch over weeds in lacy, interwoven strands. Females lay up to 200,000 eggs annually, the young emerging in 10–14 days as $\frac{1}{8}$ in. (3 mm) long fry. Growth is fairly slow, most ruffe being only 3½ in. (90 mm) in length after two years, when they reach maturity. Few live longer than six years.

Superficially, the ruffe resembles its close relative the perch, but it is shorter and stouter and its two dorsal fins are joined. It also differs from other members of the perch family in having no scales on its head.

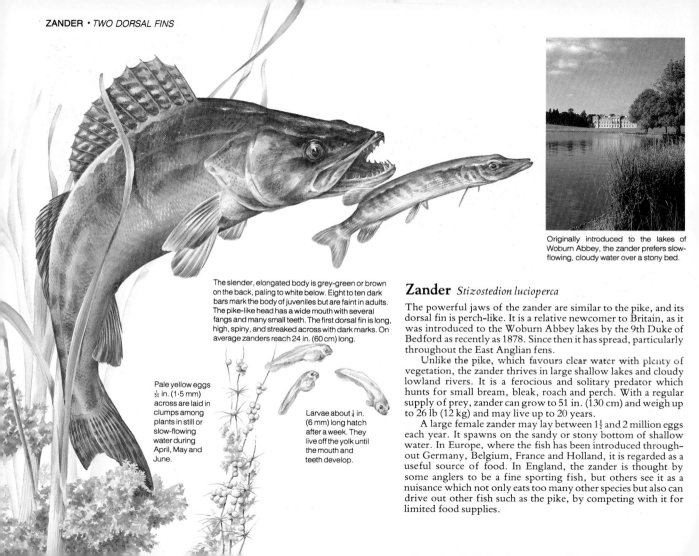

Originally introduced to the lakes of Woburn Abbey, the zander prefers slow-flowing, cloudy water over a stony bed.

The slender, elongated body is grey-green or brown on the back, paling to white below. Eight to ten dark bars mark the body of juveniles but are faint in adults. The pike-like head has a wide mouth with several fangs and many small teeth. The first dorsal fin is long, high, spiny, and streaked across with dark marks. On average zanders reach 24 in. (60 cm) long.

Pale yellow eggs $\frac{1}{16}$ in. (1·5 mm) across are laid in clumps among plants in still or slow-flowing water during April, May and June.

Larvae about $\frac{1}{4}$ in. (6 mm) long hatch after a week. They live off the yolk until the mouth and teeth develop.

Zander *Stizostedion lucioperca*

The powerful jaws of the zander are similar to the pike, and its dorsal fin is perch-like. It is a relative newcomer to Britain, as it was introduced to the Woburn Abbey lakes by the 9th Duke of Bedford as recently as 1878. Since then it has spread, particularly throughout the East Anglian fens.

Unlike the pike, which favours clear water with plenty of vegetation, the zander thrives in large shallow lakes and cloudy lowland rivers. It is a ferocious and solitary predator which hunts for small bream, bleak, roach and perch. With a regular supply of prey, zander can grow to 51 in. (130 cm) and weigh up to 26 lb (12 kg) and may live up to 20 years.

A large female zander may lay between 1½ and 2 million eggs each year. It spawns on the sandy or stony bottom of shallow water. In Europe, where the fish has been introduced throughout Germany, Belgium, France and Holland, it is regarded as a useful source of food. In England, the zander is thought by some anglers to be a fine sporting fish, but others see it as a nuisance which not only eats too many other species but also can drive out other fish such as the pike, by competing with it for limited food supplies.

Flat head

Spines

Thick lips

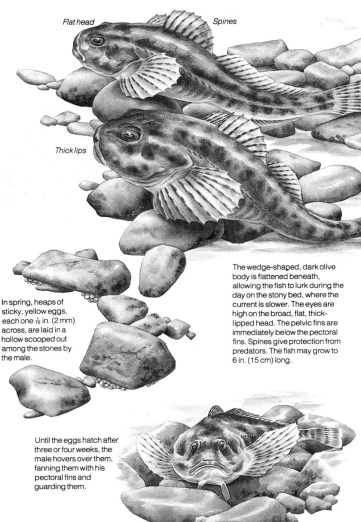

In spring, heaps of sticky, yellow eggs, each one 1/16 in. (2 mm) across, are laid in a hollow scooped out among the stones by the male.

Until the eggs hatch after three or four weeks, the male hovers over them, fanning them with his pectoral fins and guarding them.

The wedge-shaped, dark olive body is flattened beneath, allowing the fish to lurk during the day on the stony bed, where the current is slower. The eyes are high on the broad, flat, thick-lipped head. The pelvic fins are immediately below the pectoral fins. Spines give protection from predators. The fish may grow to 6 in. (15 cm) long.

The stony shallows of lakes, such as Windermere, suit the bullhead, as well as cool, clean, fast streams.

Bullhead *Cottus gobio*

The common name is derived from the broad, heavy head, which also gives rise to the fish's other name, miller's thumb, since millers were supposed to develop broad thumbs from rubbing grain between their fingers. The fish is common throughout England and Wales, but is rare in Scotland and unknown in Ireland. It lives in shallow rivers with stony bottoms, and is also found in the shallows of gravelly lakes and canals, where during the day it lies under stones and in crevices. At night, the bullhead darts out to feed on insects and crustaceans. Sometimes fish eggs are eaten, or even very small fish.

Although relatively small in size – the bullhead is usually no more than 3–4 in. (7·5–10 cm) long – it is well-protected against predators. The dorsal fin is spiny, there are spines associated with the pelvic fins, and backward-curved spines project from either side of the head. These, combined with the very broad head, make it a very difficult fish to swallow.

The female lays up to 250 eggs during March and April. Like the male stickleback, the male bullhead is an attentive parent and guards the eggs until they hatch. However, he then abandons the young, and may even eat them if they swim too close to him.

Young sticklebacks often swim in shoals of many hundreds in shallow water.

In spring the male develops spawning colours – a red throat and belly and bright blue eyes – and performs a zigzag courtship dance.

Trachurus

Leiurus

Semi-armatus

There are bony plates on the sides, which vary in the three different forms of the fish. The leiurus form is the most common.

The dance entices the female, or more than one, to the nest the male has built on the bed.

The male guards the nest and threatens any fish that comes too near by taking up a vertical posture.

Three heavy spines spaced out along the back distinguish this torpedo-shaped fish, which is dark browny-green above and silver below. In sea water, its back may be bluish. Each pelvic fin consists of a long spine and one soft ray. The fish may grow to 4 in. (10 cm) long.

Sticklebacks live in ditches, ponds, lakes and rivers, as well as in estuaries and seashore pools.

Three-spined stickleback *Gasterosteus aculeatus*

The sticklebacks are the smallest of the freshwater fish in the British Isles. Commonest is the three-spined stickleback, which is found in every type of water body from ditches to estuaries. It is one of the hardiest of fish, surviving in waters too polluted for other fish species. Sticklebacks are the only British freshwater fish to build a nest from vegetation and care for the young during their first few weeks of life.

There are three distinct races of three-spined stickleback, each with its own arrangement of scutes, or plates. The leiurus form, the common freshwater stickleback, normally has between two and nine plates. The trachurus, or most heavily plated, form is essentially migratory, spawning in rivers and passing the winter in the sea. The semi-armatus form has a plate arrangement intermediate between the other two and is much less common. It lives in brackish water, especially along the east coast. The temperature and salinity they prefer probably account for the different, but overlapping, distribution of the three forms. None of the sticklebacks is a strong swimmer and all shun fast-flowing water. Their diet is varied and includes small molluscs, crustaceans and larval insects.

The male's zigzagging courtship dance attracts the female to follow and be shown the nest built a few inches off the bed.

The slender body has short spines along the olive-green back. The tiny pelvic fins consist of one spine and one soft ray. In spring the male's throat is black, and eyes and pelvic fins bluish. Usually less than 2 in. (50 mm) long.

The female enters the nest and lays eggs there. The male fertilises them and guards the nest until the young sticklebacks leave it.

Young nine-spined sticklebacks swim in shoals, but the adults tend to be solitary.

Weed-filled estuaries and standing fresh waters are likely habitats but distribution is very localised.

Nine-spined stickleback *Pungitius pungitius*

Much less well known than its three-spined stickleback cousin, this species has in the past been called both the nine-spined and the ten-spined stickleback. The number of dorsal spines is normally nine, although as few as seven and as many as 12 have been recorded. At first sight, this species can be distinguished from the three-spined by its much slimmer, more elongated outline. The colour change in males during the spring spawning is much less dramatic than in the three-spined species, the brown-green throat turning black.

The male is responsible for nest building and care of the young. Normally the nest is tubular and about 1½ in. (40 mm) long. It is made from thread-like algae and willowmoss (*Fontinalis*), and built in the vegetation, near the bottom.

Their small size makes all sticklebacks vulnerable to many predatory fish and birds. The nine-spined species is not well protected by its small spines and therefore spends much time among weeds and other cover, in which it is well camouflaged. It feeds on a variety of small invertebrates – mainly crustaceans. Sticklebacks are relatively short-lived; the life-span of nine-spined and three-spined species is between three and four years.

53

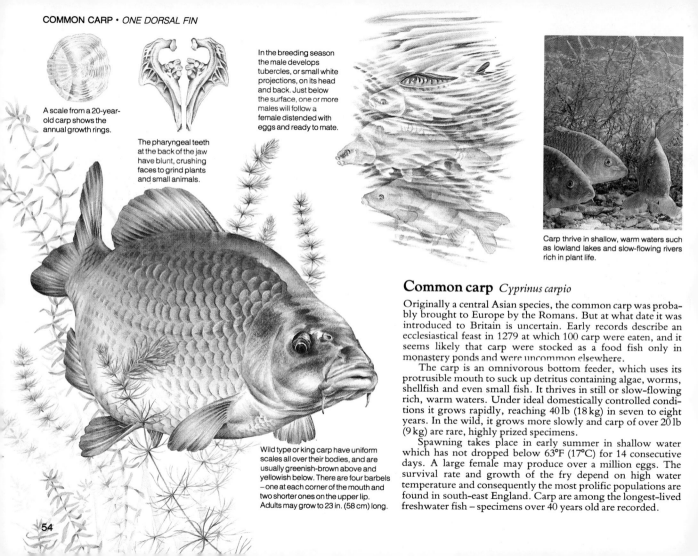

A scale from a 20-year-old carp shows the annual growth rings.

The pharyngeal teeth at the back of the jaw have blunt, crushing faces to grind plants and small animals.

In the breeding season the male develops tubercles, or small white projections, on its head and back. Just below the surface, one or more males will follow a female distended with eggs and ready to mate.

Carp thrive in shallow, warm waters such as lowland lakes and slow-flowing rivers rich in plant life.

Wild type or king carp have uniform scales all over their bodies, and are usually greenish-brown above and yellowish below. There are four barbels – one at each corner of the mouth and two shorter ones on the upper lip. Adults may grow to 23 in. (58 cm) long.

Common carp *Cyprinus carpio*

Originally a central Asian species, the common carp was probably brought to Europe by the Romans. But at what date it was introduced to Britain is uncertain. Early records describe an ecclesiastical feast in 1279 at which 100 carp were eaten, and it seems likely that carp were stocked as a food fish only in monastery ponds and were uncommon elsewhere.

The carp is an omnivorous bottom feeder, which uses its protrusible mouth to suck up detritus containing algae, worms, shellfish and even small fish. It thrives in still or slow-flowing rich, warm waters. Under ideal domestically controlled conditions it grows rapidly, reaching 40 lb (18 kg) in seven to eight years. In the wild, it grows more slowly and carp of over 20 lb (9 kg) are rare, highly prized specimens.

Spawning takes place in early summer in shallow water which has not dropped below 63°F (17°C) for 14 consecutive days. A large female may produce over a million eggs. The survival rate and growth of the fry depend on high water temperature and consequently the most prolific populations are found in south-east England. Carp are among the longest-lived freshwater fish – specimens over 40 years old are recorded.

Common varieties of carp

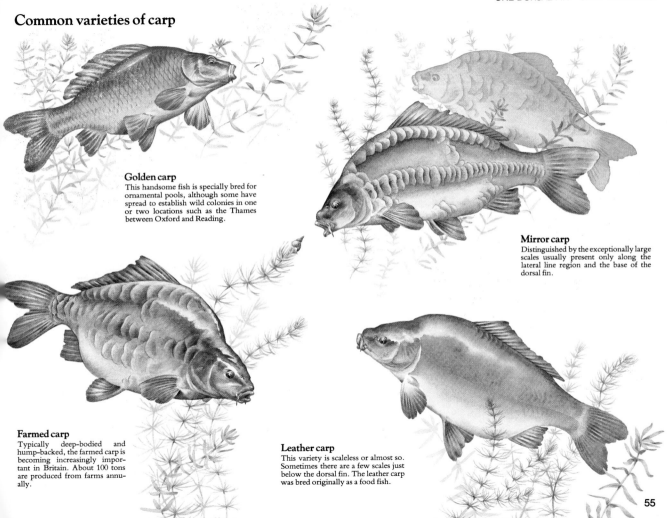

Golden carp
This handsome fish is specially bred for ornamental pools, although some have spread to establish wild colonies in one or two locations such as the Thames between Oxford and Reading.

Mirror carp
Distinguished by the exceptionally large scales usually present only along the lateral line region and the base of the dorsal fin.

Farmed carp
Typically deep-bodied and hump-backed, the farmed carp is becoming increasingly important in Britain. About 100 tons are produced from farms annually.

Leather carp
This variety is scaleless or almost so. Sometimes there are a few scales just below the dorsal fin. The leather carp was bred originally as a food fish.

55

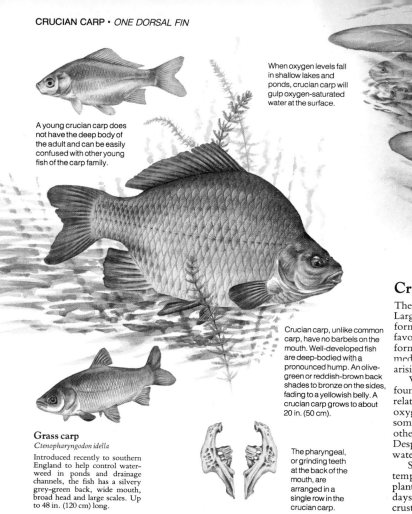

A young crucian carp does not have the deep body of the adult and can be easily confused with other young fish of the carp family.

When oxygen levels fall in shallow lakes and ponds, crucian carp will gulp oxygen-saturated water at the surface.

Small, richly vegetated ponds and gravel pits in southern England are typical habitats of crucian carp.

Crucian carp, unlike common carp, have no barbels on the mouth. Well-developed fish are deep-bodied with a pronounced hump. An olive-green or reddish-brown back shades to bronze on the sides, fading to a yellowish belly. A crucian carp grows to about 20 in. (50 cm).

Grass carp
Ctenopharyngodon idella
Introduced recently to southern England to help control water-weed in ponds and drainage channels, the fish has a silvery grey-green back, wide mouth, broad head and large scales. Up to 48 in. (120 cm) long.

The pharyngeal, or grinding teeth at the back of the mouth, are arranged in a single row in the crucian carp.

Crucian carp *Carassius carassius*

The crucian carp is a very variable species in shape and size. Large lakes where food is abundant often support a deep-bodied form which develops a characteristic humped profile; in less favourable environments it occurs as a relatively slim fish which formerly was thought to be a separate species. Every inter-mediate form is found, as well as occasional sterile hybrids arising from a cross with the common carp.

Very tolerant of unfavourable conditions, crucian carp are found in places where most other fish cannot survive, such as relatively acid waters and ponds that become temporarily de-oxygenated or freeze over in winter. For this reason it is sometimes used to stock angling waters which are unsuitable for other fish, although such crucian carp rarely grow to a good size. Despite its ability to withstand cold it is essentially a warm-water species, limited in Britain to the east and south.

Spawning takes place from May to July, and requires a water temperature of at least 56°F (13°C). The eggs adhere to water plants, and when they hatch the young remain attached for two days, absorbing the yolk sac, before beginning to feed on small crustaceans and, when older, plants and insect larvae.

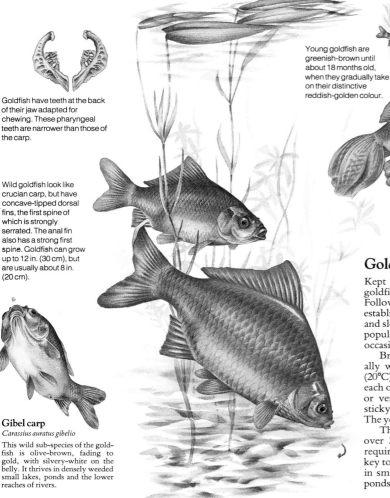

Goldfish have teeth at the back of their jaw adapted for chewing. These pharyngeal teeth are narrower than those of the carp.

Wild goldfish look like crucian carp, but have concave-tipped dorsal fins, the first spine of which is strongly serrated. The anal fin also has a strong first spine. Goldfish can grow up to 12 in. (30 cm), but are usually about 8 in. (20 cm).

Gibel carp
Carassius auratus gibelio
This wild sub-species of the goldfish is olive-brown, fading to gold, with silvery-white on the belly. It thrives in densely weeded small lakes, ponds and the lower reaches of rivers.

Young goldfish are greenish-brown until about 18 months old, when they gradually take on their distinctive reddish-golden colour.

Many varieties of goldfish, such as the veil tail, have been bred in captivity; released into the wild, few survive for long.

Normally seen in ornamental ponds, released goldfish now thrive in warm waters throughout Britain.

Goldfish   *Carassius auratus*

Kept as a domestic species in China for over 1,200 years, the goldfish was introduced to western Europe in the 17th century. Following escapes and random introductions it has become established in the wild, and viable populations now exist in lakes and slow-flowing rivers in Britain and continental Europe. Such populations tend to revert to olive-green or brown, and only occasionally produce a traditionally coloured specimen.

Breeding takes place in June and July, but only in exceptionally warm seasons when the water temperature exceeds 68°F (20°C) for some days. Several hundred thousand yellow eggs, each one $\frac{1}{20}$ in. (1·5 mm) in diameter, are laid by the female in still or very slow-flowing, densely weeded waters; the eggs are sticky and adhere to the vegetation, hatching in about a week. The young are very similar in appearance to crucian carp young.

They are long-lived fish, individuals having survived for over 30 years in captivity; in view of their special breeding requirements in European waters this longevity is probably the key to their survival in the wild. The fish is also capable of living in small, oxygen-deficient waters, making it ideal for garden ponds and aquariums.

57

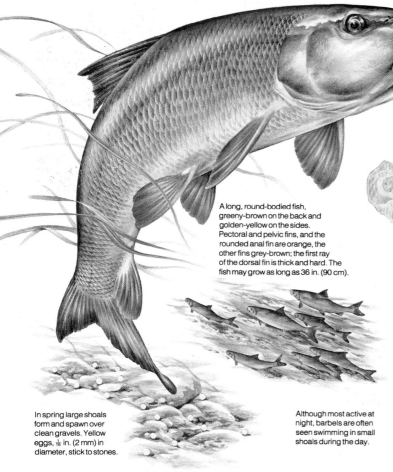

The upper jaw juts beyond the lower and the mouth has fleshy lips. Four barbels, or fleshy filaments, on the upper jaw, are well supplied with taste cells.

Small, smooth-edged oval scales are set deep in the skin.

Barbels live in the middle reaches of clear rivers with sandy or stony beds such as the Thames and its tributaries.

A long, round-bodied fish, greeny-brown on the back and golden-yellow on the sides. Pectoral and pelvic fins, and the rounded anal fin are orange, the other fins grey-brown; the first ray of the dorsal fin is thick and hard. The fish may grow as long as 36 in. (90 cm).

In spring large shoals form and spawn over clean gravels. Yellow eggs, 1/16 in. (2 mm) in diameter, stick to stones.

Although most active at night, barbels are often seen swimming in small shoals during the day.

Barbel *Barbus barbus*

The only British fish in the world-wide genus *Barbus*, which includes more than 200 species, is the barbel. Originally it was found only in the Thames and Trent river systems. It has now been introduced more widely for anglers, who know it as a fighting fish, but it is still absent from Ireland. The Thames and its tributaries, the Kennet in Berkshire, and the Lea in Hertfordshire are the best-known barbel rivers. Several hundred pounds of fish – some weighing up to 10 lb (4·5 kg) – have been caught in them in a few hours. The barbel is not now considered edible as its eggs are said to be poisonous.

Barbel populations vary in colour from almost yellow to brown-black, depending on the colour of the bed and water where they live. With its barbels and low mouth, the fish is well suited to bottom-feeding, sometimes uprooting plants while searching for caddis larvae, molluscs and other food.

Before spawning, between May and July, there may be an upstream migration as the fish search for suitable gravels. It is at this time that large shoals are seen, the males often with spawning tubercles on the head. After spawning, groups of exhausted fish can be seen resting in still water under the banks.

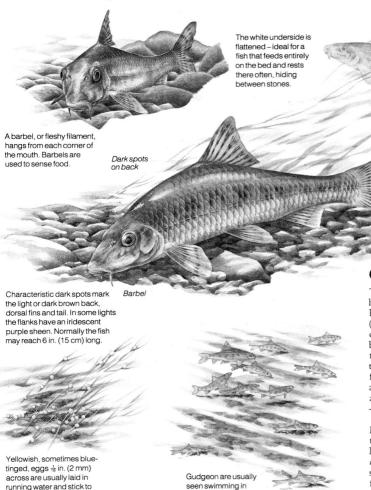

The white underside is flattened – ideal for a fish that feeds entirely on the bed and rests there often, hiding between stones.

A barbel, or fleshy filament, hangs from each corner of the mouth. Barbels are used to sense food.

Dark spots on back

Characteristic dark spots mark the light or dark brown back, dorsal fins and tail. In some lights the flanks have an iridescent purple sheen. Normally the fish may reach 6 in. (15 cm) long.

Barbel

Yellowish, sometimes blue-tinged, eggs ⅛ in. (2 mm) across are usually laid in running water and stick to stones and weeds.

Gudgeon are usually seen swimming in close-packed shoals.

Gudgeon live in rich rivers, lakes, canals and ponds, but grow especially large in some gravel-pit lakes.

Gudgeon *Gobio gobio*

The gravel-pit lakes that are now a feature of many river valleys have proved an ideal habitat for gudgeon and support considerable numbers. It is there that the fish are largest, often about 8 in. (20 cm) long and 4 oz (115 g) in weight. Despite the scanty flesh on this slender member of the carp family, it has historically been considered a culinary delicacy; in the 19th century, invalids taking the waters at Bath were given gudgeon to eat because they were 'easy of digestion'. They were plentiful enough for fishermen in southern England to catch them in cast-nets. They are found now in England and Ireland, but not in most of Wales and Scotland. In central Europe they are still caught for the table – in such numbers in places that the excess is fed to pigs.

Spawning occurs over a prolonged period, but is heaviest in May. The eggs, large for such a small fish, vary from off-white to deep yellow, and some are tinged with blue. A female usually lays between 1,000 and 3,000 each year. They hatch in about ten days, and the young live in shoals at the bottom near the spawning place. The fish mature at two or three years, and live for four or six. They feed entirely on the river-bed, eating tiny creatures, some plants and, occasionally, fish eggs.

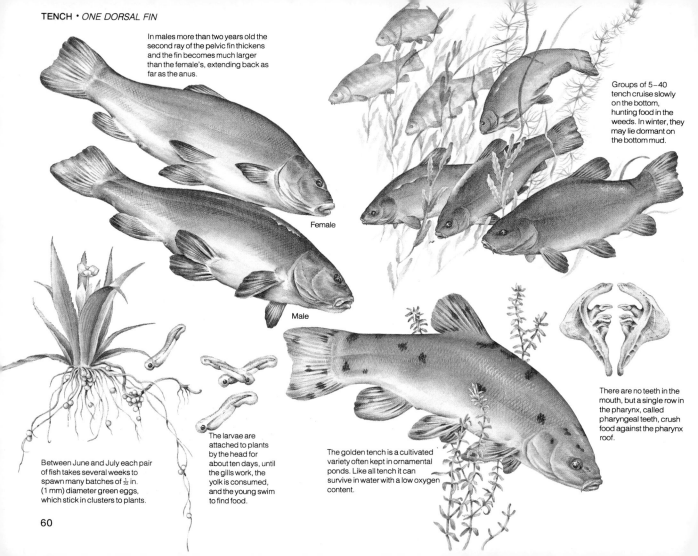

In males more than two years old the second ray of the pelvic fin thickens and the fin becomes much larger than the female's, extending back as far as the anus.

Groups of 5–40 tench cruise slowly on the bottom, hunting food in the weeds. In winter, they may lie dormant on the bottom mud.

Female

Male

Between June and July each pair of fish takes several weeks to spawn many batches of $\frac{1}{32}$ in. (1 mm) diameter green eggs, which stick in clusters to plants.

The larvae are attached to plants by the head for about ten days, until the gills work, the yolk is consumed, and the young swim to find food.

The golden tench is a cultivated variety often kept in ornamental ponds. Like all tench it can survive in water with a low oxygen content.

There are no teeth in the mouth, but a single row in the pharynx, called pharyngeal teeth, crush food against the pharynx roof.

The fish lives in shallow lakes, gravel pits and slow rivers where there is dense plant growth on the soft, muddy bed.

Thickset and deep-bodied, the tench is dark bronze-green or brown above with a yellow sheen below. The eye is red. All the fins are large and rounded. The deep-based tail fin shows little forking. The scales are small and well embedded in thick, smooth, slimy skin. Very large specimens may grow to 28 in. (70 cm) long.

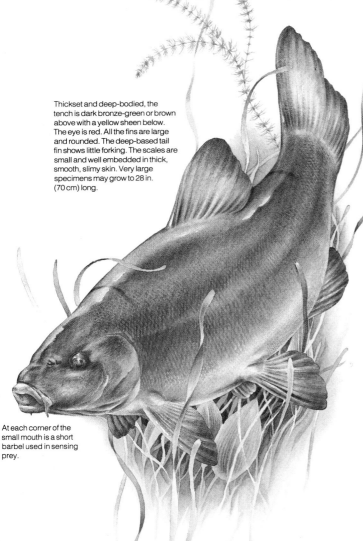

At each corner of the small mouth is a short barbel used in sensing prey.

Tench *Tinca tinca*

The unmistakably coloured tench, probably introduced into Britain from Europe hundreds of years ago, has a patchy spread here, and is found mainly in the south-east. Fish 24 in. (60 cm) long and weighing 4 lb (1·8 kg) are not uncommon and anglers occasionally catch fish twice this size. Tench are raised commercially in mainland Europe and on a few fish farms in southern England, chiefly to restock fisheries.

Males mature at three years old, females at four. Of all the carp family tench are latest in their spawning time, July and early August being normal. Before they will spawn, water temperature must be at least 64°F (18°C) for two weeks. For this reason, in the British climate, they do not spawn every year in all lakes, nor will some deep or heavily shaded waters support a self-sustaining population. Large females may lay up to 500,000 eggs, but lakes seldom become overpopulated because enormous numbers of tench die in their first year.

'Doctor fish' and 'physician of fishes' are titles given to the tench because it was claimed that wounded fish rub against it, presumably to aid healing. In medieval days headache, toothache, jaundice and other ailments were treated by applying tench slime.

61

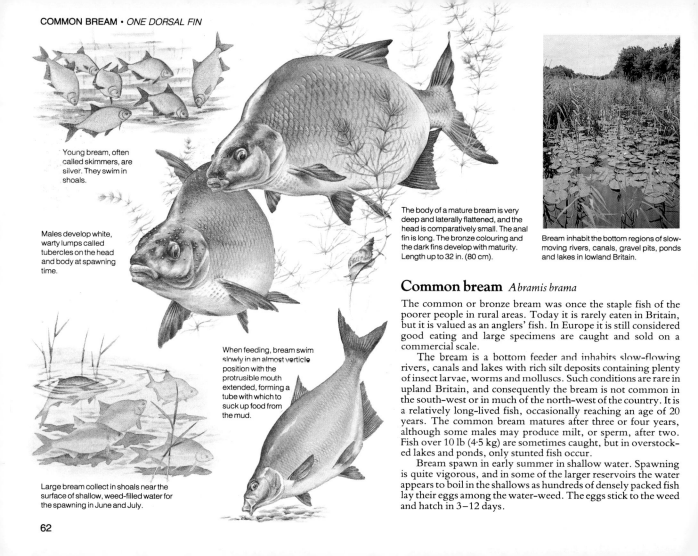

Young bream, often called skimmers, are silver. They swim in shoals.

Males develop white, warty lumps called tubercles on the head and body at spawning time.

When feeding, bream swim slowly in an almost verticle position with the protrusible mouth extended, forming a tube with which to suck up food from the mud.

Large bream collect in shoals near the surface of shallow, weed-filled water for the spawning in June and July.

The body of a mature bream is very deep and laterally flattened, and the head is comparatively small. The anal fin is long. The bronze colouring and the dark fins develop with maturity. Length up to 32 in. (80 cm).

Bream inhabit the bottom regions of slow-moving rivers, canals, gravel pits, ponds and lakes in lowland Britain.

Common bream *Abramis brama*

The common or bronze bream was once the staple fish of the poorer people in rural areas. Today it is rarely eaten in Britain, but it is valued as an anglers' fish. In Europe it is still considered good eating and large specimens are caught and sold on a commercial scale.

The bream is a bottom feeder and inhabits slow-flowing rivers, canals and lakes with rich silt deposits containing plenty of insect larvae, worms and molluscs. Such conditions are rare in upland Britain, and consequently the bream is not common in the south-west or in much of the north-west of the country. It is a relatively long-lived fish, occasionally reaching an age of 20 years. The common bream matures after three or four years, although some males may produce milt, or sperm, after two. Fish over 10 lb (4·5 kg) are sometimes caught, but in overstocked lakes and ponds, only stunted fish occur.

Bream spawn in early summer in shallow water. Spawning is quite vigorous, and in some of the larger reservoirs the water appears to boil in the shallows as hundreds of densely packed fish lay their eggs among the water-weed. The eggs stick to the weed and hatch in 3–12 days.

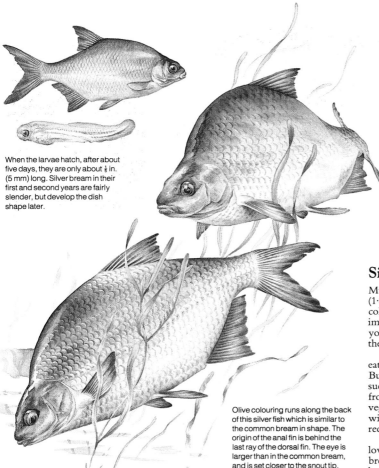

When the larvae hatch, after about five days, they are only about ⅕ in. (5 mm) long. Silver bream in their first and second years are fairly slender, but develop the dish shape later.

The light yellow eggs are laid in shallow, weed-filled water and stick to the plants.

The silver bream inhabits lowland lakes, slow-flowing rivers and canals in southeast England and the Midlands.

Olive colouring runs along the back of this silver fish which is similar to the common bream in shape. The origin of the anal fin is behind the last ray of the dorsal fin. The eye is larger than in the common bream, and is set closer to the snout tip. There is a double row of teeth in the pharynx. Silver bream grow to 14 in. (36 cm) long.

Silver bream *Blicca bjoerkna*

Much smaller than the common bream, rarely exceeding 4 lb (1·8 kg) in weight, the silver bream derives its name from its colour, which never changes from the silvery-white found in immature fish of both species. It can be distinguished from the young common bream by the larger eye and shorter anal fin, and the reddish coloration of the pectoral and pelvic fins.

It is similar in its feeding habits to the common bream and eats insect larvae, small plants, snails, crustaceans and worms. But it lacks the protrusible mouth which makes the larger fish such an efficient bottom feeder, and spends some time away from the river or lake bed feeding in mid-water among the vegetation. Small silver bream are often seen in mixed shoals with roach, rudd and dace; most large specimens have been recorded in slow-running water rather than in lakes.

Communal spawning occurs in May and June, in the shallow, vegetated waters which are also favoured by the common bream. The similarity in the life histories and habits of the two bream has led to speculation about how the silver bream is able to co-exist with the larger common bream in some waters. It is not a good sporting fish, and is discouraged in fisheries.

63

The male's belly and paired fins become red-flushed prior to spawning time, while the back darkens and white lumps called tubercles develop on the head. Minnows gather in large shoals for spawning.

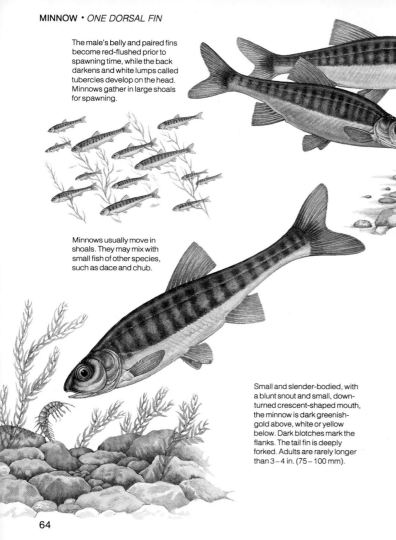

Minnows usually move in shoals. They may mix with small fish of other species, such as dace and chub.

Small and slender-bodied, with a blunt snout and small, down-turned crescent-shaped mouth, the minnow is dark greenish-gold above, white or yellow below. Dark blotches mark the flanks. The tail fin is deeply forked. Adults are rarely longer than 3–4 in. (75–100 mm).

Females produce up to 1,000 yellow eggs, 1/32 in. (1 mm) across. They are laid among the stones of the river bed in small groups.

Minnows are found in the upper reaches of rivers where clean, cool shallows run over stony bottoms, and in stony lakes.

Minnow *Phoxinus phoxinus*

The minnow belongs to the cyprinid, or carp family and is the smallest member of that family in Britain. It is a pretty, darting fish, often seen in shoals of up to 100 strong. Its size makes it easy prey for all the carnivorous fish, kingfishers, dippers, and even carnivorous water beetles. Minnows rely on their speed and close shoaling habits to save them from predators. During the summer months they often swim in mixed shoals with small bleak and roach and other fish from which it is difficult to distinguish them. They feed mainly on freshwater insects and shrimps, but also eat plants and algae.

Anglers no longer fish for minnows, except to use as bait. In the past, however, people living in the countryside caught them for eating, sometimes in an egg dish – Izaak Walton in his treatise *The Compleat Angler* (1655) advised that 'a minnow-tansy is a dainty dish of meat'. Other accounts tell of 7 gallons of minnows being served at a banquet given in 1394 by William of Wykeham, Bishop of Winchester and Richard II's chancellor.

The presence of minnows indicates well-oxygenated, clean water, so their reappearance in the lower Thames in the 1970s reflected the marked improvement in pollution control.

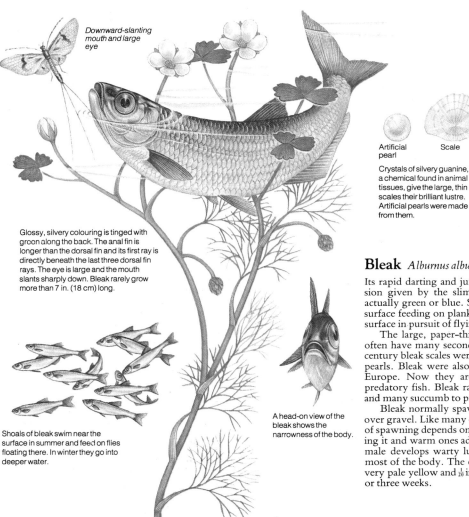

Downward-slanting mouth and large eye

Glossy, silvery colouring is tinged with green along the back. The anal fin is longer than the dorsal fin and its first ray is directly beneath the last three dorsal fin rays. The eye is large and the mouth slants sharply down. Bleak rarely grow more than 7 in. (18 cm) long.

Shoals of bleak swim near the surface in summer and feed on flies floating there. In winter they go into deeper water.

A head-on view of the bleak shows the narrowness of the body.

Artificial pearl

Scale

Crystals of silvery guanine, a chemical found in animal tissues, give the large, thin scales their brilliant lustre. Artificial pearls were made from them.

The fish occurs in slow-flowing lowland rivers and in some lakes in England and in parts of Wales.

Bleak *Alburnus alburnus*

Its rapid darting and jumping enhance the quicksilver impression given by the slim, delicate bleak, although its back is actually green or blue. Shoals swim within a foot or two of the surface feeding on plankton, and individual fish often break the surface in pursuit of flying insects.

The large, paper-thin scales come off easily and older fish often have many secondary, or replacement scales. In the 19th century bleak scales were used commercially in making artificial pearls. Bleak were also caught for eating in most of western Europe. Now they are used mainly by anglers as bait for predatory fish. Bleak rarely live longer than six or seven years and many succumb to pike or perch.

Bleak normally spawn between April and June in shallows over gravel. Like many other species of the carp family, the time of spawning depends on water temperature, cold springs delaying it and warm ones advancing it. Around spawning time, the male develops warty lumps called tubercles, sometimes over most of the body. The eggs, laid singly or in small groups, are very pale yellow and $\frac{1}{20}$ in. (1·5 mm) across; they hatch after two or three weeks.

65

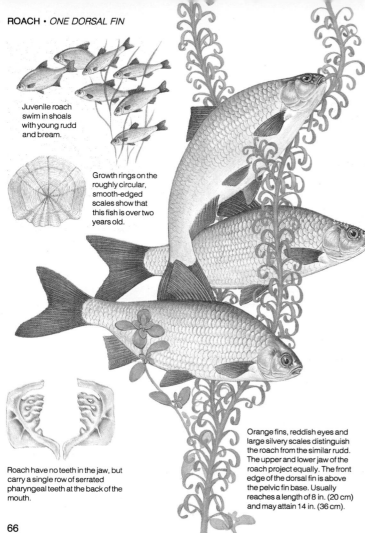

Juvenile roach swim in shoals with young rudd and bream.

Growth rings on the roughly circular, smooth-edged scales show that this fish is over two years old.

Roach have no teeth in the jaw, but carry a single row of serrated pharyngeal teeth at the back of the mouth.

Orange fins, reddish eyes and large silvery scales distinguish the roach from the similar rudd. The upper and lower jaw of the roach project equally. The front edge of the dorsal fin is above the pelvic fin base. Usually reaches a length of 8 in. (20 cm) and may attain 14 in. (36 cm).

Yellowish eggs, ½ in. (1 mm) across, are laid in shallow water in May and June. They stick to plants and stones.

Grey-white tubercles – small, hard, rounded projections – may form on the adult male's head in spring.

Man-made waters, like the Kennet and Avon Canal, have extended the habitats available to the adaptable roach.

Roach *Rutilus rutilus*

The coarse fish caught most in Great Britain is the highly adaptable roach, which inhabits most fresh waters except the fastest flowing streams. The extensive lowland stretches of the Thames, Severn and Trent support numerous roach. They are often the main food of pike, perch and zander.

In different environments the roach varies from silver to nearly black. Its body depth also varies, from slim to very deep. Roach weighing more than 3 lb (1·4 kg) are rare and mainly from reservoirs and gravel pits in the south of England and the Midlands. Growth is seasonal and regular, so that it is one of the easiest fish to age by the pattern of growth rings on its scales.

Spawning usually occurs in late May, but is delayed by a cold spring and encouraged by a warm one. Like many other males of the carp family, the male roach develops warty lumps called tubercles on the head and sometimes on the body before spawning. Each is the size of a pinhead. They may develop as early as January, but disappear by July. Tubercles form patterns on the head and above the eyes. When they occur on the body, there is only one on each scale. Only when it bears tubercles is a male roach distinguishable from a female.

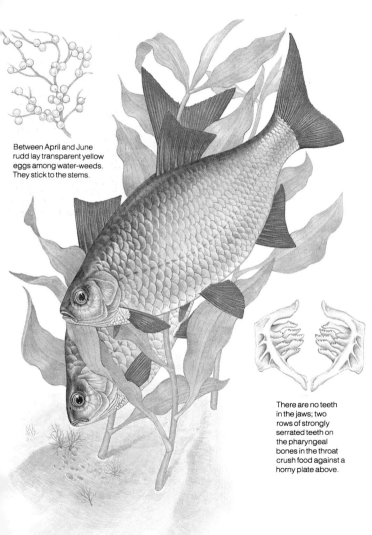

Between April and June rudd lay transparent yellow eggs among water-weeds. They stick to the stems.

There are no teeth in the jaws; two rows of strongly serrated teeth on the pharyngeal bones in the throat crush food against a horny plate above.

All the fins are reddish, with those on the underside bright red. The dorsal fin starts well behind the pelvic fin base. The mouth slants steeply down in the small head and the eye is orange-yellow. The deep body, dark olive on the back, bronze-yellow below, has a sharp ridge between the pelvic and anal fins. Adults usually reach a length of about 12 in. (30 cm).

Rudd thrive among abundant plant growth in the slow-flowing and still waters of lakes, backwaters and canals.

Rudd *Scardinius erythrophthalmus*

Izaak Walton referred to the rudd as 'a kind of bastard roach', a fish with which it is often confused. It is differentiated by its protruding lower jaw, redder fins, greener and deeper body, and the position of the dorsal fin, which is nearer the tail. Like the roach, the rudd is a popular fish among anglers as it takes a bait readily. Unfortunately, the red-eye, as it is sometimes known, is also just as poor as a culinary fish.

The rudd is found throughout lowland England and in Ireland but it is absent from Scotland and Wales. In waters shared by roach and rudd, roach are normally more frequent, except in ponds and lakes in soft-water areas where rudd are often the dominant fish, though their growth is stunted.

Rudd feed off small insects, crustaceans and plant material at middle depths or on the surface of the water. They have great difficulty in feeding off the bottom because of the position of their mouths. Larger specimens, of over 2 lb (1 kg), may take small fish. Rudd are normally slower growing than roach; in some waters they may take four years to reach maturity, when they will be about 5 in. (12·5 cm) long. Where conditions are good, they are commonly 8–12 in. (20–30 cm) long.

The omnivorous chub has a double row of curved teeth in the back of the throat to crush its food (minnows and other small fish, insects and water plants) against a hard plate at the top of the pharynx.

Young and medium-sized chub are usually seen in small, open shoals. Large fish are sometimes solitary.

The head is slightly broader than that of the dace, and the mouth is wide. The eye is large and golden.

Strong-flowing, middle reaches of large rivers in England and southern Scotland are typical chub habitats.

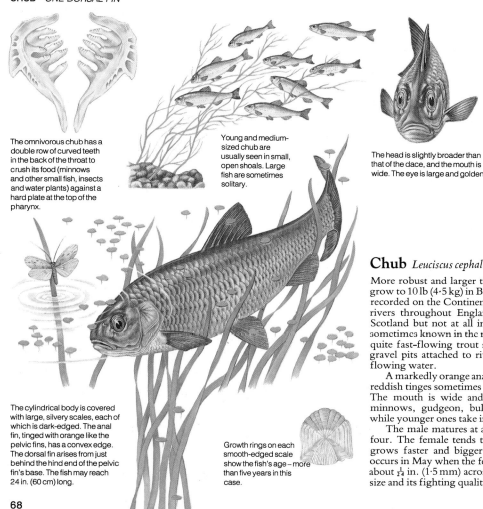

The cylindrical body is covered with large, silvery scales, each of which is dark-edged. The anal fin, tinged with orange like the pelvic fins, has a convex edge. The dorsal fin arises from just behind the hind end of the pelvic fin's base. The fish may reach 24 in. (60 cm) long.

Growth rings on each smooth-edged scale show the fish's age – more than five years in this case.

Chub *Leuciscus cephalus*

More robust and larger than the roach or dace, the chub may grow to 10 lb (4·5 kg) in Britain, while 18 lb (8 kg) fish have been recorded on the Continent. It is found mainly in slow-flowing rivers throughout England, parts of Wales and in southern Scotland but not at all in Ireland. The chub, or skelly as it is sometimes known in the north of England, can also be found in quite fast-flowing trout streams and, more rarely, in lakes or gravel pits attached to rivers. However, it will breed only in flowing water.

A markedly orange anal fin is characteristic of the species and reddish tinges sometimes appear on the pectoral and pelvic fins. The mouth is wide and larger chub eat small fish such as minnows, gudgeon, bullheads, and young dace and roach, while younger ones take insects, fish eggs and worms.

The male matures at about three years, the female at about four. The female tends to live longer – up to 12 years – and grows faster and bigger than the male. Spawning normally occurs in May when the female will lay up to 100,000 eggs, each about $\frac{1}{14}$ in. (1·5 mm) across. The fish is prized by anglers for its size and its fighting qualities but it makes poor eating.

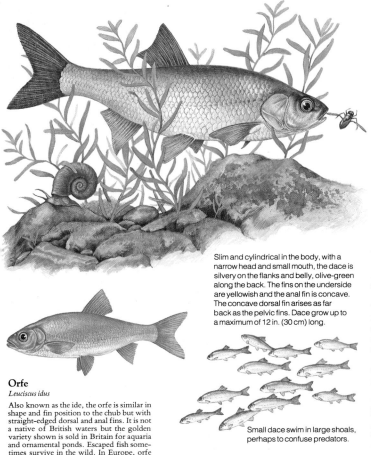

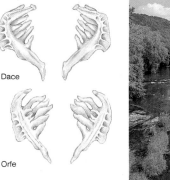

Dace

Orfe

Dace and orfe both have a double row of pharyngeal teeth. The orfe's are longer.

Middle reaches of clean, fairly fast rivers are the dace's commonest habitat. It occurs occasionally in gravel pits.

Slim and cylindrical in the body, with a narrow head and small mouth, the dace is silvery on the flanks and belly, olive-green along the back. The fins on the underside are yellowish and the anal fin is concave. The concave dorsal fin arises as far back as the pelvic fins. Dace grow up to a maximum of 12 in. (30 cm) long.

Orfe
Leuciscus idus

Also known as the ide, the orfe is similar in shape and fin position to the chub but with straight-edged dorsal and anal fins. It is not a native of British waters but the golden variety shown is sold in Britain for aquaria and ornamental ponds. Escaped fish sometimes survive in the wild. In Europe, orfe inhabit lowland lakes, large rivers and brackish estuaries, where they grow to a maximum length of 3 ft (1 m).

Small dace swim in large shoals, perhaps to confuse predators.

Dace *Leuciscus leuciscus*

Although found over most of Europe including England, the dare or dart, as the dace is sometimes known, is absent from Scotland, most of Ireland and large parts of Wales and Cornwall. It is similar to the roach and chub but can be identified by its yellow eyes which have no red flecks and its yellow or off-white anal fin with a concave edge. It is a slim-bodied species, preferring fairly fast rivers with moderate currents; less commonly, it is found in gravel pits. The male is sexually mature after about one year, the female after two years. The dace rarely grows longer than 10 in. (25 cm) or lives more than six years.

The spawning season is from February to May. During this period the male develops small swellings, called tubercles, on its head. The female lays up to 27,000 pale orange eggs among the gravel on the stream bed; they take up to 25 days to hatch. The dace feeds on larval and adult flying insects, as well as crustaceans and some vegetation.

The dace is a nuisance to anglers as it may compete with young salmon and trout for food and will also take those lures designed to catch the adult forms of these game fish.

69

Male

Female

The male is slimmer than
the female and has
slightly larger, more
pointed pectoral fins.

Yellow eggs, less than
½ in. (1 mm) across, stick
to stones and weeds.

Because it is a night feeder the
stone loach relies on its barbels
to detect prey. The flattened
underside is suited to a fish that
lies on the bottom.

The fish is common in stony-bottomed,
clean rivers throughout most of England,
Wales and Ireland.

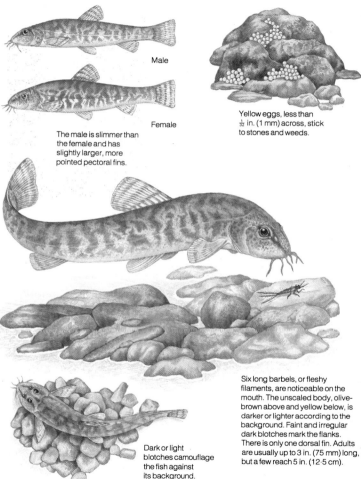

Six long barbels, or fleshy
filaments, are noticeable on the
mouth. The unscaled body, olive-
brown above and yellow below, is
darker or lighter according to the
background. Faint and irregular
dark blotches mark the flanks.
There is only one dorsal fin. Adults
are usually up to 3 in. (75 mm) long,
but a few reach 5 in. (12·5 cm).

Dark or light
blotches camouflage
the fish against
its background.

Stone loach *Noemacheilus barbatulus*

Stone loach spend the daylight hours hiding under and between
small rocks in stony rivers, but they can also be found on sandy
or muddy bottoms. On rare occasions they have been found in
lakes. At night they emerge to feed on bottom-dwelling invertebrates such as midge larvae, small crustaceans like ostracods
and copepods, and shrimps. They are clean-water fish and
disappear quickly from even mildly polluted streams. They are
common in most of Britain except the north of Scotland and the
extreme south-west of England.

The six barbels round its mouth are the source of the name
'beardie' which is sometimes given to the stone loach. It is fairly
short-lived, five years being the maximum age except in rare
cases. Assessing the age is difficult as the fish has no scales with
their informative growth rings. Eggs, laid between April and
June among stones or water-weeds, hatch after about 14 days,
depending on temperature. Growth to 3 in. (75 mm) is rapid and
fish are usually mature at a year old.

Apart from being used as bait to catch pike, perch and trout,
stone loach are now of little interest to anglers, but formerly
they were prized as a delicate dish.

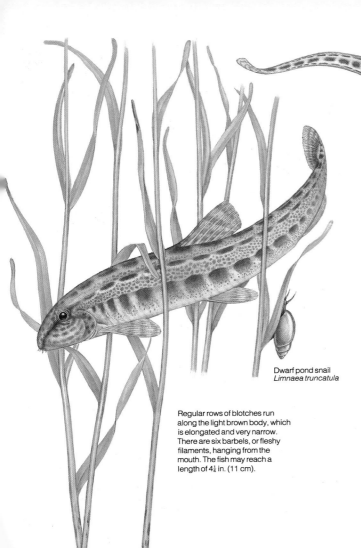

In a pocket under each eye is a retractable two-pointed spine, from which the fish is named.

The upper jaw juts beyond the lower. The six barbels are short and all of the same length.

The fish, found in a few areas of the east Midlands, is becoming popular as a hardy and attractive aquarium fish.

Dwarf pond snail
Limnaea truncatula

Regular rows of blotches run along the light brown body, which is elongated and very narrow. There are six barbels, or fleshy filaments, hanging from the mouth. The fish may reach a length of 4¼ in. (11 cm).

Spined loach *Cobitis taenia*

Because of its habit of burying itself in the sandy or muddy river-bed with only its head exposed, the spined loach was formerly known as the groundling. It also hides in the reeds and weeds of the densely vegetated habitats it prefers. Unlike the stone loach, this fish is able to live in water that has relatively little oxygen, so it occurs in slow-flowing rivers, canals, stagnant ponds and ditches, and lakes or reservoirs connected to slow rivers. Where oxygen is very low, it rises to the surface to gulp in air, from which oxygen is absorbed in the gut. Little scientific study has been made of the spined loach. It is thought to be similar to the stone loach in its eating and breeding, and to be active at night or where the light is poor. The spines under the eyes may be for defence against predators such as the pike and perch.

This loach is rare and found only in the east Midlands and East Anglia. One fish expert, Tate Regan, dismissed it in 1911 as having little value as food and 'no value for any other purposes, so that its scarcity in our waters need not concern us greatly'. But now it is gaining some popularity as a hardy and attractive, but not very active, aquarium fish.

71

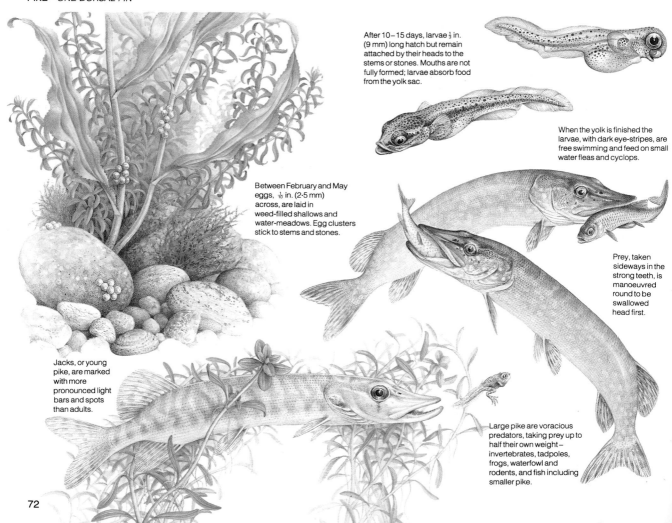

After 10–15 days, larvae ⅓ in. (9 mm) long hatch but remain attached by their heads to the stems or stones. Mouths are not fully formed; larvae absorb food from the yolk sac.

When the yolk is finished the larvae, with dark eye-stripes, are free swimming and feed on small water fleas and cyclops.

Between February and May eggs, 1/10 in. (2·5 mm) across, are laid in weed-filled shallows and water-meadows. Egg clusters stick to stems and stones.

Prey, taken sideways in the strong teeth, is manoeuvred round to be swallowed head first.

Jacks, or young pike, are marked with more pronounced light bars and spots than adults.

Large pike are voracious predators, taking prey up to half their own weight – invertebrates, tadpoles, frogs, waterfowl and rodents, and fish including smaller pike.

Teeth in roof of mouth and lower jaw.

Golden-green bands and spots mark the greenish-brown, long body of the pike. Below, it is yellow. The dorsal fin is as far back as the anal fin; both are near the tail. The head is broad, the snout pointed. Adults may reach 42 in. (107 cm) long.

Pike live in slow-flowing rivers, and in canals, reservoirs, lakes and ponds all over Britain. They lurk motionless among the weeds waiting for prey.

Pike *Esox lucius*

Its good camouflage and large size both contribute to the pike's success in different habitats. Even quite small ponds will hold pike as long as there is some weedy corner for shelter, but they grow large only where food is plentiful. A monster of 72 lb (32·7 kg) was caught in Loch Ken, Kirkcudbright, in the 18th century.

Although recognised as beneficial in many coarse fisheries, the pike has always been regarded as a pest in salmon and trout rivers. It is well known that pike eat most fish species, but water rats, water birds and frogs are taken when available. Several accounts in Europe blame pike for attacks on people, dogs and mules, but these are likely to have been exaggerated.

Spawning usually takes place in quiet shallows. Smaller fish lay their eggs first, the larger ones not being ripe until later. Between 10,000 and 20,000 eggs are produced for each pound of the female's weight, so a large one may lay 480,000 eggs. Cannibalism and competition for food take their toll of young fish, but once a pike has attained several pounds in weight, it has few enemies except man, and may live many years. Adults 17 years old have been caught in the Thames in recent years.

73

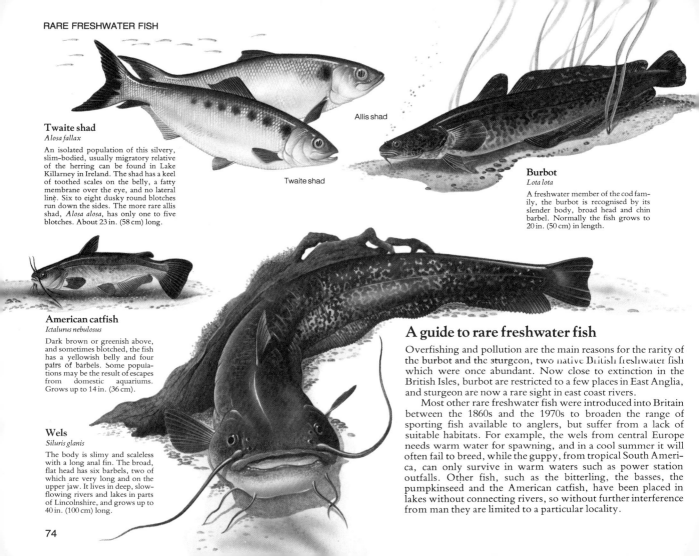

Twaite shad
Alosa fallax

An isolated population of this silvery, slim-bodied, usually migratory relative of the herring can be found in Lake Killarney in Ireland. The shad has a keel of toothed scales on the belly, a fatty membrane over the eye, and no lateral line. Six to eight dusky round blotches run down the sides. The more rare allis shad, *Alosa alosa*, has only one to five blotches. About 23 in. (58 cm) long.

Allis shad

Twaite shad

Burbot
Lota lota

A freshwater member of the cod family, the burbot is recognised by its slender body, broad head and chin barbel. Normally the fish grows to 20 in. (50 cm) in length.

American catfish
Ictalurus nebulosus

Dark brown or greenish above, and sometimes blotched, the fish has a yellowish belly and four pairs of barbels. Some populations may be the result of escapes from domestic aquariums. Grows up to 14 in. (36 cm).

Wels
Siluris glanis

The body is slimy and scaleless with a long anal fin. The broad, flat head has six barbels, two of which are very long and on the upper jaw. It lives in deep, slow-flowing rivers and lakes in parts of Lincolnshire, and grows up to 40 in. (100 cm) long.

A guide to rare freshwater fish

Overfishing and pollution are the main reasons for the rarity of the burbot and the sturgeon, two native British freshwater fish which were once abundant. Now close to extinction in the British Isles, burbot are restricted to a few places in East Anglia, and sturgeon are now a rare sight in east coast rivers.

Most other rare freshwater fish were introduced into Britain between the 1860s and the 1970s to broaden the range of sporting fish available to anglers, but suffer from a lack of suitable habitats. For example, the wels from central Europe needs warm water for spawning, and in a cool summer it will often fail to breed, while the guppy, from tropical South America, can only survive in warm waters such as power station outfalls. Other fish, such as the bitterling, the basses, the pumpkinseed and the American catfish, have been placed in lakes without connecting rivers, so without further interference from man they are limited to a particular locality.

Rock bass
Ambloplites rupestris

The dull, greenish-brown body is deep and has lines of slightly darker colour running along it. There is a dark spot on the gill cover. The fish grows up to 12 in. (30 cm) long, and is only found locally in the south.

Bitterling
Rhodeus sericeus

The lateral line is very short and there is a blue-green stripe from the middle of each flank to the base of the tail fin. The female grows a long tube, which it uses to deposit the eggs inside a mussel for protection. Found in only a few lakes in north-west England, the fish grows to 3½ in. (90 mm).

Pumpkinseed
Lepomis gibbosus

This fish is confined to a few lakes in south-east England. The back is blue and green, the flanks are spotted with orange or gold, and there is a black patch on the gill cover. It grows to 10 in. (25 cm).

Wild form

Guppy
Poecilia reticulata

Unlike the brightly coloured aquarium varieties from which it has probably derived, the wild guppy is olive-green to brown. This small fish needs warm water and has become established in some warm effluents in canals and rivers. It may grow 2 in. (50 mm) long.

Aquarium variety

Largemouth bass
Micropterus salmoides

Of the same family as the pumpkinseed and rock bass, this species and its relative the smallmouth bass (*Micropterus dolomieui*) are slimmer, longer and bronze-green in colour. It is found in Dorset and usually grows to about 21 in. (53 cm).

Sturgeon
Acipenser sturio

This very rare fish has four barbels, huge bony plates instead of scales, and an upturned tail. Females may grow to 11½ ft (3·5 m) long.

75

Marsh frog
Rana ridibunda

The largest frog in Europe, this frog is usually browny-green with black spots. Males have a pea-like vocal sac at each shoulder and produce raucously loud calls, especially when many mass at one spot. Females may be 5 in. (12.5 cm) long; males are smaller.

Edible frog
Rana esculenta

This species closely resembles the marsh frog but typically has a yellow, central stripe on its back. It is bigger and greener than the common frog and usually stays close to water. The plump hind legs are the edible part.

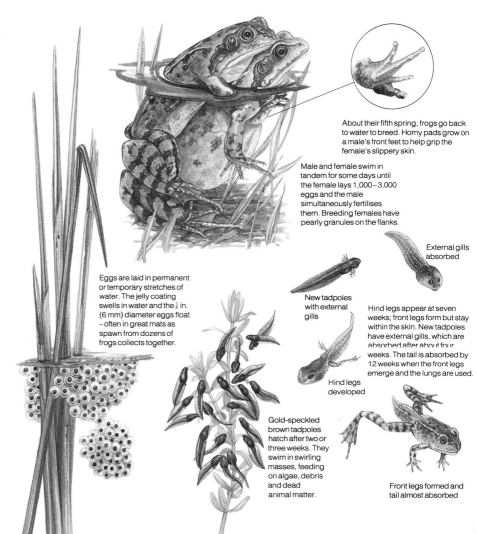

About their fifth spring, frogs go back to water to breed. Horny pads grow on a male's front feet to help grip the female's slippery skin.

Male and female swim in tandem for some days until the female lays 1,000–3,000 eggs and the male simultaneously fertilises them. Breeding females have pearly granules on the flanks.

Eggs are laid in permanent or temporary stretches of water. The jelly coating swells in water and the ¼ in. (6 mm) diameter eggs float – often in great mats as spawn from dozens of frogs collects together.

External gills absorbed

New tadpoles with external gills

Hind legs appear at seven weeks; front legs form but stay within the skin. New tadpoles have external gills, which are absorbed after about four weeks. The tail is absorbed by 12 weeks when the front legs emerge and the lungs are used.

Hind legs developed

Gold-speckled brown tadpoles hatch after two or three weeks. They swim in swirling masses, feeding on algae, debris and dead animal matter.

Front legs formed and tail almost absorbed

Frogs have a smooth, moist skin and move by springy leaps, whereas toads are rough-skinned, drier, warty and crawl. The common frog is mottled in shades of green, yellow or brown with a dark patch behind the eye. Females grow to 3 in. (75 mm) in length; males are smaller.

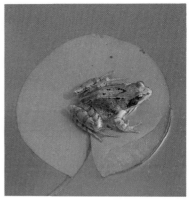

Found all over Britain, common frogs gather in ponds between January and March to spawn. The rest of the year is spent on land in damp places.

Common frog *Rana temporaria*

Contrary to popular belief, common frogs spend much of their lives on land; but between January and March they gather in shallow ponds to spawn, when quiet, buzzing croaks can be heard from the males as they pair up with females. The pairs remain in the spawning posture for days, even weeks, before the spawn is laid suddenly. All the eggs are shed in one batch, usually early in the morning. The males fight to win females and many frogs die at this time from exhaustion and subsequent capture by predators. Survivors often hide in deeper water for some weeks before spending the summer on land.

Newly hatched tadpoles swarm round the spawn clumps, but as they grow they become more secretive and hide from preying fish, newts and insects among the weeds. Even so, few survive to leave the pond as froglets in June or July.

Edible frogs and marsh frogs, introduced into Britain in the last 150 years, are patchily distributed mainly in east and south-east England. In summer they bask in bright sun on the banks of ponds and ditches and dive in with a loud splash when approached. They breed in late May and June and the froglets are fully formed by September or October.

On damp nights in June large groups of young toads leave the water for woods and fields. From October they hibernate until the spring.

Spawning, in March and early April, occurs under water. The male grasps the female from behind and fertilises the eggs as they are laid, croaking loudly if attacked by other males.

Great diving beetle

Few tadpoles survive their many predators, which include great diving beetles, water boatmen, dragonflies and great crested newts.

Natterjack toad
Bufo calamita

The prominent yellow stripe down its back distinguishes the natterjack. It is rarer than the common toad, faster moving, and usually inhabits sandy places where it spends most of the day in its burrow. Breeding males keep up a noisy, protracted croaking during May nights.

Natterjack eggs form a single row in a strand of spawn. Young leave the water fully formed after only six weeks.

The double row of black eggs lies within a continuous strand of clear jelly. Up to 10 ft (3 m) long, the strand twines round stems and leaves.

Tadpoles are black with a rounded tail tip.

During the daytime, toads remain concealed in a hole, cavity or shady spot among herbage, but in the late evening or early night they move out to hunt for food.

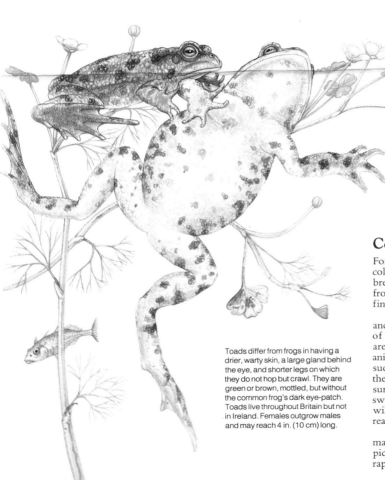

Toads differ from frogs in having a drier, warty skin, a large gland behind the eye, and shorter legs on which they do not hop but crawl. They are green or brown, mottled, but without the common frog's dark eye-patch. Toads live throughout Britain but not in Ireland. Females outgrow males and may reach 4 in. (10 cm) long.

Common toad *Bufo bufo*

For a short time each year common toads congregate – often in colonies of thousands – to lay their eggs. They are 'explosive' breeders, with many hundreds of sexually mature individuals from a district migrating to their chosen breeding ground, finding mates and spawning within a week or so.

Tadpoles hatch after two to three weeks, and through May and June swim freely in open water, sometimes in shoals of tens of thousands – behaviour rare in frog tadpoles. Toad tadpoles are omnivorous, feeding on algae, rotting plants and dead animals. Although their skins are distasteful to many predators such as fish, tadpoles are greedily eaten by various insects. Of the 1,000–4,000 eggs laid by a female only 5 per cent are likely to survive. Even so, on damp nights in June or July, pond banks swarm with tiny toads dispersing into the countryside. They will not return, except for a soak in a dry spell, until they are ready to breed two or three years later.

The slightly smaller, now rare natterjack toad is found mainly in sandy coastal areas of east and north-west England. It picks shallower ponds in which to spawn and the tadpoles grow rapidly. It has been legally protected in Britain since 1975.

79

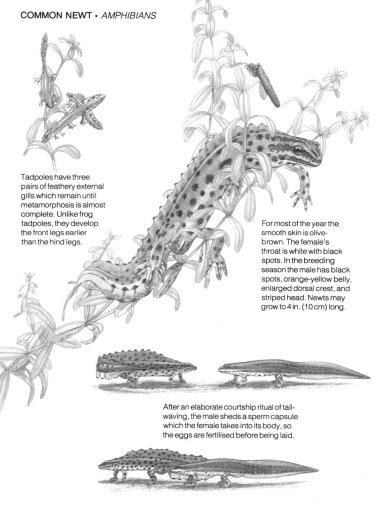

Tadpoles have three pairs of feathery external gills which remain until metamorphosis is almost complete. Unlike frog tadpoles, they develop the front legs earlier than the hind legs.

For most of the year the smooth skin is olive-brown. The female's throat is white with black spots. In the breeding season the male has black spots, orange-yellow belly, enlarged dorsal crest, and striped head. Newts may grow to 4 in. (10 cm) long.

After an elaborate courtship ritual of tail-waving, the male sheds a sperm capsule which the female takes into its body, so the eggs are fertilised before being laid.

The female wraps each egg individually in a leaf and secretes a fluid that seals it up.

Newts breed throughout Britain in small bodies of water with abundant plant growth – even garden ponds.

Common or smooth newt *Triturus vulgaris*

As its name suggests, the common newt is the most numerous of Britain's three native newt species. It is found throughout Britain, but less often in mountainous areas or heathlands. Despite their wide distribution, newts are a rare sight during most of the year. They live on land, hiding throughout the day and emerging at night to feed. In winter they hibernate under stones and in crevices. Between February and June is the time to see them, for that is when they move to the water and breed; during this period the male is displaying its bright breeding colours. For breeding, newts will occupy any small body of water, including an ornamental garden pond, provided that it has an abundant growth of water-weeds.

The spring courtship and breeding behaviour of the common newt is very similar to that of the palmate and great crested newts. The tadpoles hatch after about a week, and are entirely carnivorous, feeding on tiny animals such as *Daphnia*. They are slim, reaching 1¼–1½ in. (30–40 mm) long, and less conspicuous than those of frogs and toads, staying hidden in pond-weed to avoid predators. The young newts leave the water in August or September, but some overwinter still in tadpole form.

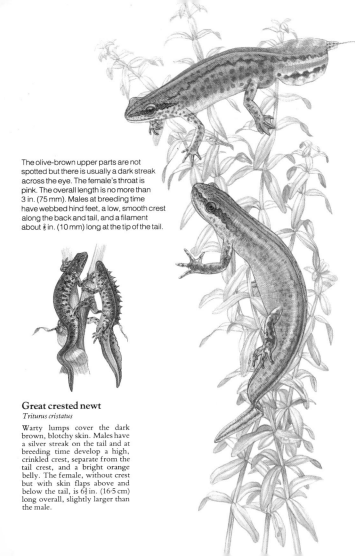

The olive-brown upper parts are not spotted but there is usually a dark streak across the eye. The female's throat is pink. The overall length is no more than 3 in. (75 mm). Males at breeding time have webbed hind feet, a low, smooth crest along the back and tail, and a filament about ⅜ in. (10 mm) long at the tip of the tail.

Great crested newt
Triturus cristatus

Warty lumps cover the dark brown, blotchy skin. Males have a silver streak on the tail and at breeding time develop a high, crinkled crest, separate from the tail crest, and a bright orange belly. The female, without crest but with skin flaps above and below the tail, is 6½ in. (16·5 cm) long overall, slightly larger than the male.

Tadpoles are practically identical to those of the common newt. Only close comparison shows their eyes to be set slightly more forward. The tadpoles hide from predators among pond-weed for much of the time.

Palmate newts move from hibernation nooks on land to breed in small water bodies, including high mountain tarns.

Palmate newt *Triturus helveticus*

This smallest of our three native newt species may be found throughout Britain but is rare in eastern England. It is most common in the moorlands and heathy places that common newts shun. In spring palmate newts leave their hibernation nooks on land to breed in small water bodies, including high mountain tarns. Our largest newt, the great crested newt, spends more time in water, mainly in lowland England; it breeds in large ponds. It is now rare, protected by law since 1981.

The spring behaviour of all three newt species is similar. Males court females by swimming before them vibrating their tails rapidly. If the female is receptive the male deposits a small bag of sperm on the bed. The female moves over it and takes it up into its cloaca. Fertilisation occurs internally. The female climbs among water plants laying up to 300 eggs, singly, and wrapping each in a leaf to protect it from predators. In behaviour and development tadpoles of great crested and palmate newts are like those of common newts. Adults feed greedily in the ponds, snapping up tadpoles and small aquatic insects. They surface frequently to breathe. In early summer they leave the pond but a few may return to hibernate there.

81

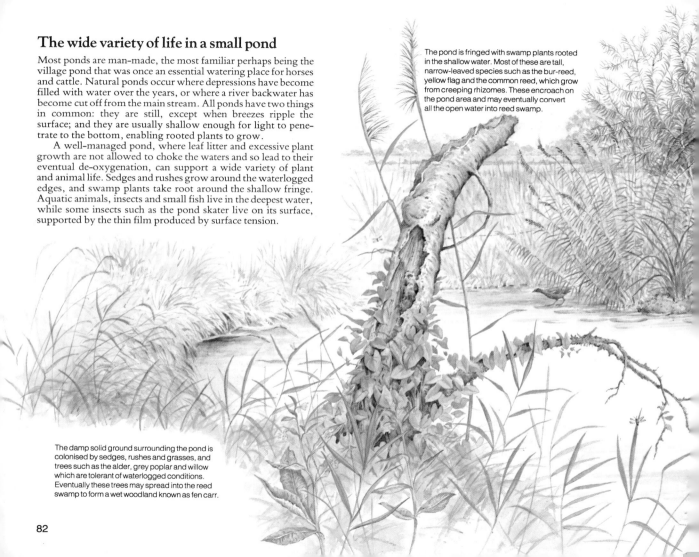

The wide variety of life in a small pond

Most ponds are man-made, the most familiar perhaps being the village pond that was once an essential watering place for horses and cattle. Natural ponds occur where depressions have become filled with water over the years, or where a river backwater has become cut off from the main stream. All ponds have two things in common: they are still, except when breezes ripple the surface; and they are usually shallow enough for light to penetrate to the bottom, enabling rooted plants to grow.

A well-managed pond, where leaf litter and excessive plant growth are not allowed to choke the waters and so lead to their eventual de-oxygenation, can support a wide variety of plant and animal life. Sedges and rushes grow around the waterlogged edges, and swamp plants take root around the shallow fringe. Aquatic animals, insects and small fish live in the deepest water, while some insects such as the pond skater live on its surface, supported by the thin film produced by surface tension.

The pond is fringed with swamp plants rooted in the shallow water. Most of these are tall, narrow-leaved species such as the bur-reed, yellow flag and the common reed, which grow from creeping rhizomes. These encroach on the pond area and may eventually convert all the open water into reed swamp.

The damp solid ground surrounding the pond is colonised by sedges, rushes and grasses, and trees such as the alder, grey poplar and willow which are tolerant of waterlogged conditions. Eventually these trees may spread into the reed swamp to form a wet woodland known as fen carr.

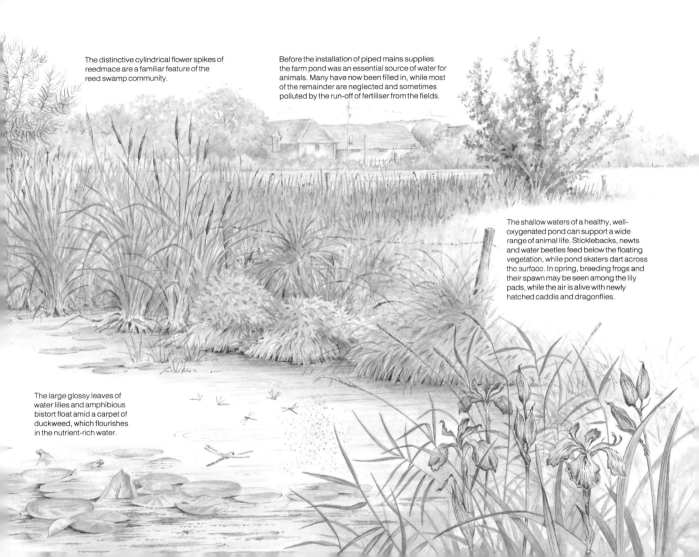

The distinctive cylindrical flower spikes of reedmace are a familiar feature of the reed swamp community.

Before the installation of piped mains supplies the farm pond was an essential source of water for animals. Many have now been filled in, while most of the remainder are neglected and sometimes polluted by the run-off of fertiliser from the fields.

The shallow waters of a healthy, well-oxygenated pond can support a wide range of animal life. Sticklebacks, newts and water beetles feed below the floating vegetation, while pond skaters dart across the surface. In spring, breeding frogs and their spawn may be seen among the lily pads, while the air is alive with newly hatched caddis and dragonflies.

The large glossy leaves of water lilies and amphibious bistort float amid a carpet of duckweed, which flourishes in the nutrient-rich water.

A female lays 200–500 cigar-shaped eggs, which are attached at one end to the surface of a leaf or stem of a water plant such as a reed.

The larvae are less than $\frac{1}{16}$ in. (2 mm) long when they hatch. After they emerge they drop into the water.

The adult alderfly, seen here on an alder leaf, has two pairs of dark, smoky-brown wings, which are folded like a roof over the back when the insect is at rest.

Alderflies are common in most types of fresh water; larvae have even been found at considerable depths in deep lakes.

Alderfly *Sialis lutaria*

There are two species of alderfly in Britain, of which the dark brown *Sialis lutaria* is the commonest. In early summer large numbers may be found on the vegetation near rivers, ponds and lakes. Although active at dusk, they are reluctant flyers, and during the day they spend most of their time crawling on plants and stones near the water's edge.

The female alderfly lays its eggs in neatly arranged batches on plants near the water. The larva which emerges from each egg is an aquatic carnivore, equipped with vicious pincer-like jaws, which preys on caddisfly larvae, mayfly nymphs and other invertebrates. It can tolerate low levels of dissolved oxygen, undulating its body to increase the flow of water over the seven pairs of feathery gills on the abdomen.

The larva remains in the water for one or two years, moulting nine times as it grows. When fully grown in the spring it crawls out onto the bank to pupate in a cavity which it digs in the soft earth near the water. After two or three weeks' pupation it emerges as an adult alderfly, which breeds and dies within a very short period. Anglers use the larva as bait and make imitations of the adult for dry fly fishing.

The single-tailed larva is distinctively marked with yellow and brown. It has large, powerful jaws and seven pairs of slender, jointed filaments along the abdomen. About 1½ in. (40 mm) long.

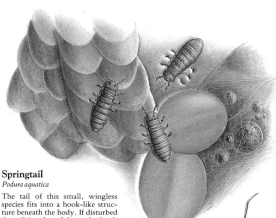

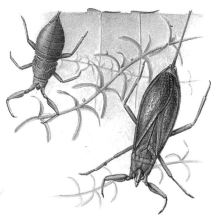

Water scorpion
Nepa cinerea

The front limbs of this greenish-brown insect are adapted solely for grabbing prey such as tadpoles, other insects and even small fish. It is found in weedy, stagnant ponds and lakes. The adult, which has a longer tail than the nymph, grows up to 1⅓ in. (35 mm) long.

Springtail
Podura aquatica

The tail of this small, wingless species fits into a hook-like structure beneath the body. If disturbed the tail is released from the hook and acts like a spring, driving the insect into the air. This species is dusky black and grows to about 1/16 in. (1·5 mm) long.

Water stick insect
Ranatra linearis

The front pair of limbs are adapted for grasping prey such as water mites and tadpoles. It lives in southern England in shallow ponds with abundant vegetation and plant debris. The brown, narrow body reaches about 2 in. (50 mm) in length.

Springtail
Isotoma palustris

Less common than *Podura aquatica*, this springtail has longer antennae though its body is smaller at about 1/25 in. (1 mm) long.

Pond insects with one tail

One of the wingless springtails, *Podura aquatica* is common in a few places where it lives on the surface of stagnant waters, often in masses several inches across. *Isotoma* is another smaller pond-dwelling springtail. The other two insects, *Nepa* and *Ranatra*, are bugs; their mouths are adapted for sucking the juices of other animals. The long tail is a breathing tube used to break the water surface and obtain oxygen. In spring the females insert their eggs into the stems and leaves of plants just below the surface.

Some stoneflies, such as *Perla microcephala*, produce short-winged adults incapable of flight.

After three years in the water, the nymph crawls out for its final moult. The wing-buds are black.

The female *Perla* runs across the surface of the water and releases a mass of black eggs from her abdomen.

Stoneflies live in well-oxygenated, unpolluted rivers and streams, particularly those with beds of loose stones.

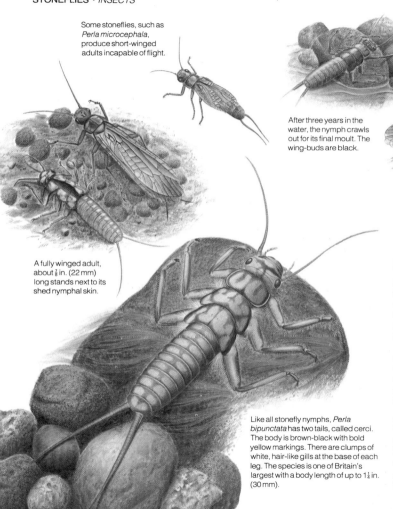

A fully winged adult, about ⅞ in. (22 mm) long stands next to its shed nymphal skin.

Like all stonefly nymphs, *Perla bipunctata* has two tails, called cerci. The body is brown-black with bold yellow markings. There are clumps of white, hair-like gills at the base of each leg. The species is one of Britain's largest with a body length of up to 1⅛ in. (30 mm).

Stonefly *Perla bipunctata*

Found crawling among stones in fast-flowing rivers and streams and on rocky, unsilted, lake shores, stonefly nymphs are slender, flattened insects similar to mayfly nymphs. They can be distinguished by their sluggish habits, and the presence of only two jointed tails, or cerci, which *Perla bipunctata* retains throughout the adult stage.

Small stonefly nymphs feed on algae, diatoms and organic debris, but *Perla bipunctata* and some of the larger species (called creepers by anglers) are voracious predators, taking worms, caddis and fly larvae and even the nymphs of mayflies and other stoneflies. The prey is located using the long, jointed antennae, then attacked and eaten quickly.

After moulting over 30 times during its three years in the water the nymph climbs out onto the bank for the final moult. The skin splits down the back and the winged adult emerges, pale and soft at first but hardening and darkening after a few hours. The adults live on for less than a month, for their mouthparts are much reduced and many species cannot eat at all. After mating they fly off to lay their eggs on the surface of the water; the eggs sink to the bottom and hatch in three weeks.

Six common stonefly nymphs

Leuctra hippopus
A yellowish, slender-bodied species found under stones. The wing-buds lie parallel to the body. Adult *Leuctra* are called needleflies. About $\frac{1}{5}-\frac{7}{16}$ in. (5–8 mm) long.

Brachyptera risi
This slender nymph is brown with lighter markings and a lighter underside. The wing-buds are set at an angle to the body. $\frac{5}{16}-\frac{2}{5}$ in. (7–10 mm) long, excluding tails.

Amphinemura sulcicollis has distinctive white gills.

Isoperla grammatica
Also called yellow sally, this is one of the commonest of the Perlodid family of stoneflies. It is found mainly in fast-flowing streams and has a body length of $\frac{7}{16}-\frac{5}{8}$ in. (11–16 mm).

Chloroperla torrentium
A yellowish-brown species, distinguished by the almost circular outline of its wing-buds and the hairiness of its tails. The body is $\frac{1}{10}-\frac{2}{5}$ in. (7–10 mm) long.

Amphinemura sulcicollis
The chocolate-brown nymphs are often camouflaged by debris which sticks to the body hairs. This species is distinguished by the two tufts of white gills under the chin. Grows $\frac{1}{6}-\frac{1}{4}$ in. (4–6 mm) long.

Nemoura erratica
Like other species of the stonefly family Nemouridae, *Nemoura erratica* is dark brown with stout wing-buds set at an angle to the body. Body length is $\frac{1}{4}-\frac{2}{5}$ in. (6–10 mm).

A rich growth of algae, mosses and liverworts covers the rocks in the splash zone near the waterfall. Small eddies at the margins of the stream may support the surface dwelling water cricket, which feeds on flying insects that fall into the water, and the springtail. An occasional bullhead may be seen, sheltering among the stones on the stream bed.

Adult stoneflies emerge in early summer. They fly very little, and are usually found on streamside rocks and vegetation. The rocks also make convenient perches for the dipper, a small bird which dives beneath the surface to forage for food on the bed of the stream.

Life in an upland stream

In the wet upland climate the rainwater flows off the hills into steep, swift watercourses, broken up by rapids and waterfalls. Any suspended rock particles are swept along in the current, leaving the bed of the stream bare of sediment.

Few water plants can grow in these conditions, and the insect life in the main part of the stream is restricted to species which are adapted to the fast-flowing water. The flattened nymphs of mayflies and stoneflies cling between stones, and carnivorous caddisfly larvae anchor themselves to the bottom in silken nets placed to trap small waterborne invertebrates. The water which drains off the rocks of upland areas is usually too poor in calcium to support many shelled molluscs, which need calcium for shell building.

The characteristic fish of upland streams is the brown trout, which thrives in the well-oxygenated conditions created by the constant splashing and turbulence of the water. Swimming against the current to keep station, it feeds on the insect larvae which drift downstream.

Melting snow and mountain rainstorms may swell the stream to a torrent, which scours out the bed and strips surrounding vegetation and soil off the underlying rocks. During a dry summer the flow may be a mere trickle between the stones on the bed – but it will be enough for insect larvae to survive.

89

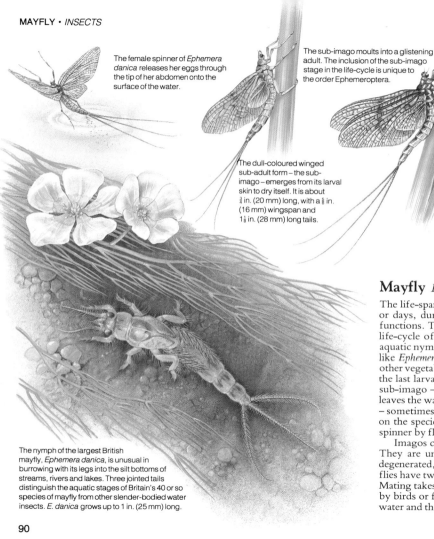

The female spinner of *Ephemera danica* releases her eggs through the tip of her abdomen onto the surface of the water.

The sub-imago moults into a glistening adult. The inclusion of the sub-imago stage in the life-cycle is unique to the order Ephemeroptera.

The dull-coloured winged sub-adult form – the sub-imago – emerges from its larval skin to dry itself. It is about ¾ in. (20 mm) long, with a ⅝ in. (16 mm) wingspan and 1⅛ in. (28 mm) long tails.

The nymph of the largest British mayfly, *Ephemera danica*, is unusual in burrowing with its legs into the silt bottoms of streams, rivers and lakes. Three jointed tails distinguish the aquatic stages of Britain's 40 or so species of mayfly from other slender-bodied water insects. *E. danica* grows up to 1 in. (25 mm) long.

The River Lambourn in Berkshire is among the many chalk streams where mayfly nymphs are abundant.

Mayfly *Ephemera danica*

The life-span of the adult mayfly is brief – perhaps a few hours or days, during which mating and egg laying are its primary functions. This, however, is but one stage of several in the full life-cycle of the mayfly. The first and longest stage is that of aquatic nymph. In most species it lasts one year; in some others, like *Ephemera danica*, two or three. Nymphs graze on algae and other vegetable matter in the water. They moult many times and the last larval skin to split reveals a dull-coloured sub-adult, or sub-imago – also known as a dun – which, flying laboriously, leaves the water to rest on a stone or plant. There it moults again – sometimes in minutes, sometimes in up to 30 hours depending on the species – to become a shining adult, or imago, called a spinner by fly fishermen.

Imagos collect in swarms over water during May and June. They are unable to feed because their mouths have by now degenerated, although some species are known to drink. Mayflies have two pairs of wings, the front pair larger than the rear. Mating takes place in flight, after which the male dies or is eaten by birds or fish. The female lays her eggs on the surface of the water and then she too dies or is eaten.

Mayfly nymphs

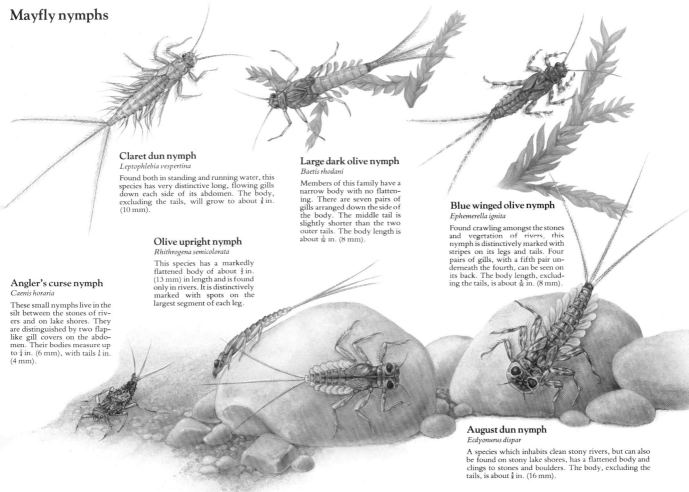

Claret dun nymph
Leptophlebia vespertina

Found both in standing and running water, this species has very distinctive long, flowing gills down each side of its abdomen. The body, excluding the tails, will grow to about ⅜ in. (10 mm).

Large dark olive nymph
Baetis rhodani

Members of this family have a narrow body with no flattening. There are seven pairs of gills arranged down the side of the body. The middle tail is slightly shorter than the two outer tails. The body length is about 5/16 in. (8 mm).

Olive upright nymph
Rhithrogena semicolorata

This species has a markedly flattened body of about ½ in. (13 mm) in length and is found only in rivers. It is distinctively marked with spots on the largest segment of each leg.

Blue winged olive nymph
Ephemerella ignita

Found crawling amongst the stones and vegetation of rivers, this nymph is distinctively marked with stripes on its legs and tails. Four pairs of gills, with a fifth pair underneath the fourth, can be seen on its back. The body length, excluding the tails, is about 5/16 in. (8 mm).

Angler's curse nymph
Caenis horaria

These small nymphs live in the silt between the stones of rivers and on lake shores. They are distinguished by two flap-like gill covers on the abdomen. Their bodies measure up to ¼ in. (6 mm), with tails ⅛ in. (4 mm).

August dun nymph
Ecdyonurus dispar

A species which inhabits clean stony rivers, but can also be found on stony lake shores, has a flattened body and clings to stones and boulders. The body, excluding the tails, is about ⅝ in. (16 mm).

Tail claspers of the male

The slender greenish nymph has a pale line down its back, and usually two narrow dark bands on the three tails. About 1¼ in. (30 mm) long.

During mating, the male grips the female behind the head with a pair of claspers at the tip of his abdomen, while she bends her abdomen forwards to permit fertilisation.

The recently hatched adult stays by the shed larval skin until its body and wings have hardened.

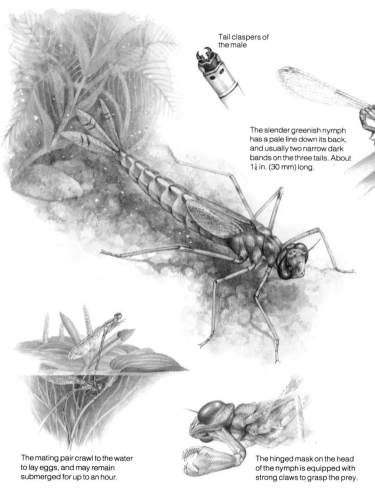

The mating pair crawl to the water to lay eggs, and may remain submerged for up to an hour.

The hinged mask on the head of the nymph is equipped with strong claws to grasp the prey.

Common blue damselfly *Enallagma cyathigerum*

Compared to the closely related dragonflies, adult damselflies are slender, delicate insects with widely spaced eyes and a slow fluttering flight. At rest the wings are folded along the body or held partly open, unlike those of the larger dragonflies which are permanently outstretched.

The common blue damselfly is widely distributed throughout the British Isles, and often found in large numbers round shallow lakes, rivers and ponds which contain plenty of water plants. The eggs are laid in plant stems, and hatch into aquatic larvae which immediately moult to emerge as second-stage nymphs. Like all damselfly nymphs they have long, slim bodies, and three leaf-like tails, or caudal lamellae, which appear to function as gills. In contrast the three tails of mayfly nymphs, which may occur in the same habitat, are narrow and segmented.

A very efficient predator, the nymph feeds on a variety of small animals, catching them by rapid extension of the hinged jaws, or mask. It grows quickly, moulting about ten times until it reaches the final stage. It then crawls out of the water and sheds the larval skin, emerging as an adult damselfly.

Six damselflies and their nymphs

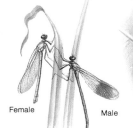

Female   Male

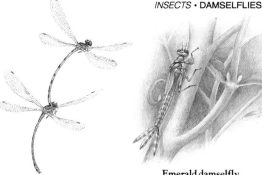

Banded demoiselle
Calopteryx splendens

Muddy yellow, with two pale tail bars, the nymphs have antennae with long first segments. The two outer tails are triangular in section. About 1½ in. (40 mm) long.

The wing-bands of the adult male vary from light purple-brown to deep blue depending on age.

Azure damselfly
Coenagrion puella

Found in weedy ponds, lakes and canals, the nymph is bright green with spots on the head. About 1 in. (25 mm) long.

Adult azure damselflies may be seen flying in tandem.

Emerald damselfly
Lestes sponsa

The grey or yellowish nymph is similar to that of the banded demoiselle, but the antennae do not have long first segments. All three tails are leaf-like. Grows 1 in. (25 mm) long.

The slender green *Lestes* often rests with its wings half open.

Large red damselfly
Pyrrhosoma nymphula

The nymph is stout-bodied, with dark-patterned tails and an almost rectangular head. About ¾ in. (20 mm) long.

The red body of the adult is distinctive.

Red-eyed damselfly
Erythromma najas

The large nymph can be identified by the three dark tail-bands. May grow over 1¼ in. (30 mm) long.

The adult has prominent red eyes.

Probably the commonest British damselfly, it is on the wing from end of May to end of August.

Blue-tailed damselfly
Ischnura elegans

Green or brown in colour, with long usually sharply pointed tails, the nymph may grow to 1 in. (25 mm).

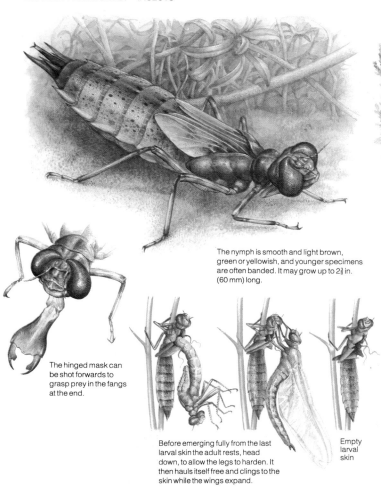

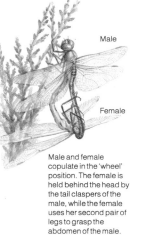

Male

Female

Male and female copulate in the 'wheel' position. The female is held behind the head by the tail claspers of the male, while the female uses her second pair of legs to grasp the abdomen of the male.

Common in the Midlands and southern England, the emperor dragonfly lays its eggs in weedy ponds and lakes.

The nymph is smooth and light brown, green or yellowish, and younger specimens are often banded. It may grow up to 2⅜ in. (60 mm) long.

The hinged mask can be shot forwards to grasp prey in the fangs at the end.

Before emerging fully from the last larval skin the adult rests, head down, to allow the legs to harden. It then hauls itself free and clings to the skin while the wings expand.

Empty larval skin

Emperor dragonfly *Anax imperator*

Adult dragonflies are large, heavily built, powerful predators capable of very fast flight. They feed on other insects, which they catch and often consume on the wing. In comparison with the smaller damselflies, adult dragonflies have very large eyes, and their gauze-like wings remain stiffly outstretched when the insect is at rest.

The emperor is the largest of the dragonflies, and is on the wing from the end of May to the end of August in southern Britain. The eggs are anchored into incisions made in water plants such as broad-leaved pondweed, and hatch into aquatic predatory nymphs which live among the water-weed. Sluggish by habit, the nymph lurks in wait for its prey, and captures it with a swift movement of the hinged, strongly clawed mask. A well-grown nymph will take insects, tadpoles and small fish. It obtains oxygen by drawing water into the rectum, which is lined with gills. By rapidly ejecting the water the nymph can propel itself forwards away from danger.

After two years or more the nymph crawls out of the water for the final moult. The larval skin splits down the back, enabling the fully developed insect to haul itself free.

Six dragonflies and their nymphs

The large adult is easily identified by the yellow banding.

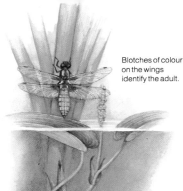

Blotches of colour on the wings identify the adult.

The swollen tail of the adult male gives the species its name.

Golden-ringed dragonfly
Cordulegaster boltonii

This brown, hairy nymph with black markings, small eyes and an angular head may lie buried in mud. About 1½ in. (40 mm) long.

Four-spotted chaser or libellula
Libellula quadrimaculata

A mud-dwelling nymph found in shallow waters, dark brown with a spine on the mid-line of segments four to eight. It grows to 1 in. (25 mm) long.

Club-tailed dragonfly
Gomphus vulgatissimus

Flattened, yellow-brown and hairy, the nymph lies buried in river mud. It grows to 1¼ in. (30 mm) long.

The adult is common in heath and moorland areas.

An alert, fast flyer, the adult is common in the south-east.

Mating adults may be seen flying in tandem.

Common hawker or aeshna
Aeshna juncea

A long and narrow, greenish-brown, hairless nymph, it has short projections from abdominal segments seven to nine. Up to 1¾ in. (45 mm) long.

Downy emerald
Cordulia aenea

The short-bodied nymph has small eyes and a dome-shaped mask. It is usually found clambering among weed. Up to 1 in. (25 mm) long.

Common darter
Sympetrum striolatum

Light to dark brown according to its surroundings, the nymph has spines on the abdominal segments. It may be ¾ in. (18 mm) long.

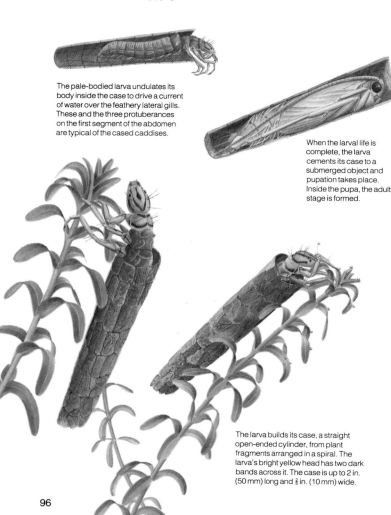

The pale-bodied larva undulates its body inside the case to drive a current of water over the feathery lateral gills. These and the three protuberances on the first segment of the abdomen are typical of the cased caddises.

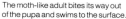

When the larval life is complete, the larva cements its case to a submerged object and pupation takes place. Inside the pupa, the adult stage is formed.

The moth-like adult bites its way out of the pupa and swims to the surface.

The species is widely distributed throughout Britain, where it is found in slow-flowing rivers and ponds.

The larva builds its case, a straight open-ended cylinder, from plant fragments arranged in a spiral. The larva's bright yellow head has two dark bands across it. The case is up to 2 in. (50 mm) long and ⅜ in. (10 mm) wide.

Great red sedge *Phryganea grandis*

Caddis flies or sedges are moth-like insects with hairy wings, generally dull grey or brown in colour, and active by night. The larvae, with one exception (*Enoicyla pusilla*), live in fresh water. Many species use plant fragments, sand grains and other material to build protective cases for their soft bodies.

The eggs of the largest British species, the great red sedge or murragh, are laid in a mass of jelly on water plants. After hatching, the larva begins to spin a sheath of silk around its abdomen, and to this it attaches pieces of plant material, all cut by its biting jaws to the same length and cemented together in a spiral pattern. As the larva grows it extends the tube by adding material to the front end. The case is retained by a pair of hooks at the end of the body which engage with a silken plug closing off the tube. The plug is porous, enabling water to circulate through the case and over the gills on the abdomen.

The insect lives as a larva for a year, feeding on plant matter and insect larvae, and withdrawing into the case if danger threatens. After pupating within the case it breaks free and emerges as a winged adult, which reproduces and dies within a few weeks.

Caddis larvae and their cases

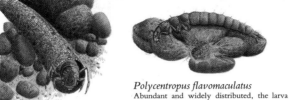

Limnephilus rhombicus
The case, built mainly from vegetable matter arranged tangentially, is characteristic of this caddis species. It is usually found in still water where there is plenty of vegetation, and is about 1¼–1½ in. (30–40 mm) long.

Sericostoma personatum
The curved case, made of sand grains, is sealed with a secretion at the back end, except for a small hole in the centre of the seal. The larva is found in stony fast-flowing streams, and sometimes on stony lake shores. The case is ⅗ in. (15 mm) long and ¹⁄₁₆–⅛ in. (2–3 mm) wide.

Polycentropus flavomaculatus
Abundant and widely distributed, the larva spins a net slung between stones to trap prey. Found in streams and the edges of lakes, it grows ⁹⁄₁₆ in. (14 mm) long.

Rhyacophila species
This free-living form does not build a case. It is light green, preys on small invertebrates in stony streams, and grows up to 1 in. (25 mm) long.

Agapetus species
The case, made of small stones, is flat below and rounded on top. Found only in rivers and on stony lake shores. The larva and case are about ¹⁄₁₆ in. (7 mm) long.

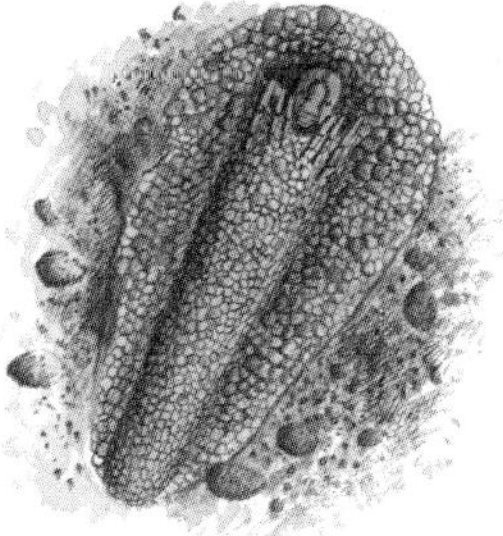

Molanna angustata
The case, made of sand grains, consists of a central tube with a shield-shaped extension round it. Found in lakes, pools and slow-flowing rivers and streams where there are patches of sand and fine gravel. The larva is ¾ in. (18 mm) long, the case 1 in. (25 mm).

Anabolia nervosa
The yellow head has black markings. The straight, tubular case of sand grains has one or more long twigs attached. The cases are found in fairly fast-flowing streams – each one is 1¼–1½ in. (30–40 mm) long without twigs.

Grey sedge, or grey flag
Hydropsyche angustipennis

This grey-brown or yellow larva builds no case. It lives in a roughly elongate silken shelter, incorporating small stones and debris, which it spins under stones in swift streams to catch prey. It grows ¾ in. (18 mm) long.

Hydroptila species
The flattened case consists of sand grains and is lined with silk. About ⅕ in. (5 mm) long and ⅛ in. (3 mm) wide.

Goera pilosa
The straight-sided case of sand grains incorporates a few larger stones, which form wing-like extensions – possibly acting as ballast to prevent its being swept away. Found in fast-flowing streams, the case is ⅗ in. (15 mm) long.

Agapetus species have cases with both openings on the underside.

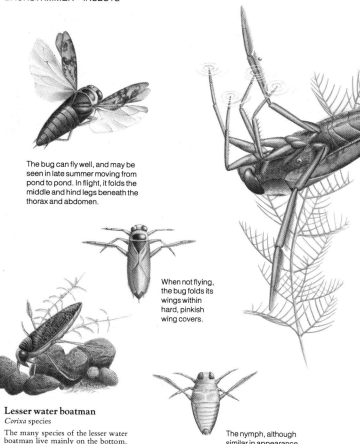

The bug can fly well, and may be seen in late summer moving from pond to pond. In flight, it folds the middle and hind legs beneath the thorax and abdomen.

When not flying, the bug folds its wings within hard, pinkish wing covers.

Lesser water boatman
Corixa species

The many species of the lesser water boatman live mainly on the bottom. The shovel-like end of the fore-legs is used to scrabble up food. Smaller than the backswimmer, they have a series of stripes on the wing covers. Up to ½ in. (13 mm) long.

The nymph, although similar in appearance to the adult, has no wings.

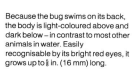

Because the bug swims on its back, the body is light-coloured above and dark below – in contrast to most other animals in water. Easily recognisable by its bright red eyes, it grows up to ⅝ in. (16 mm) long.

Ponds rich in plant life are among the many places that the backswimmer can be seen during the summer.

Backswimmer *Notonecta glauca*

Often numerous in the still or slow-flowing water of ponds, canals and ditches, the backswimmer is one of the commonest British species of water bug. Like all the water bugs it has fore-wings modified into hard protective covers for the thin, delicate hind-wings, and mouthparts which are extended into a piercing beak, or rostrum.

The backswimmer is a predator. Motionless, it hangs upside-down just below the surface of the water, with the first two pairs of legs and the tip of the abdomen just touching the surface film, sensitive to the slightest vibration. Quickly attracted to any disturbance it pursues its prey using the powerful hind legs to propel itself in a series of jerks. It will attack insects, small fish and tadpoles, injecting digestive chemicals called enzymes through the rostrum and sucking out the juices.

Mating occurs in the spring, and the oval eggs are embedded lengthwise into plant stems. The nymphs moult five times before they achieve the adult state in August.

The lesser water boatmen, of which there are about 30 species in Britain, are bottom-living bugs which use the smaller, weaker rostrum to suck up organic debris and algae.

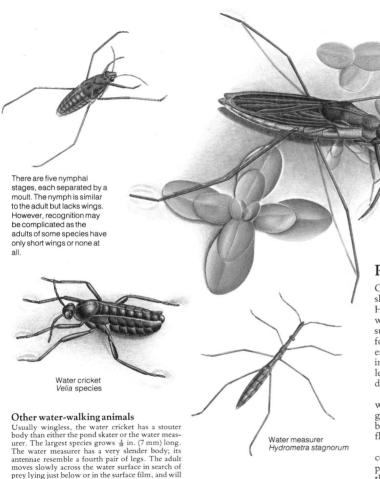

There are five nymphal stages, each separated by a moult. The nymph is similar to the adult but lacks wings. However, recognition may be complicated as the adults of some species have only short wings or none at all.

Water cricket
Velia species

The narrow body is mainly brown or black. The middle pair of legs is used for propulsion, the rear for steering, and the short front legs for gripping prey. At rest, the wings lie folded across the back. About ½–⅝ in. (13–16 mm) long.

Pond skaters are frequently found under man-made structures, such as jetties and boathouses, on rivers and lakes.

Other water-walking animals

Usually wingless, the water cricket has a stouter body than either the pond skater or the water measurer. The largest species grows ⁷⁄₂₄ in. (7 mm) long. The water measurer has a very slender body; its antennae resemble a fourth pair of legs. The adult moves slowly across the water surface in search of prey lying just below or in the surface film, and will jump if disturbed. It grows up to about ½ in. (13 mm) long.

Water measurer
Hydrometra stagnorum

Pond skater *Gerris lacustris*

Common on lakes, ponds and slow-flowing streams, the pond skater is a predatory water bug belonging to the insect order Hemiptera – a group that includes the backswimmer and the water scorpion. It lives entirely above water, supported on the surface by dense pads of hair on the underside and on the ends of four of the six legs. The pads trap air and are not wetted. This enables the insect to propel itself over the water without breaking the surface film. Normally, it uses the long second pair of legs with a rowing action and steers with the third pair. If disturbed it may move rapidly away in a series of hops.

The pond skater normally feeds on dead and dying insects which fall on to the water. Attracted by the disturbance, it grasps the prey using its short, sturdy front legs, pierces the body with its proboscis, or feeding tube, and sucks out the body fluids.

The female lays her eggs in the spring. Protected by a covering of jelly, they are laid in small groups on the submerged parts of water plants. They hatch into nymphs which resemble the adults in appearance and habits, moulting five times before achieving maturity.

The sausage-shaped eggs are laid in slits made in the stems of submerged plants.

Sucker discs on the male's forelimbs allow it to grip the female during mating.

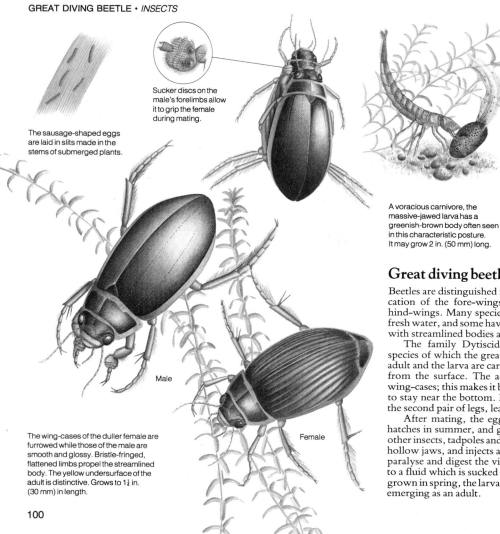

Male

The wing-cases of the duller female are furrowed while those of the male are smooth and glossy. Bristle-fringed, flattened limbs propel the streamlined body. The yellow undersurface of the adult is distinctive. Grows to 1¼ in. (30 mm) in length.

Female

A voracious carnivore, the massive-jawed larva has a greenish-brown body often seen in this characteristic posture. It may grow 2 in. (50 mm) long.

The great diving beetle larva, found in still, weedy waters, breathes through holes in its tail appendages.

Great diving beetle *Dytiscus marginalis*

Beetles are distinguished from most other insects by the modification of the fore-wings into hard, protective cases for the hind-wings. Many species spend all or part of their life-cycle in fresh water, and some have become well adapted for swimming, with streamlined bodies and flattened oar-like limbs.

The family Dytiscidae contains over a hundred British species of which the great diving beetle is the largest. Both the adult and the larva are carnivores, and both breathe air obtained from the surface. The adult beetle stores the air beneath the wing-cases; this makes it buoyant, and it has to swim vigorously to stay near the bottom. It often hangs on to water plants with the second pair of legs, leaving the first pair free to catch food.

After mating, the eggs are laid in water plants. The larva hatches in summer, and grows rapidly on a diet which includes other insects, tadpoles and small fish. It seizes the prey with huge hollow jaws, and injects a mixture of enzymes. These chemicals paralyse and digest the victim's internal organs, reducing them to a fluid which is sucked back up through the jaws. When fully grown in spring, the larva pupates at the edge of the water before emerging as an adult.

Water beetles and their larvae

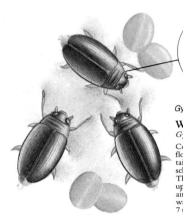

Gyrinus natator eye

Whirligig beetle
Gyrinus natator

Common in still or slow-flowing waters throughout Britain, where it congregates in schools gyrating on the surface. The eye is divided in two, the upper part adapted for vision in air, and the lower for vision in water. About $\frac{1}{5}$–$\frac{7}{16}$ in. (5–7 mm) long.

Gyrinus natator larva

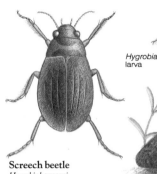

Screech beetle
Hygrobia hermanni

Recognisable by its habit of screeching when alarmed, and found in muddy ponds where it feeds on worms. It is rounded and reddish-yellow with black markings. Grows up to $\frac{1}{2}$ in. (13 mm) long.

Hygrobia larva

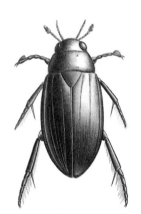

Hydrophilus piceus larva

Silver water beetle
Hydrophilus piceus

The largest British beetle, with a high-domed back and velvet-like hairs underneath which trap air bubbles. It swims upside-down, exposing the silvery, shining abdomen. The larva is carnivorous, and up to 3 in. (75 mm) long.

Elmis aenea larva

Elmis aenea

One of a group of beetles which do not swim, but crawl slowly over stones and water plants in fast-flowing streams. They absorb oxygen from the water into a bubble of air. Adults and larvae are herbivorous. The larvae are flat with a hard covering. $\frac{1}{16}$–$\frac{1}{10}$ in. (1·5–2·5 mm) long.

Overgrown with birch, crack willow and alder, and inaccessible to anglers, the older sections of the gravel pit provide excellent cover for nesting water birds such as herons, coots and moorhens. The flooded pit attracts many species, including the distinctive black-necked Canada goose, the mute swan, the great crested grebe and the greylag goose, originally a native and recently re-introduced to southern Britain.

How gravel pits return to nature

Disused gravel pits provide a perfect example of how nature takes over when man, having intruded on nature in the first place, moves on. A freshly abandoned gravel pit is a desolate place, with little or no vegetation and barely a living creature in sight. But soon the hole will fill with water – during gravel extraction the pit is pumped continuously – and among its first occupants are the algae which are present in almost all water. Then come the larger plants, their seeds carried on the wind or by birds. Their progress is sometimes slow, for the fast-growing algae may filter out the sunlight which has to penetrate the water if the shoots are to develop. But the spread of algae is soon checked by the tiny plant-eating creatures known as zoo-plankton, and as the seasons pass the water plants become established and provide an environment for dragonflies, aquatic beetles and pond snails. Soon frogs, newts and fish appear, and water birds such as herons and great crested grebes. Eventually the gravel pit becomes a thriving community, a self-contained piece of nature where the lives of all members are closely interlinked in the competition to survive.

A submerged dead tree creates a backwater, encouraging the growth of water plants such as Canadian pondweed and water milfoil. The weed harbours pond snails and crustaceans; these provide food for fish such as roach which prefer still water. Carp, perch and often pike, inhabit the deeper, open water in the centre of the lake.

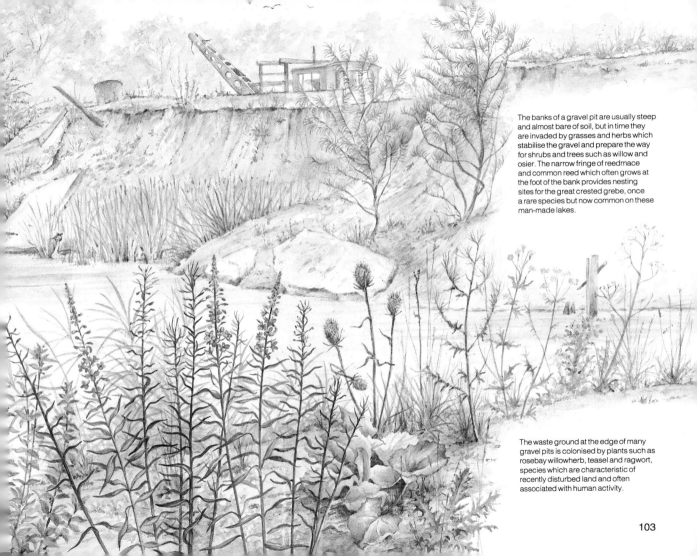

The banks of a gravel pit are usually steep and almost bare of soil, but in time they are invaded by grasses and herbs which stabilise the gravel and prepare the way for shrubs and trees such as willow and osier. The narrow fringe of reedmace and common reed which often grows at the foot of the bank provides nesting sites for the great crested grebe, once a rare species but now common on these man-made lakes.

The waste ground at the edge of many gravel pits is colonised by plants such as rosebay willowherb, teasel and ragwort, species which are characteristic of recently disturbed land and often associated with human activity.

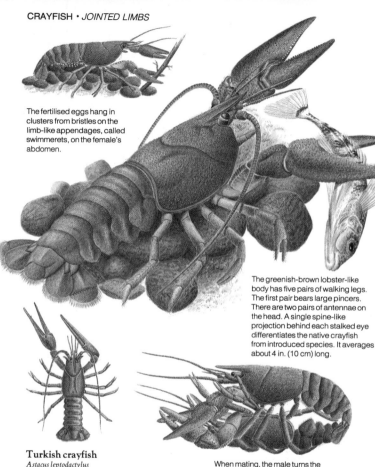

The fertilised eggs hang in clusters from bristles on the limb-like appendages, called swimmerets, on the female's abdomen.

The greenish-brown lobster-like body has five pairs of walking legs. The first pair bears large pincers. There are two pairs of antennae on the head. A single spine-like projection behind each stalked eye differentiates the native crayfish from introduced species. It averages about 4 in. (10 cm) long.

Turkish crayfish
Astacus leptodactylus

The colour is brownish or bluish, and the claws of the male are longer and more slender than those of the native crayfish. Up to 6 in. (15 cm) long.

When mating, the male turns the female on to her back and fixes adhesive packets of sperm on to her abdomen. Shortly after this, the eggs are laid and are fertilised.

Newly hatched crayfish look similar to the adult form but have a domed shell over the front of the body within which yolk is stored. The young cling to the female for about ten days until their first moult.

The native crayfish lives in rivers, streams and lakes where the water is hard. It is not found in soft water.

Crayfish *Austropotamobius pallipes*

The freshwater crayfish is related to the marine lobster and the crawfish, both of which it resembles. In the British Isles the native species is also known as the white-footed crayfish. Its typical habitat is clear, shallow, fast-flowing water, though it also occurs in lakes and ponds. It lives in holes in the banks or under large stones. The Turkish crayfish and the signal crayfish (*Pacifastacus leniusculus*) are both introduced species.

The crayfish is both a predator and scavenger, feeding mainly at night. It moves about by walking but it can also swim quickly backwards by a flick of its muscular tail. Its prey – snails, insect larvae and dead or dying fish – is caught, held and sometimes torn apart by the powerful pincers before being transferred to the mouth and ground up by the jaws. The crayfish itself is preyed on by many animals, particularly eels, otters and herons, and is very vulnerable when it is moulting.

Mating takes place in autumn, and the young hatch in the following May or June. From birth they are voracious predators and cannibals, especially when other young crayfish are moulting their shells. Moulting takes place many times during the three to four year growing period, after which it is annual.

The predatory water spider is common in weedy ponds and ditches, particularly in peat-stained moorland waters.

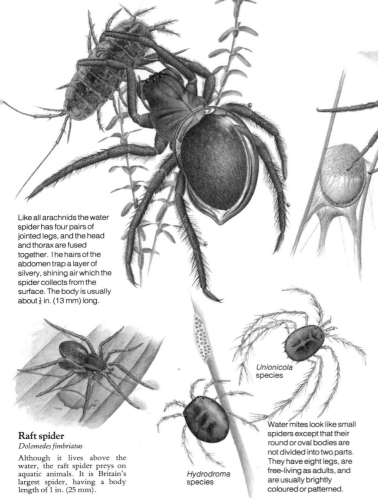

Like all arachnids the water spider has four pairs of jointed legs, and the head and thorax are fused together. The hairs of the abdomen trap a layer of silvery, shining air which the spider collects from the surface. The body is usually about ½ in. (13 mm) long.

The water spider visits the surface to collect more air bubbles on its abdomen and so add to the store of air inside the bell-shaped silk net.

Raft spider
Dolomedes fimbriatus

Although it lives above the water, the raft spider preys on aquatic animals. It is Britain's largest spider, having a body length of 1 in. (25 mm).

Hydrodroma species

Unionicola species

Water mites look like small spiders except that their round or oval bodies are not divided into two parts. They have eight legs, are free-living as adults, and are usually brightly coloured or patterned.

Water spider *Argyroneta aquatica*

The only truly aquatic species of spider in Britain, the water spider is physically very similar to the familiar land spiders. It has to breathe air, so collects it from the surface and stores it within bell-shaped webs attached to submerged plants. Once the air-bells are fully stocked, the oxygen used from each bell by the spider is replaced by diffusion from the oxygen dissolved in the water. Further journeys to the surface are rarely necessary.

When swimming, the water spider carries its air supply trapped in the fine hairs on its abdomen. Using its poison fangs it immobilises prey such as small crustaceans, insect larvae, or even small fish, carrying them back to an air-bell to be eaten. The female lays 20 to 100 eggs within a specially constructed bell. They hatch after two or three weeks into miniatures of the adult and shelter inside the bell for some weeks.

Water mites are much smaller than water spiders. They are carnivorous arachnids – the class which also includes spiders, scorpions and harvestmen – and are able to absorb oxygen directly from the water. The eggs of many species are laid in jelly masses on submerged stones or plants, and the six-legged larvae become parasitic on aquatic insects such as diving beetles.

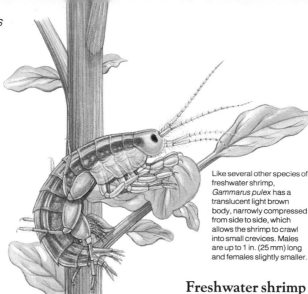

At breeding time the male grips the female from behind and the pair swim in tandem for several days before mating.

Like several other species of freshwater shrimp, *Gammarus pulex* has a translucent light brown body, narrowly compressed from side to side, which allows the shrimp to crawl into small crevices. Males are up to 1 in. (25 mm) long and females slightly smaller.

Gammarus pulex lives where water plants and leaf litter are abundant; a cress farm is ideal.

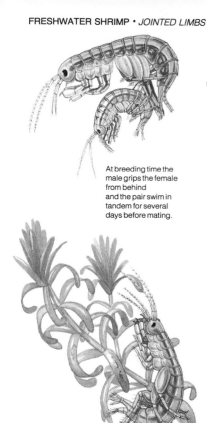

Crangonyx pseudogracilis
Introduced from America, this species is similar to *Gammarus* but has a pale blue-grey sheen and is slightly smaller. It swims upright, not on its side. The rear body segments are smooth, with no hairs or bristles.

Freshwater shrimps of the *Gammarus* genus swim on their sides.

Eggs are incubated in a brood pouch under the female's body and hatch into miniature adults, not into larvae.

Freshwater shrimp *Gammarus pulex*

Freshwater shrimps are in fact not shrimps at all but amphipods, members of a group that also includes the marine sandhoppers. They occur widely in a variety of well-oxygenated waters. The various species are difficult to distinguish without the aid of a microscope, for they all have curved, laterally compressed bodies and many limbs adapted for different functions. *Gammarus pulex* is one of the most common species.

The first two pairs of limbs are used for grasping, while the next five are modified for swimming. Plate-like gills, attached to the inner surfaces of the swimming legs, are supplied with fresh oxygenated water by the vibrations of three pairs of feathery structures on the abdomen. The animals are also equipped with two pairs of long antennae, which assist in locating the decomposing plant and animal material on which they feed.

The male is larger than the female, and mating pairs are often found swimming together until the female moults and loses its shell, after which the male deposits its sperm and breaks away. The fertilised eggs are retained in a brood pouch between the female's leg bases, where the eggs develop into small replicas of the adult.

Its colouring gives the water slater effective camouflage in its habitual activity of crawling over plants or along the muddy bottom.

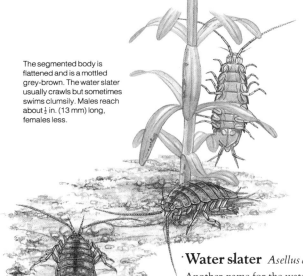

The segmented body is flattened and is a mottled grey-brown. The water slater usually crawls but sometimes swims clumsily. Males reach about ½ in. (13 mm) long, females less.

Moderately polluted rivers, canals and weedy ponds suit water slaters, which can tolerate low oxygen levels.

Asellus meridianus
This closely related species is similar to *A. aquaticus* but has different head markings, is smaller, and has shorter antennae – about three-quarters of the body length.

Prior to the spring mating, male and female swim in tandem. The female incubates the fertilised eggs in a brood pouch under the body.

Water slater *Asellus aquaticus*

Another name for the water slater is the water hog louse. This freshwater crustacean is related to the familiar woodlouse and there are two common species in Britain: *Asellus aquaticus* and the smaller *A. meridianus*. Both have segmented bodies, seven pairs of jointed legs and two small appendages called uropods on the abdomen. There are five broad gills beneath the body. A third species, *A. cavaticus*, is white and blind and lives in wells and caves in southern England and South Wales.

Common where the current is not too strong and where there is plenty of dead organic material for it to feed on, the water slater can live in poorly oxygenated conditions. It is often abundant where there is moderate organic pollution.

Prior to mating – which occurs mainly in spring – males and females pair off and the male, which is larger, carries the female around until the eggs are laid. This is done into a brood pouch beneath the body of the female, where the young remain protected for several days after hatching. Similar in appearance to the adults, but transparent and lacking one pair of legs, they undergo a series of moults at intervals of a few weeks before achieving the adult stage.

107

Canals – wildlife havens amid industry's scars

In the 17th and 18th centuries, Britain's canals provided a link between mining, industrial and commercial centres, but their heyday was short-lived. Today they provide havens for wildlife in the heart of cities and industrial areas.

Canals differ from rivers in that they are made to a uniform width and depth, are slow-moving and have a virtually constant water level. The natural history of a canal is, therefore, very similar throughout its length. The banks often support luxuriant plant growth and are inhabited by many birds and animals, and in the shallow water's edge grow reeds and rushes. But in the deep central channel, plants are cut back so as not to hinder navigation. This is the angler's water – the domain of bream, roach and tench.

The canal towpath often becomes overgrown with a rich flora of grasses, sedges and tall marsh plants such as meadowsweet, purple loosestrife and the exotic orange balsam, merging with wasteland plants such as willowherb, brambles, nettles and buddleia. The tangle of vegetation provides cover for nesting birds, mice and water voles.

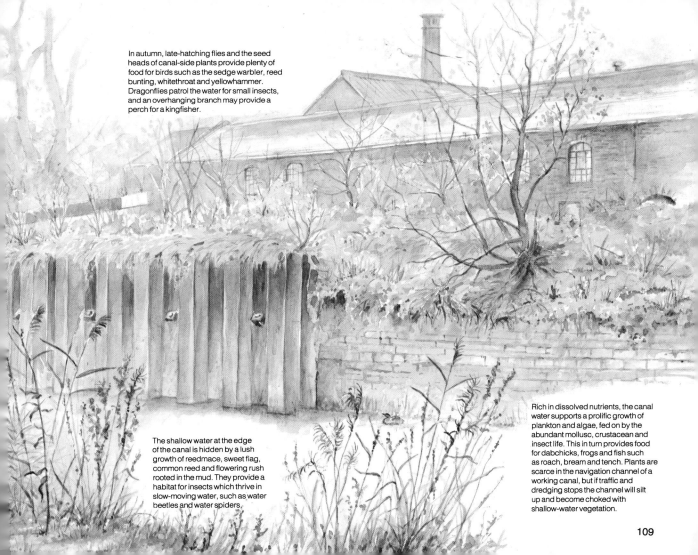

In autumn, late-hatching flies and the seed heads of canal-side plants provide plenty of food for birds such as the sedge warbler, reed bunting, whitethroat and yellowhammer. Dragonflies patrol the water for small insects, and an overhanging branch may provide a perch for a kingfisher.

The shallow water at the edge of the canal is hidden by a lush growth of reedmace, sweet flag, common reed and flowering rush rooted in the mud. They provide a habitat for insects which thrive in slow-moving water, such as water beetles and water spiders.

Rich in dissolved nutrients, the canal water supports a prolific growth of plankton and algae, fed on by the abundant mollusc, crustacean and insect life. This in turn provides food for dabchicks, frogs and fish such as roach, bream and tench. Plants are scarce in the navigation channel of a working canal, but if traffic and dredging stops the channel will silt up and become choked with shallow-water vegetation.

109

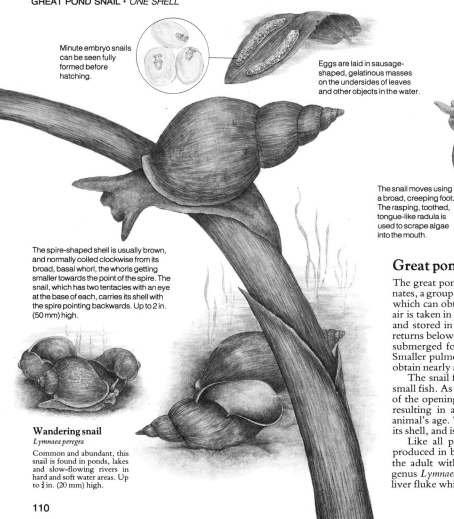

Minute embryo snails can be seen fully formed before hatching.

Eggs are laid in sausage-shaped, gelatinous masses on the undersides of leaves and other objects in the water.

The snail moves using a broad, creeping foot. The rasping, toothed, tongue-like radula is used to scrape algae into the mouth.

Found in southern and central England, the snail inhabits ponds and slow-moving waters rich in plant life.

The spire-shaped shell is usually brown, and normally coiled clockwise from its broad, basal whorl, the whorls getting smaller towards the point of the spire. The snail, which has two tentacles with an eye at the base of each, carries its shell with the spire pointing backwards. Up to 2 in. (50 mm) high.

Wandering snail
Lymnaea peregra

Common and abundant, this snail is found in ponds, lakes and slow-flowing rivers in hard and soft water areas. Up to $\frac{3}{4}$ in. (20 mm) high.

Great pond snail *Lymnaea stagnalis*

The great pond snail is the largest of British freshwater pulmonates, a group of single-shelled molluscs related to the land snails which can obtain oxygen from the air using a simple lung. The air is taken in at the water surface through a breathing aperture, and stored in a cavity lined with blood vessels while the snail returns below to feed. In well-oxygenated waters it may remain submerged for long periods, absorbing oxygen by diffusion. Smaller pulmonates such as limpets and some ramshorn species obtain nearly all their oxygen this way.

The snail feeds on organic debris, algae, fish eggs and even small fish. As it grows the shell is enlarged by adding to the edge of the opening. Growth is fast in summer but slow in winter, resulting in a pattern of growth marks which relate to the animal's age. The snail needs a good supply of calcium to build its shell, and is common only in hard water areas.

Like all pulmonates it is a hermaphrodite. The eggs are produced in batches of up to 500, and hatch into miniatures of the adult without going through a larval stage. Snails of the genus *Lymnaea* are the hosts in one stage of the life-cycle of the liver fluke which infests sheep and cattle in wet pastures.

Identifying other freshwater snails

Of the snails illustrated, two – the ramshorn and the bladder snail – are pulmonates, breathing by means of a simple lung. The others are operculates and have a tough, limy or horny trap-door structure – the operculum – which closes the entrance to the shell once the snail has withdrawn inside. Operculate snails obtain oxygen by passing water over their gills. Most water snails lay eggs but Jenkins' spire shell and the viviparous snail give birth to already formed young.

Nerite
Theodoxus fluviatilis

Fairly common in hard water areas, the nerite has a solid shell with a half-moon shaped aperture. The colour is variable. The background may be yellow, brown or black with purple-pink or white markings. Height $\frac{1}{3}$–$\frac{1}{2}$ in. (9–10 mm).

White ramshorn
Planorbis albus

Common in ponds, lakes and streams, this species has a dull white shell shaped like a flattened coil. It bears spiral marks as well as transverse growth lines. Diameter $\frac{1}{16}$ in. (7 mm).

Valve snail
Valvata piscinalis

The greenish-yellow, slightly flattened shell, often used by the caddis larva *Limnephilus flavicornis* for case making, is sometimes banded. The snail is common in gently flowing streams but rare in still water. Height $\frac{1}{5}$ in. (5 mm).

River snail
Viviparus viviparus

The greenish-brown shell is marked by three dark bands. The operculum is patterned with concentric rings. Though common in southern England, it is not found north of Yorkshire. Height 1$\frac{1}{4}$–1$\frac{1}{2}$ in. (30–40 mm).

Jenkins' spire shell
Hydrobia jenkinsi

This small snail is very numerous in rivers and still waters on stones or mud. The shell is dark brown or black, the body grey. Height up to $\frac{1}{5}$ in. (5 mm).

The eggs of water snails, such as those of the great ramshorn, are laid in jelly-covered masses on water plants or stones.

Common bithynia
Bithynia tentaculata

The glossy yellow to reddish shell has five or six whorls. The hard, limy operculum, which is patterned with concentric rings, cannot be drawn into the aperture. Height $\frac{1}{3}$–$\frac{2}{3}$ in. (9–15 mm).

Fountain bladder snail
Physa fontinalis

The horn-coloured shell has three or four whorls – which wind anticlockwise from the very large basal whorl. Finger-like projections from the body partly cover the shell when the animal is extended. It is common in clear, running water with abundant vegetation. About $\frac{1}{2}$ in. (13 mm) high.

Great ramshorn
Planorbarius corneus

This massive brown shell is the largest ramshorn. It is fairly common in large bodies of hard water. Diameter 1$\frac{1}{4}$ in. (30 mm), height $\frac{1}{2}$ in. (13 mm).

River limpets are found in standing and flowing silt-free waters throughout the British Isles.

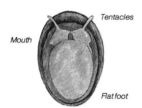

The river limpet has a cone-shaped shell with a curved top. The shell is thin and dark in colour, often black. A side view shows how the outline of the shell fits the contours of the rock to which it is attached. It is usually about ¼ in. (6 mm) long.

Mouth

Tentacles

Flat foot

An examination of the underside shows the oval, flat foot, the dark mouth and the two long tentacles.

Lake limpet
Acroloxus lacustris

Despite its name, the lake limpet is also found in rivers. It is slightly smaller than the river limpet and has a narrower shell.

River limpet *Ancylus fluviatilis*

Although only distantly related to the marine limpet the fresh-water limpet has evolved a very similar mode of life, clinging to rocks and water plants with its foot and clamping down its thin, conical shell for protection. The margin of the shell is soft and easily accommodates irregularities in the surface, forming a strong seal which gives the animal great tenacity.

The river limpet will not tolerate silt. For this reason, it is usually found in fast-flowing, stony rivers. Running water is not essential, however, and it is common on the stony margins of large lakes where there is enough water movement to prevent silting. The limpet's relatively low calcium requirement for shell-building enables it to live in the soft, silt-free waters of the Lake District. The lake limpet (*Acroloxus lacustris*), by contrast, is tolerant of the silty water of ponds, lowland lakes and slow rivers, where it is normally found clinging to reeds and submerged vegetation.

Both species feed by grazing on the diatoms and blue-green algae which grow on rocks and water plants. Like the pond snail, river and lake limpets have no gills but obtain oxygen from an air supply stored in a simple lung.

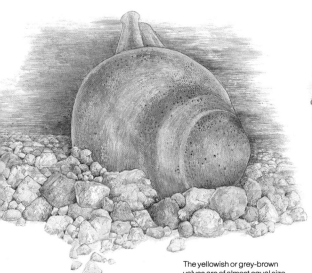

Orb shells may be easily overlooked because they so closely match the size, shape and colour of the gravel in which they live.

Found in mud, gravel and among the roots of water plants, *Sphaerium* inhabits a wide variety of fresh waters.

The yellowish or grey-brown valves are of almost equal size, and the umbo, or conical knob, is near the centre of the upper valve. When alive the cockle can be distinguished from *Pisidium* by its two siphons. Fully grown, the shell is about ¾ in. (20 mm) long.

Pea shell cockle
Pisidium species

Sixteen species of pea shell cockles are found in Britain, in standing and flowing water, and also in the deep regions of lakes. Each cockle has a single siphon, and the whitish-yellow or brown shell rarely exceeds ⅜ in. (10 mm).

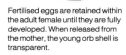

Fertilised eggs are retained within the adult female until they are fully developed. When released from the mother, the young orb shell is transparent.

An orb shell can climb the stems of water plants with the aid of its foot.

Orb shell cockle *Sphaerium corneum*

Of the four species of orb shell cockle that occur in Britain, *Sphaerium corneum* is the commonest. A wide variety of fish and water birds feed on *Sphaerium* and the cockles are often found adhering to the feet of waders, which may explain their wide distribution.

Quite active, the orb shell moves about using the slimy, tongue-shaped foot which projects between the shell valves. The extended foot adheres to the rock or plant surface, and when it contracts the shell is hauled after it. The orb shell feeds by drawing water into an aperture or siphon at the rear of the shell. Food particles are sieved out by the gills, oxygen is absorbed, and the waste is expelled through the upper siphon.

The orb shell is hermaphrodite, but it cross-fertilises by releasing sperm into the water, which is taken in through the inhalant tubes of other individuals. Once fertilised, the eggs are retained inside the shell in brood pouches formed in the gill area. After many months the young emerge as miniature adults.

In Britain 16 species of the pea shell cockle (*Pisidium*) occur. In appearance and way of life they are similar to the orb shell, but *Pisidium* has only one siphon and its shell is flatter.

113

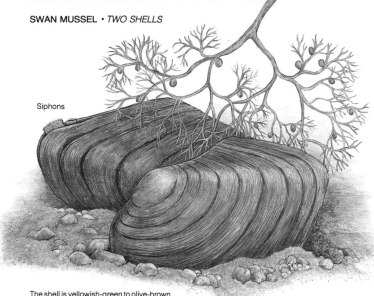

Siphons

The larvae, called glochidia (meaning 'arrowheads' in Latin), swim by snapping their tiny valves together.

Using the muscular foot that protrudes from its shell, the mussel can crawl slowly or half bury itself in the bed.

The swan mussel is common in muddy-bottomed, slow-moving or still waters that are fairly hard.

The shell is yellowish-green to olive-brown and an elongated oval in shape with a sharply angled posterior end. Distinct growth lines mark the surface. The front part of the shell normally lies buried but the siphons at the posterior end are always exposed. It is the largest bivalve found in British waters and may grow up to 9 in. (23 cm) long.

Duck mussel
Anodonta anatina

This species has a more oval and swollen shell than the swan mussel, and is usually darker in colour. About 4 in. (10 cm) long.

Anodonta anatina

Anodonta cygnea

Teeth   Byssus

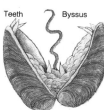

A coiled, sticky thread, the byssus, arises from the central muscle of the larva. The larva uses it and the teeth and spines to attach itself to the gills (or sometimes the skin) of fish to begin its parasitic stage of development.

Swan mussel *Anodonta cygnea*

One of the commonest large freshwater bivalves found in Britain, the swan mussel is notable for its long life-span – probably over 11 years under normal conditions. The age of the animal is related to the number of concentric lines on either of the two equal sized valves, or shells. The mussel is not eaten.

The animal breathes and feeds by drawing a current of water into the gill area within the valves, using an inhalant tube, or siphon, which is fringed with finger-like outgrowths. The water is passed through the gills, allowing oxygen to be extracted and trapping small animals and algae for food. It is then expelled through a second siphon.

The female produces up to a million eggs, which are retained within a brood pouch, called the marsupium, in the gill region. Here, over a period of several months, they develop into larval forms called glochidia. In spring these are ejected into the water, where the majority fall victim to predators. The survivors attach themselves to fish, and become encapsulated as the skin of the fish grows over them to form a cyst. Within the cyst the larva feeds parasitically on its host until it changes into a young mussel; it then breaks out and sinks to the river-bed.

Four common freshwater mussels

In Roman times British rivers were famous for freshwater pearls obtained from the pearl mussel, which is restricted to fairly fast-flowing rivers. Small pearls are also produced occasionally by the painter's mussel. Both these species and the swollen river mussel are parasitic on fish in their larval stages, but the zebra mussel produces free-swimming larvae like the marine molluscs. None of the mussels are eaten.

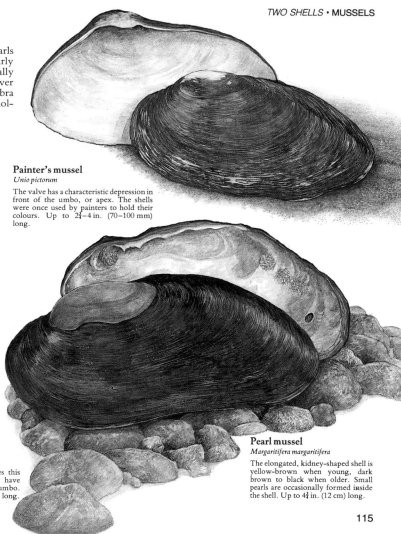

Painter's mussel
Unio pictorum

The valve has a characteristic depression in front of the umbo, or apex. The shells were once used by painters to hold their colours. Up to 2¾–4 in. (70–100 mm) long.

Zebra mussel
Dreissena polymorpha

Conspicuous, dark, zigzag stripes mark the two valves, which are shaped like those of the common marine mussel. The zebra mussel attaches itself to hard objects such as stones and posts by sticky byssus threads. Often found in clusters in slow rivers, canals, lakes and reservoirs, it grows to ⅝–¾ in. (15–18 mm) high and 1⅛–1½ in. (28–40 mm) long.

Swollen river mussel
Unio tumidus

The inflated umbo, or apex, gives this species its name. The shell may have greenish stripes radiating from the umbo. It grows up to 2½–3½ in. (64–90 mm) long.

Pearl mussel
Margaritifera margaritifera

The elongated, kidney-shaped shell is yellow-brown when young, dark brown to black when older. Small pearls are occasionally formed inside the shell. Up to 4¾ in. (12 cm) long.

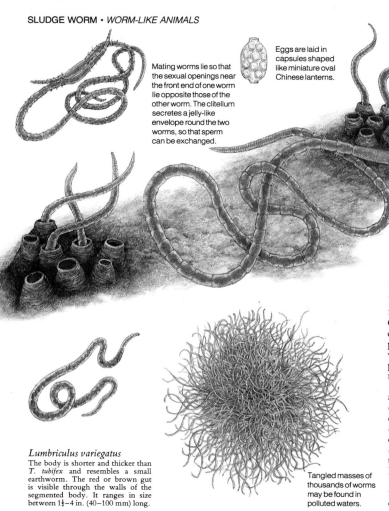

Mating worms lie so that the sexual openings near the front end of one worm lie opposite those of the other worm. The clitellum secretes a jelly-like envelope round the two worms, so that sperm can be exchanged.

Eggs are laid in capsules shaped like miniature oval Chinese lanterns.

The long, thin, red worm lives in a vertical burrow, cemented with slime, in sand or mud. The front end, which has bristles on its segments, is fixed in the burrow and the hind end projects. Up to 3½ in. (90 mm) long, but is usually about 1½ in. (40 mm) long.

The tails of a mass of sludge worms, protruding from mud burrows, clearly show the red pigment in the blood.

Lumbriculus variegatus
The body is shorter and thicker than *T. tubifex* and resembles a small earthworm. The red or brown gut is visible through the walls of the segmented body. It ranges in size between 1½–4 in. (40–100 mm) long.

Tangled masses of thousands of worms may be found in polluted waters.

Sludge worm *Tubifex tubifex*

When sludge worms are found as the most numerous animal in a river or pond, it is a good indication that the water is polluted. Once, they provided the main source of food for ducks that lived on or near the Thames, other food-species having found the polluted waters impossible to live in. Large numbers of sludge worms live on the filter beds of sewage farms, and even where a pond or river is fairly clean, they are most abundant near the muddy banks and in slow-moving water.

Tangled red masses of these worms will be familiar to aquarium owners, who buy them as food for their fish. In a pond or stream, they look like a red smudge on the mud. This red colouring, caused by the blood pigment haemoglobin, gives the clue to the animal's ability to live in organically polluted water containing little dissolved oxygen. Haemoglobin is an excellent carrier of oxygen and maximises the worm's ability to extract the gas from its surroundings. The worm's tail stirs the water to help oxygen absorption.

Lumbriculus variegatus also lives in mud tubes in standing waters. It is common in woodland ponds and looks like a small earthworm.

Identifying other freshwater worms

There are three main groups of freshwater worms – annelid, round and flat. The annelid worms, which include the sludge worm, *Stylaria* and *Eiseniella*, have segmented bodies. The roundworms such as *Dorylaimus*, and the similar horsehair worms, are unsegmented. The flatworms, which prey on small larvae and dead animals, have unsegmented, ribbon-like bodies, normally with distinct heads and eyes.

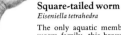

Square-tailed worm
Eiseniella tetrahedra

The only aquatic member of the earth-worm family, this brownish, segmented worm has a body square in cross-section in the hind region. Found near the banks of streams or edges of lakes, it grows up to 2 in. (50 mm) long.

Dendrocoelum lacteum

A large, milky-white flatworm with a semi-transparent body through which the brownish gut shows. It has two eyes set in a squared-off head. Common in streams and lakes. Up to 1 in. (25 mm) long.

Dugesia lugubris

The two eyes are set fairly close together in a spade-shaped head. The body is black or sometimes grey-brown. A very common flatworm in standing and slow-flowing water, it grows up to ½ in. (20 mm) long.

Dorylaimus stagnalis

This long, thin, colourless roundworm is unsegmented and pointed at both ends. It moves by S-shaped flexing of the body. Found among plant roots and in mud in ponds and slow-flowing water, it is one of the largest freshwater roundworms, growing to about ⅜ in. (8 mm) long.

Polycelis nigra

The unsegmented body is usually dark grey or black, the head is squared off. Many small eye-spots dot the edges of the front end. A common flatworm in all waters. Up to ½ in. (13 mm) long.

Stylaria lacustris

The yellow to brown transparent segmented body has bristles, eye-spots and a long, thin proboscis. The worm is common in fresh and brackish water, especially in standing water. It may be seen in chains. Up to ¾ in. (18 mm) long.

Horsehair worm
Gordius species

Yellow, brown or black in colour, these very thin, hair-like worms live as adults on vegetation in still water. The larvae are parasitic on insects and other arthropods. The worms grow up to 32 in. (80 cm) long.

117

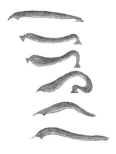

The medicinal leech swims by undulating its body. It can also crawl like a caterpillar by arching its back and using front and rear suckers to push and pull.

The underside is green or yellowish-green and is sometimes marked with irregular black spots. The sucker at the tail end attaches the leech while it feeds.

The elongated, slightly flattened body is olive-green with reddish-yellow or reddish-brown stripes down the back. Strong teeth in the jaw leave a Y-shaped cut in the prey's skin. It grows 4–6 in. (10–15 cm) long and ⅖–⅗ in. (10–15 mm) wide.

Cocoons of eggs are laid in a damp place above the water line. Each cocoon contains five to fifteen eggs and is ⅖ in. (10 mm) long.

The medicinal leech was once common in village ponds where horses were watered in centuries gone by.

Medicinal leech *Hirudo medicinalis*

The only British leech with teeth that can bite through the human skin is the medicinal leech. As a full-grown specimen can take up to five times its own weight of blood at one time, it was once widely used for blood-letting. This treatment was used for such a wide range of ills that by the 19th century the demand for medicinal leeches almost wiped out British stocks.

The medicinal leech, like other British leeches, lives in fresh water where horses, cattle and other mammals – its main food source – drink. The demise of horse transport, the advent of piped water supplies for livestock, the draining of ponds and the reduction in the numbers of frogs (on which young *Hirudo* feed) have all contributed to the near extinction of this species.

Leeches such as the fish leech feed by means of an extendible tube, or proboscis, to penetrate the host. Others, like the medicinal leech, have toothed jaws that bite into the prey. And some, like *Erpobdella*, have no teeth or proboscis and swallow their prey – gnat larvae, small insects and small crustaceans – whole. The eggs of leeches are laid in cocoons, which some species carry until they hatch. With other species, the parent broods over the eggs by covering the cocoon with the body.

118

Six common leeches

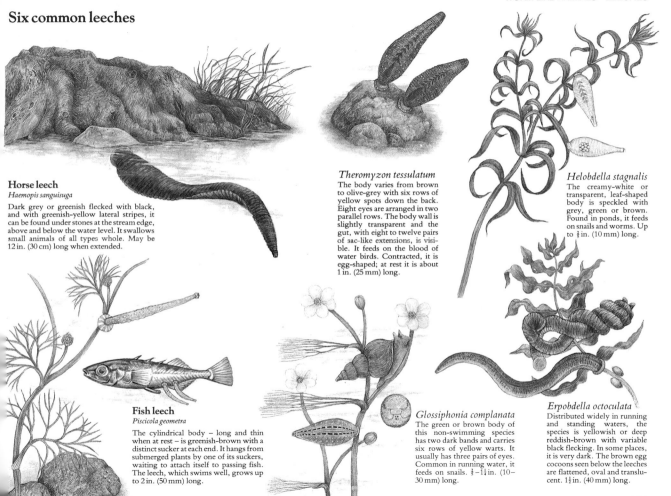

Horse leech
Haemopis sanguisuga

Dark grey or greenish flecked with black, and with greenish-yellow lateral stripes, it can be found under stones at the stream edge, above and below the water level. It swallows small animals of all types whole. May be 12 in. (30 cm) long when extended.

Fish leech
Piscicola geometra

The cylindrical body – long and thin when at rest – is greenish-brown with a distinct sucker at each end. It hangs from submerged plants by one of its suckers, waiting to attach itself to passing fish. The leech, which swims well, grows up to 2 in. (50 mm) long.

Theromyzon tessulatum
The body varies from brown to olive-grey with six rows of yellow spots down the back. Eight eyes are arranged in two parallel rows. The body wall is slightly transparent and the gut, with eight to twelve pairs of sac-like extensions, is visible. It feeds on the blood of water birds. Contracted, it is egg-shaped; at rest it is about 1 in. (25 mm) long.

Glossiphonia complanata
The green or brown body of this non-swimming species has two dark bands and carries six rows of yellow warts. It usually has three pairs of eyes. Common in running water, it feeds on snails. ¾–1¼ in. (10–30 mm) long.

Helobdella stagnalis
The creamy-white or transparent, leaf-shaped body is speckled with grey, green or brown. Found in ponds, it feeds on snails and worms. Up to ⅜ in. (10 mm) long.

Erpobdella octoculata
Distributed widely in running and standing waters, the species is yellowish or deep reddish-brown with variable black flecking. In some places, it is very dark. The brown egg cocoons seen below the leeches are flattened, oval and translucent. 1½ in. (40 mm) long.

119

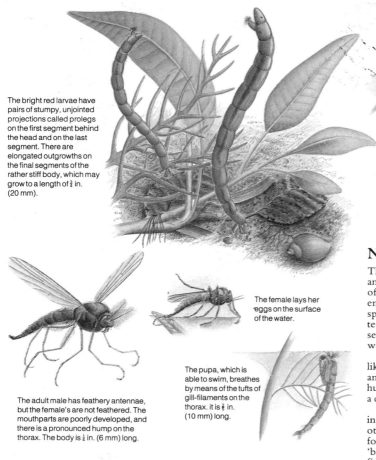

The bright red larvae have pairs of stumpy, unjointed projections called prolegs on the first segment behind the head and on the last segment. There are elongated outgrowths on the final segments of the rather stiff body, which may grow to a length of ¾ in. (20 mm).

Contained in a mass of jelly, the eggs are anchored to water plants.

Chironomidae, one of which is seen emerging from its pupa, are often abundant in organically polluted waters.

The female lays her eggs on the surface of the water.

The adult male has feathery antennae, but the female's are not feathered. The mouthparts are poorly developed, and there is a pronounced hump on the thorax. The body is ¼ in. (6 mm) long.

The pupa, which is able to swim, breathes by means of the tufts of gill-filaments on the thorax. it is ⅜ in. (10 mm) long.

Non-biting midge *Chironomus plumosus*

The non-biting midges are often mistaken for the biting gnats and mosquitoes, but they have very weak mouthparts and most of them cannot even feed, mating and dying within a few days of emergence from the pupa. The larvae of one of the largest species, *Chironomus plumosus*, can be seen from April to September in still and slow-flowing waters. Rivers polluted with sewage may contain vast numbers of these bright red 'bloodworms', living in densities of thousands per square yard.

Sometimes, on warm summer evenings, what at first looks like a dense cloud of smoke can be seen hovering around trees and buildings near water. It is in fact a mating swarm of hundreds of thousands of male midges hovering up and down in a courtship dance to attract the females.

Enormous numbers of eggs are laid in a spiral rope contained in a mass of jelly. This is attached to plants under water or some other submerged object. The tiny larvae that hatch out are ideal food for many species of young fish. Anglers also collect the 'bloodworms' and use them for bait. After many moults the final larval stage pupates. The pupa is mobile and eventually splits to release the adult midge.

Identifying other legless larvae

Over 500 species of Diptera, two-winged flies, occur in Britain, many of which have aquatic, legless larvae. The adults are usually seen during the summer, for most species, such as the phantom midges, craneflies and hoverflies, overwinter as larvae. They can be found in a wide variety of freshwater habitats, but blackfly larvae only live in rivers and streams, while mosquito and hoverfly larvae are most numerous in still waters. Some species of beetle also have legless larvae.

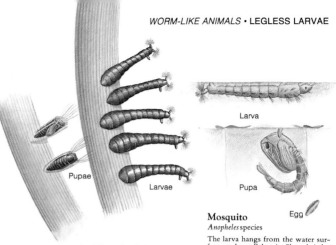

Larva

Phantom midge
Chaoborus species

The larva is transparent with two shiny black air sacs. Prey is caught by hook-like antennae. It grows to ⅝ in. (15 mm) long.

Pupae

Larvae

Blackfly or buffalo gnat
Simulium species

Two fans of hairs on the head of the larva filter food from running water. Both larva and pupa are attached to vegetation or stones. Up to ⅝ in. (15 mm) long.

Pupa

Egg

Mosquito
Anopheles species

The larva hangs from the water surface, and parallel to it. Short bristles on the head create a feeding current. Up to ⅜ in. (10 mm) long.

Cranefly
Dicranota species

This mud-dwelling larva with a very small head and five pairs of prolegs is fairly common in ponds and streams. About ¾ in. (20 mm) long.

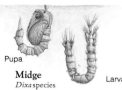

Pupa

Midge
Dixa species

Larva

The body segments of the larva are all similar in size, unlike those of mosquito larvae. The U-shaped posture is characteristic. Up to ⅜ in. (10 mm) long.

Hoverfly
Eristalis species

The rat-tailed maggot inhabits polluted water and breathes through a telescopic tube. Up to ¾ in. (20 mm).

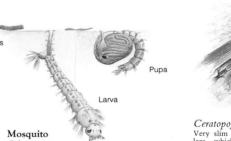

Eggs

Pupa

Larva

Mosquito
Culex species

Similar to *Anopheles*, but this larva hangs from the surface at an angle. The eggs float in rafts. Up to ⅜ in. (10 mm) long.

Ceratopogon species

Very slim larvae without prolegs, which swim by lashing backwards and forwards. They are normally ½ in. (13 mm) long.

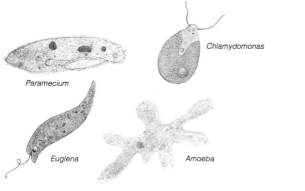

Paramecium

Chlamydomonas

Euglena

Amoeba

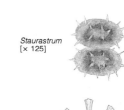

Volvox
[× 75]

Staurastrum
[× 125]

Small floating algae
The algae include desmids such as *Staurastrum*, colonial green algae such as *Volvox*, and diatoms such as *Asterionella*.

Asterionella
[× 175]

Single-celled organisms
Euglena and *Chlamydomonas* contain green chlorophyll and can manufacture their own food by photosynthesis. *Paramecium* swims by vibrating a fringe of hair-like cilia, while *Amoeba* moves by extending and contracting its jelly-like body. Both feed on minute particles of organic matter.

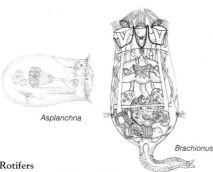

Asplanchna

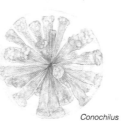

Conochilus

Brachionus

Keratella

Rotifers
Among the most abundant of freshwater animals, rotifers are characterised by a ring of hair-like cilia at the front end, used for propulsion and feeding on bacteria and floating organic particles. Most are solitary, but *Conochilus* is colonial.

Minute floating animals and plants

Rivers, lakes and ponds contain an abundance of small floating, or planktonic, animals and plants. Among them are a variety of microscopic, single-celled organisms. They include plants such as the algae and diatoms, which can convert simple chemicals into organic food by photosynthesis, and animals such as *Amoeba* which cannot do this and must feed on other organisms. There are also organisms which cannot be placed in either category, such as *Euglena* and *Chlamydomonas*. It is thought that they can behave as animals or plants, undergoing photosynthesis in the light but able to feed on organic material in low light conditions.

The small multicellular animals such as the rotifers feed on microscopic life. Although more complex, some of the rotifers are smaller than the largest single-celled organisms.

Much larger are the many crustaceans which float in the plankton. In ideal conditions of high temperatures and abundant food many of them, such as the water fleas, can multiply very quickly to form dense populations. They are a major food of many freshwater fish, particularly the fry stage, when they are very small.

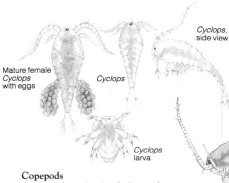

Mature female
Cyclops
with eggs

Cyclops

Cyclops,
side view

Cyclops
larva

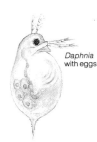

Female
Diaptomus
with eggs

Male
Diaptomus

Female *Diaptomus,*
side view

Copepods

Common, free-swimming freshwater forms include *Cyclops* species, $\frac{1}{12}$ in. (1 mm) long, so-named because of the single eye, and *Diaptomus* species, $\frac{1}{8}$ in. (3 mm) long, which have very long balancing antennae. The females carry egg sacs; *Cyclops* has two, *Diaptomus* has one.

Fish louse

Argulus species

Although it feeds by attaching itself to fish and sucking their blood, *Argulus* is often found in open water. Up to $\frac{1}{5}$ in. (5 mm) long.

Daphnia
with eggs

*Simocephalus
vetulis*

Water fleas

A very numerous and diverse group of small crustaceans which feed by filtering small organisms out of the water. Among the commonest are *Daphnia* species and *Simocephalus vetulis.* Both are about $\frac{1}{8}$ in. (3 mm) long.

Fairy shrimp

Chirocephalus species

A filter-feeding relative of the water fleas, found in muddy pools in southern England. Up to $\frac{2}{5}$ in. (10 mm) long.

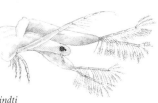

Leptodora kindti

Related to *Daphnia,* with very long antennae and no outer shell, *Leptodora* is a predator on other small crustaceans. Up to $\frac{2}{5}$ in. (10 mm) long.

Ostracod

Candona species

Similar to bivalve molluscs in appearance, ostracods are crustaceans with jointed legs used for swimming and feeding on organic particles. Up to $\frac{1}{12}$ in. (2 mm) long.

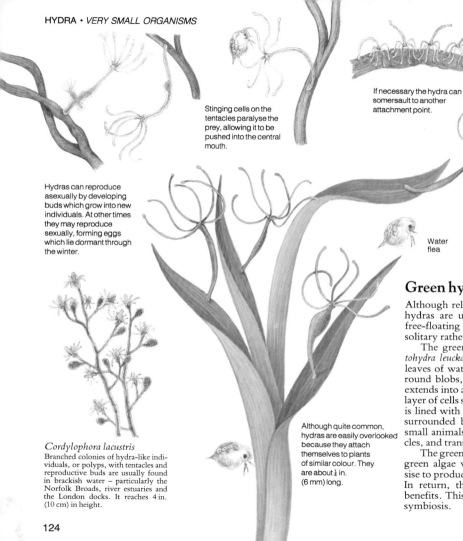

Stinging cells on the tentacles paralyse the prey, allowing it to be pushed into the central mouth.

If necessary the hydra can somersault to another attachment point.

Hydras can reproduce asexually by developing buds which grow into new individuals. At other times they may reproduce sexually, forming eggs which lie dormant through the winter.

Water flea

Cordylophora lacustris
Branched colonies of hydra-like individuals, or polyps, with tentacles and reproductive buds are usually found in brackish water – particularly the Norfolk Broads, river estuaries and the London docks. It reaches 4 in. (10 cm) in height.

Although quite common, hydras are easily overlooked because they attach themselves to plants of similar colour. They are about ¼ in. (6 mm) long.

The green hydra occurs in well-vegetated ponds, lakes and streams throughout the British Isles.

Green hydra *Hydra viridissima*

Although related to the sea anemones, jellyfish and corals, the hydras are unusual among this group in that they have no free-floating medusa stage in their life-cycle, and they live solitary rather than colonial lives.

The green hydra and the less common brown hydra *Protohydra leuckarti* are normally found attached to the stems and leaves of water plants. When alarmed they contract into small round blobs, but if left undisturbed the body of each animal extends into a long, hollow tube formed from an inner and outer layer of cells separated by a jelly-like material. The central cavity is lined with digestive cells, and terminates in a single aperture surrounded by tentacles. When feeding, the hydra captures small animals by using adhesive and stinging cells on its tentacles, and transfers the prey to its body cavity for digestion.

The green colour of *Hydra viridissima* is due to the presence of green algae within its cells. These small plants photosynthesise to produce sugars, some of which can be used by the hydra. In return, the algae obtain protection and, possibly, other benefits. This mutually beneficial relationship is an example of symbiosis.

124

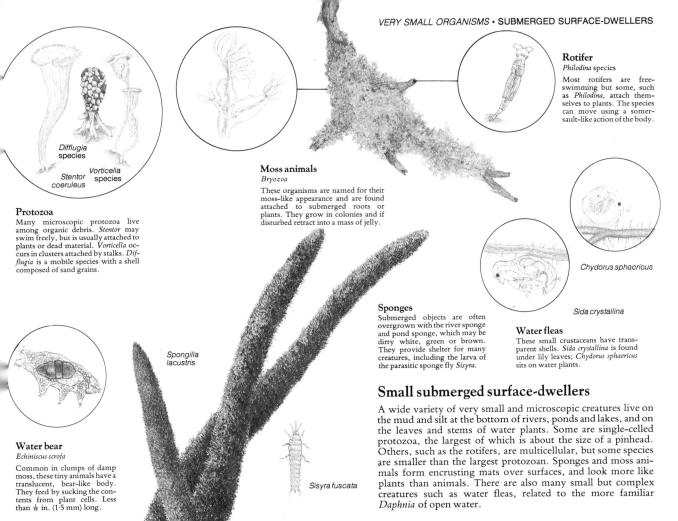

Rotifer
Philodina species

Most rotifers are free-swimming but some, such as *Philodina*, attach themselves to plants. The species can move using a somersault-like action of the body.

Diffugia
species

Stentor
coerulea

Vorticella
species

Protozoa
Many microscopic protozoa live among organic debris. *Stentor* may swim freely, but is usually attached to plants or dead material. *Vorticella* occurs in clusters attached by stalks. *Difflugia* is a mobile species with a shell composed of sand grains.

Moss animals
Bryozoa

These organisms are named for their moss-like appearance and are found attached to submerged roots or plants. They grow in colonies and if disturbed retract into a mass of jelly.

Chydorus sphaericus

Sida crystallina

Sponges
Submerged objects are often overgrown with the river sponge and pond sponge, which may be dirty white, green or brown. They provide shelter for many creatures, including the larva of the parasitic sponge fly *Sisyra*.

Water fleas
These small crustaceans have transparent shells. *Sida crystallina* is found under lily leaves; *Chydorus sphaericus* sits on water plants.

Spongilla
lacustris

Small submerged surface-dwellers

A wide variety of very small and microscopic creatures live on the mud and silt at the bottom of rivers, ponds and lakes, and on the leaves and stems of water plants. Some are single-celled protozoa, the largest of which is about the size of a pinhead. Others, such as the rotifers, are multicellular, but some species are smaller than the largest protozoan. Sponges and moss animals form encrusting mats over surfaces, and look more like plants than animals. There are also many small but complex creatures such as water fleas, related to the more familiar *Daphnia* of open water.

Water bear
Echiniscus scrofa

Common in clumps of damp moss, these tiny animals have a translucent, bear-like body. They feed by sucking the contents from plant cells. Less than ⅛ in. (1·5 mm) long.

Sisyra fuscata

Crystalwort
Riccia fluitans

The branched fronds of this liverwort form mats which float just below the surface of still water and may cover several square yards. Fronds up to 2 in. (50 mm) long.

Concephalum conicum

This liverwort, common in damp, shady places, forms lush green carpets. Branches up to 8 in. (20 cm) long but very variable.

Marchantia polymorpha aquatica

A large liverwort of marshes, streams and rivers. Wavy-edged branches, with raised cup-like structures, grow to 4 in. (10 cm) long.

Scapania undulata

A liverwort of wet mountainous areas, particularly in fast-flowing streams. Submerged shoots may reach 4 in. (10 cm).

Rhacomitrium aquaticum

A moss often found by mountain streams, growing in patches of varying size. Stems are $1\frac{1}{2}$–$4\frac{3}{4}$ in. (4–12 cm) long.

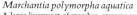

Drepanocladus fluitans

This moss grows both above and below water in bogs and marshes. Stems may be very short or up to 12 in. (30 cm) long.

Sphagnum plumilosum

A common *Sphagnum* found in bog and moorland all over Britain. Low-growing and usually tinged red.

Water ferns, mosses and liverworts

Some of the simplest forms of plant life are non-flowering species so small that individually they are invisible to the naked eye, but growing in a mass they provide much of the colour and texture in, on and near streams, ponds and lakes. Other larger non-flowering plants associated with fresh water, and illustrated here, include mosses, liverworts, water ferns, quillworts and horsetails. These plants are not only colourful but also play an important role in water life, giving cover in which small creatures can hide from predators, providing food for aquatic animals and serving as sites for egg-laying.

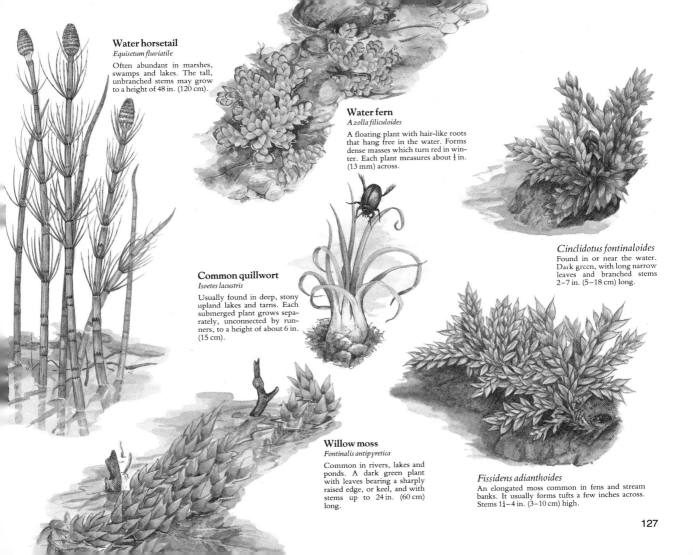

Water horsetail
Equisetum fluviatile

Often abundant in marshes, swamps and lakes. The tall, unbranched stems may grow to a height of 48 in. (120 cm).

Water fern
Azolla filiculoides

A floating plant with hair-like roots that hang free in the water. Forms dense masses which turn red in winter. Each plant measures about ½ in. (13 mm) across.

Common quillwort
Isoetes lacustris

Usually found in deep, stony upland lakes and tarns. Each submerged plant grows separately, unconnected by runners, to a height of about 6 in. (15 cm).

Cinclidotus fontinaloides
Found in or near the water. Dark green, with long narrow leaves and branched stems 2–7 in. (5–18 cm) long.

Willow moss
Fontinalis antipyretica

Common in rivers, lakes and ponds. A dark green plant with leaves bearing a sharply raised edge, or keel, and with stems up to 24 in. (60 cm) long.

Fissidens adianthoides
An elongated moss common in fens and stream banks. It usually forms tufts a few inches across. Stems 1¼–4 in. (3–10 cm) high.

Huge wintering flocks of waders such as dunlins, knots, oystercatchers and shelducks probe the rich mud for worms and shellfish. Brent geese, swans and ducks such as mallard, teal and wigeon graze on the salt-marsh grasses, and on the green seaweed and eel grass which grows on the mud-flats.

The fluctuating world of the river estuary

An estuary is the tidal mouth of a river, where fresh and salt water mix. There are more than 300 estuaries in Britain, ranging in size from the Severn, some ten miles across at its widest point, to small creeks only a few yards wide. Most estuaries are funnel-shaped, with a wide mouth flanked by sand or mud-flats and narrower upper reaches sometimes bordered by steep banks or cliffs. Within the estuary are a variety of habitats, each supporting its own particular plant and animal life. Extensive areas of mud are formed from silt brought down by the river or carried in by the tide. In quiet stretches silt, sand and algae bind together and build up into mounds which eventually become colonised by plants. The tide cuts channels through these flats and a salt-marsh develops.

Few plants and animals can tolerate the rapid changes in the saltiness, temperature and turbidity of the water during the tidal cycle, but those that do so occur in vast numbers because of the continuous influx of nutrients and organic matter supplied by river and sea. Seabirds and waders find rich pickings among the huge populations of worms, shellfish and shrimps, while for fishermen there is often a good harvest of edible shellfish such as cockles, winkles and mussels.

Herons, cormorants, grebes and divers search for gobies, smelts, sprats, sticklebacks, migrating eels and lampreys. Shore crabs, which can tolerate wide variations in salinity and temperature, are found throughout the estuary.

Hidden in the soft mud-flats exposed at low tide are vast numbers of ragworms, lugworms and molluscs such as the Baltic tellin. The tiny snail *Hydrobia* lives on the surface of the mud, feeding on organic debris, and extensive beds of oysters and mussels may develop. Around the margins of the estuary, salt-tolerant plants such as glasswort, common cord-grass and common salt-marsh grass colonise the mud, trapping silt and raising the level to form a salt-marsh. Other plants take root, including thrift, sea aster, sea lavender and marsh mallow, and provide food and nesting sites for a variety of birds including reed buntings and meadow pipits, as well as a colourful show of summer flowers.

A GUIDE TO FRESHWATER AND MARINE SHELLS

The following chart will help to identify the shells most likely to be found in British waters. It describes where each living shell can be found, the maximum size to which it will grow and, where the family is mentioned in the book, gives a page number for further information about the species itself or the family.

Freshwater – single shell

Nautilus ramshorn
Armiger (Planorbis) crista
On water plants in ponds, ditches and marshes. Widespread, 1/16 in. (2 mm) wide. Page 111

White ramshorn
Planorbis albus
In all fresh waters. Widespread, 1/4 in. (6 mm) across. Page 111

Ear pond snail
Lymnaea auricularia
Fairly common in slow-flowing rivers and lakes. Widespread, 1 3/8 in. (35 mm) high. Page 110

Nerite
Theodoxus fluviatilis
Under stones or plants in lime-rich waters. Fairly common in England. 1/2 in. (13 mm) wide. Page 111

River snail
Viviparus viviparus
In slow-flowing, well-oxygenated waters. Common in England. 1 1/2 in. (40 mm) high. Page 111

Great ramshorn
Planorbarius corneus
In lakes, rivers and marshes. Widespread (except north Scotland). 1 1/4 in. (30 mm) across. Page 111

Lister's river snail
Viviparus contectus
In slow-flowing, well-oxygenated, hard waters in central and eastern England. 1 1/2 in. (40 mm) high. Page 111

Valve snail
Valvata piscinalis
On weeds in slow-flowing, well-oxygenated waters. Widespread, 1/8 in. (7 mm) wide. Page 111

Jenkins' spire shell
Hydrobia jenkinsi
On stones, weed and mud in running fresh and brackish water. Widespread, 1/8 in. (5 mm) high. Page 111

Common bithynia
Bithynia tentaculata
On stones and weeds in lime-rich slow rivers, canals and lakes. Widespread, but rare in north and west. 3/8 in. (15 mm) high. Page 111

Great pond snail
Lymnaea stagnalis
In ponds, lakes and slow-flowing rivers. Widespread (except north Scotland). 2 in. (50 mm) high. Page 110

Bithynia leachii
Slow-flowing rivers and large lakes. Mainly south and east England. 1/4 in. (6 mm) high. Page 111

Flat valve snail
Valvata cristata
Among water plants in slow-flowing muddy waters. Fairly common and widespread. 1/4 in. (5 mm) wide. Page 111

Ramshorn
Planorbis planorbis
Common in ponds, ditches and slow-flowing rivers. Widespread, 3/4 in. (18 mm) across. Page 111

Wandering snail
Lymnaea peregra
Very common in all fresh waters. Widespread, 3/4 in. (20 mm) high. Page 110

Fountain bladder snail
Physa fontinalis
In well-vegetated rivers and streams. Widespread, 1/2 in. (13 mm) high. Page 111

Lake limpet
Acroloxus lacustris
On water plants in slow and still waters. England and Wales. Rare in Scotland. 1/16 in. (7 mm) long. Page 112

River limpet
Ancylus fluviatilis
On stones and plants in streams, rivers and lakes. Widespread, 1/3 in. (9 mm) long. Page 112

Freshwater – two shells

Zebra mussel
Dreissena polymorpha
In slow rivers, canals and
reservoirs; rare in lakes.
Devon to mid-Scotland.
1½ in. (40 mm). Page 115

Orb shell cockle
Sphaerium lacustrae
In well-oxygenated ponds
and drains. Widespread,
distinctive cap on point of
shell. ⅝ in. (15 mm).
Page 113

Orb shell cockle
Sphaerium species
In most types of fresh water.
Common and widespread.
¾ in. (20 mm). Page 113

Pearl mussel
Margaritifera margaritifera
In fast-flowing rivers with low
lime content. Widespread, 4¾ in.
(12 cm). Page 115

Painter's mussel
Unio pictorum
In lime-rich slow rivers, ponds
and lakes. Mainly Midlands and
S.E. 4 in. (10 cm). Page 115

Pea shell cockle
Pisidium amnicum
In rivers, canals and lowland lakes,
moderately hard water. Widespread,
⁷⁄₁₆ in. (11 mm). Page 113

Swollen river mussel
Unio tumidus
Slow-flowing rivers and canals.
Mainly Midlands and S.E. 3½ in.
(90 mm). Page 115

Marine – single shell

Netted dog whelk
Nassarius reticulatus
Under stones and in
crevices, lower shore and
shallows. Widespread,
1½ in. (40 mm). Page 233

Dog whelk
Nucella lapillus
Crevices and rocks,
among barnacles and
mussels, middle and
lower shore.
Widespread, 1¼ in.
(30 mm). Page 233

Duck mussel
Anodonta anatina
In rivers and canals with
sandy bottoms. Common in
England. 4 in. (10 cm).
Page 114

Swan mussel
Anodonta cygnea
In slow rivers, canals, lakes and
large ponds with muddy bottoms.
Widespread (not north Scotland).
9 in. (23 cm). Page 114

Common whelk
Buccinum undatum
Sand, mud and rocks,
low water and below.
Widespread, 4¾ in.
(12 cm). Page 232

Common necklace shell
Natica alderi

In sand, lower shore and below. Widespread, ⅔ in. (15 mm). Page 231

Violet sea snail
Janthina janthina

An open sea species: dead specimens on strandline. South-west coasts. ⅔ in. (15 mm). Page 289

Small winkle
Littorina neritoides

In crevices, upper shore. Widespread, ³⁄₁₆ in. (5 mm). Page 231

Purple top shell
Gibbula umbilicalis

On rocks, upper and middle shore. Widespread (not east coasts). ½ in. (13 mm). Page 235

Grey top shell
Gibbula cineraria

On and under stones and seaweeds, lower shore and below. Widespread, ½ in. (13 mm). Page 235

Thick-lipped dog whelk
Nassarius incrassatus

Under stones and in crevices, lower shore and shallows. Widespread, ⅔ in. (15 mm). Page 233

Large necklace shell
Natica catena

In sand, lower shore and below. Widespread, 1½ in. (40 mm). Page 231

Rough winkle
Littorina saxatilis

In crevices, on stones, upper and middle shore. Widespread, ⁵⁄₁₆ in. (8 mm). Page 231

Thick top shell
Monodonta lineata

On rocks, middle shore. South-west and west coasts. 1 in. (25 mm). Page 235.

Common winkle
Littorina littorea

On rocks and seaweeds, middle and lower shore. Widespread, 1¼ in. (30 mm). Page 230

Spindle shell
Neptunea antiqua

On sand and mud. Widely distributed except Channel Islands and south coasts. 6 in. (15 cm). Page 233

Oyster drill (sting winkle)
Ocenebra erinacea

On rocks, lower shore and below. Widespread, 2⅜ in. (60 mm). Page 233

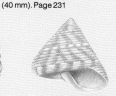

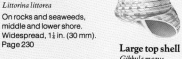

Painted top shell
Calliostoma zizyphinum

On rocks and kelp, lower shore and below. Widespread, 1 in. (25 mm). Page 235

Flat winkle
Littorina littoralis

On seaweeds, middle and lower shore. Widespread, ⅜ in. (10 mm). Page 231

Large top shell
Gibbula magus

On muddy sand and gravel, lower shore and below. South-west and west coasts. ¾ in. (20 mm). Page 235

Acteon shell
Acteon tornatilis

In sand and mud, lower shore and below. Widespread, ¾ in. (20 mm). Page 231

China limpet
Patella aspera

On rocks, low on the shore.
South-west, west and north-east
coasts. 2½ in. (60 mm).
Page 226

Common limpet
Patella vulgata

On rocks, stones and in pools
throughout the shore.
Widespread, 2½ in. (60 mm).
Page 226

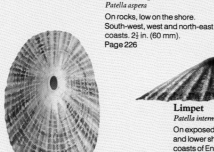

Limpet
Patella intermedia

On exposed rocks, middle
and lower shore. South-west
coasts of England and Wales.
1½ in. (40 mm). Page 226

Tortoiseshell limpet
Acmaea tessulata

On rocks, lower shore and
shallows. North coasts. 1 in.
(25 mm). Page 227

Keyhole limpet
Diodora apertura

Rocks, lower shore.
Widespread, 1½ in.
(40 mm). Page 227

White tortoiseshell limpet
Acmaea virginea

On rocks and among kelp, lower
shore and below. Widespread,
½ in. (13 mm). Page 227

Ormer
Haliotis tuberculata

Among rocks, lower shore
and shallow water. Channel
Isles. 4 in. (10 cm). Page 227

Blue-rayed limpet
Patina pellucida

On kelp plants, lower shore
and below. Widespread, 1 in.
(25 mm). Page 227

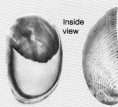

Inside
view

Side
view

Slipper limpet
Crepidula fornicata

Shallow water, in chains attached to
stones and shells. Widespread (not north
Scotland). 2 in. (50 mm). Page 227

Slit limpet
Emarginula reticulata

On and under rocks, lower shore
and below. Widespread, ¾ in. (20 mm).
Page 227

Tusk shell
Dentalium entalis

In sand and mud, offshore.
Widespread, 1⅜ in. (35 mm).
Page 237

Pelican's foot shell
Aporrhais pes-pelecani

In muddy gravel, shallow water
and below. Widespread, 1½ in.
(40 mm). Page 237

European cowrie
Trivia monacha

Among rocks and sea
squirts, lower shore and
below. Widespread, ½ in.
(13 mm). Page 234

Cowrie
Trivia arctica

Among rocks and sea
squirts, lower shore (rarely)
and below. Widespread,
⅜ in. (10 mm). Page 234

Tower shell
Turritella communis

In sand and mud, offshore.
Widespread, 2⅜ in. (60 mm).
Page 236

Banded chink shell
Lacuna vincta

On seaweed, lower shore
and shallows. Widespread,
5/16 in. (8 mm). Page 231

Needle shell
Bittium reticulatum

Among rocks and stones,
lower shore and shallows.
Widespread, ½ in. (13 mm).
Page 237

Common wendletrap
Clathrus clathrus

On rocks on lower shore (in spring),
and offshore. Widespread,
1½ in. (40 mm). Page 237

Marine –
two shells

Portuguese oyster
Crassostrea angulata

In commercial beds, in
shallow water. Mostly south
and east coasts. 7 in. (18 cm).
Page 241

Oval venus
Venus ovata

In sand and gravel,
offshore. Widespread, ¾ in.
(20 mm). Page 244

Edible oyster
Ostrea edulis

On firm bottoms in shallow
water, mostly in commercial
beds. Widespread, 4 in.
(10 cm). Page 240

Banded venus
Venus fasciata

In coarse sand or gravel,
offshore. Widespread, 1 in.
(25 mm). Page 244

Saddle oyster
Anomia ephippium

Attached to rocks and shells,
middle shore downwards.
Widespread, 2⅜ in. (60 mm).
Page 241

Dosinia species

In sand and shell gravel,
extreme lower shore and
below. Widespread, 2¼ in.
(55 mm). Page 244

Banded carpet shell
Venerupis rhomboides

In sand and gravel, extreme
lower shore and below.
Widespread, 2½ in. (60 mm).
Page 244

Pullet carpet shell
Venerupis pullastra

In hard sand and gravel, middle
shore and below. Widespread, 2 in.
(50 mm). Page 244

Common otter shell
Lutraria lutraria

In sand and sandy mud,
extreme lower shore and
below. Widespread, 4½ in.
(12 cm). Page 246

Warty venus
Venus verrucosa

In sand and gravel,
extreme lower shore and
below. South and west
coasts. 2½ in. (64 mm).
Page 244

Striped venus
Venus striatula

In sand, lower shore
downwards. Widespread,
1¾ in. (45 mm). Page 244

Callista chione
In clean sand, offshore.
South-west and west coasts
and Channel Islands. 3½ in.
(90 mm). Page 244

Cross cut carpet shell
Venerupis decussata

In sand and gravel. Mainly
south and west coasts. 3⅛ in.
(80 mm). Page 244

Common nut shell
Nucula nucleus

In gravel and coarse sand,
offshore. Widespread, ½ in.
(13 mm).

137

Dog cockle
Glycymeris glycymeris

In sand, mud or gravel, offshore. Widespread, 2¾ in. (70 mm). Page 243

Tellina fabula
In sand, extreme lower shore and below. Widespread, ¾ in. (20 mm). Page 245

Sand gaper
Mya arekaria

In mud and sand, lower shore and below. Widespread, 6 in. (15 cm). Page 246

Elliptical trough shell
Spisula elliptica

In muddy sand and gravel, lower shore and below. Widespread, 1¼ in. (30 mm).

Banded wedge shell
Donax vittatus

In clean sand, from middle shore downwards. Widespread, 1½ in. (40 mm).

Thin tellin
Tellina tenuis

In fine sand, middle shore down to shallows. Widespread, ¾ in. (20 mm). Page 245

Blunt gaper
Mya truncata

In mud or sand, middle shore and below. Widespread, 3 in. (75mm). Page 246

Baltic tellin
Macoma balthica

In mud and muddy sand, especially on estuary shores. Widespread, 1 in. (25 mm). Page 245

Cut trough shell
Spisula subtruncata
In sand and muddy sand, extreme lower shore and below. Widespread, 1 in. (25 mm).

Blunt tellin
Tellina crassa
In mud, sand and gravel, offshore. Widespread, 2½ in. (64 mm). Page 245

Rayed trough shell
Mactra corallina
In clean sand, extreme lower shore and below. Widespread, 2 in. (50 mm).

Thick trough shell
Spisula solida
In sand and gravel, extreme lower shore and below. Widespread, 1¾ in. (45 mm).

Iceland cyprina
Arctica islandica
In firm sand and muddy sand, extreme lower shore and below. Widespread, 5 in. (12·5 cm).

139

Large sunset shell
Gari depressa
In sand, extreme lower shore
and below. Widespread,
2⅜ in. (60 mm).

Peppery furrow shell
Scrobicularia plana
On muddy shores, especially
in the brackish water of
estuaries. Widespread, 2½ in.
(64 mm).

Faroe sunset shell
Gari fervensis
In sand, lower shore and
below. Widespread, 2 in.
(50 mm).

Fan mussel
Pinna fragilis
On muddy sand and gravel,
offshore. Mainly south and
west coasts. 12 in. (30 cm).
Page 239

Curved razor shell
Ensis ensis
In sand, extreme lower
shore and shallows.
Widespread, 5 in.
(12·5 cm). Page 247

Pod razor shell
Ensis siliqua
In fine sand, extreme
lower shore and
below. Widespread,
8 in. (20 cm). Page 247

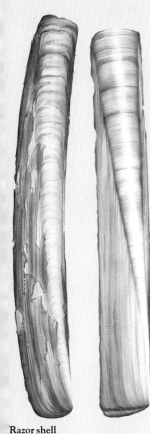

Great scallop
Pecten maximus

On sand and gravel,
offshore. Widespread, 6 in.
(15 cm). Page 248

Razor shell
Ensis arcuatus

In sand, extreme lower
shore and below. Widespread,
6 in. (15 cm). Page 247

Grooved razor shell
Solen marginatus

In sand, lower shore and
shallows. South and west
coasts. 5 in. (12·5 cm).
Page 247

Tiger scallop
Chlamys tigerina

On sand, gravel and stones.
Extreme lower shore
downwards. Widespread,
1 in. (25 mm). Page 249

Hunchback scallop
Chlamys distorta

In rock crevices and kelp
holdfasts, lower shore and
below. Widespread, 1½ in.
(40 mm). Page 249

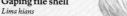

Variegated scallop
Chlamys varia

Among rocks and stones, extreme lower shore and below. Widespread, 2⅜ in. (60 mm). Page 249

Gaping file shell
Lima hians

Among stones and weed, extreme lower shore. South and west coasts. 1 in. (25 mm). Page 249

Oval piddock
Zirfaea crispata

Bores into soft rock, lower shore and shallows. Widespread, 3½ in. (90 mm). Page 251

Red nose
Hiatella arctica

Bores into soft rock, lower shore and shallows. Widespread, 1½ in. (40 mm). Page 251

White piddock
Barnea candida

Bores into soft rock and clay, lower shore and shallows. Widespread, 2½ in. (60 mm). Page 251

Queen scallop
Chlamys opercularis

On firm gravel and sand, extreme lower shore and below. Widespread, 3½ in. (90 mm). Page 249

Common piddock
Pholas dactylus

Bores into soft rock, wood or firm sand, lower shore and shallows. South and south-west coasts. 6 in. (15 cm). Page 250

Common mussel
Mytilus edulis

On rocky, wave-swept coasts and in estuaries, on the shore and below. Widespread, 4 in. (10 cm). Page 238

Horse mussel
Modiolus modiolus

Among kelp, extreme lower shore and often in offshore beds. Widespread, 8 in. (20 cm). Page 239

Edible cockle
Cerastoderma edule

In sediment, middle shore to just below low water. Widespread, 2 in. (50 mm). Page 242

Lagoon cockle
Cerastoderma glaucum

In soft sand and mud, in brackish water and lagoons. Widespread, 2 in. (50 mm). Page 243

Bearded horse mussel
Modiolus barbatus

Under rocks and in kelp holdfasts, lower shore and below. Widespread, 2 in. (50 mm). Page 239

Prickly cockle
Acanthocardia echinata

In sand offshore. Widespread, 3 in. (75 mm). Page 243

Musculus discors
Under rocks, middle shore and below. Widespread, ½ in. (13 mm). Page 239

Little cockle
Parvicardium exiguum

On sand and mud, extreme lower shore and below. Widespread, ½ in. (13 mm). Page 243

The black body has distinctive, large, white areas on the throat and belly and smaller white patches on the back, with a characteristic lens-shaped patch behind the eye. A male may grow up to 30 ft (9 m) and a female 15 ft (4·5 m).

A powerful swimmer, the killer whale pursues large prey such as porpoises and other whales. It also eats fish, penguins, seals and seabirds.

The killer whale is found in all the world's oceans and turns up frequently in British waters.

A single baby is born after a gestation of one year. It is grey and creamy-white for its first year.

The adult male, whose dorsal fin may reach 6 ft (1·8 m) high, is usually solitary but can be seen occasionally with females in small groups.

The killer whale is inquisitive and often lifts its head out of the water to look around. There are up to 50 large pointed teeth in its jaws.

Killer whale *Orcinus orca*

This fearsome-looking creature with an equally fearsome name is, in fact, a playful, intelligent animal which becomes very tame in captivity. It readily learns to perform tricks in a dolphinarium, and the act in which the trainer puts his head in a well-fed killer whale's mouth is not nearly as dangerous as it looks. In the wild, however, they will certainly attack boats.

The killer whale's formidable reputation comes from its enormous appetite for seabirds, penguins, seals and even other whales. It is reported to attack even sharks and polar bears. With its rows of pointed teeth, strong jaws and powerful body it can grab a full-grown seal and shake it as a dog would worry a rat, before tearing it apart to be eaten. In captivity it may eat as much as 100 lb (45 kg) of food a day, but in the wild it will probably consume even more.

The killer whale is widespread throughout the world's oceans and is quite common off the coasts of Britain. It is a powerful swimmer, travelling at up to 30 mph (50 kph) and leaping clear of the water in play. Small groups sometimes join together to form a pack to hunt down seals, showing a co-operation that is a mark of their intelligence.

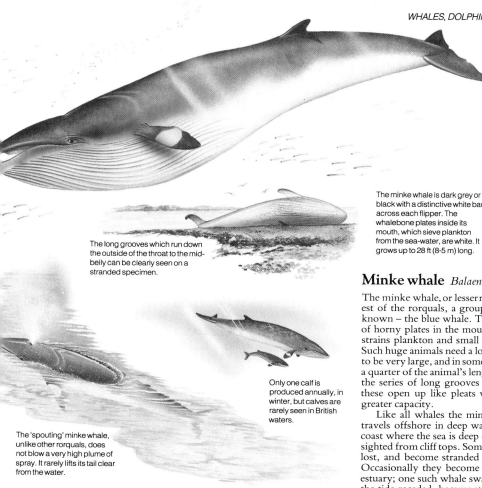

The long grooves which run down the outside of the throat to the mid-belly can be clearly seen on a stranded specimen.

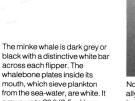

The minke whale is dark grey or black with a distinctive white bar across each flipper. The whalebone plates inside its mouth, which sieve plankton from the sea-water, are white. It grows up to 28 ft (8·5 m) long.

Normally solitary, minke whales are usually seen off north-west Scotland migrating to and from the North Sea.

Only one calf is produced annually, in winter, but calves are rarely seen in British waters.

The 'spouting' minke whale, unlike other rorquals, does not blow a very high plume of spray. It rarely lifts its tail clear from the water.

Minke whale *Balaenoptera acutorostrata*

The minke whale, or lesser rorqual, is the smallest and commonest of the rorquals, a group which includes the largest animal known – the blue whale. They are filter feeders, having a series of horny plates in the mouth each with a bristly fringe which strains plankton and small fish from great gulps of sea-water. Such huge animals need a lot of food, so this filtering system has to be very large, and in some rorquals the mouth takes up almost a quarter of the animal's length. A characteristic of all rorquals is the series of long grooves under the throat; it is thought that these open up like pleats when the mouth is open to give it greater capacity.

Like all whales the minke migrates seasonally and usually travels offshore in deep water. However, around the Scottish coast where the sea is deep close to the shore, minkes are often sighted from cliff tops. Sometimes they come too close in, or get lost, and become stranded on the shore by the outgoing tide. Occasionally they become disorientated and head into a river estuary; one such whale swam up the River Thames and, when the tide receded, became stranded at Kew, more than 50 miles (80 km) from the open sea.

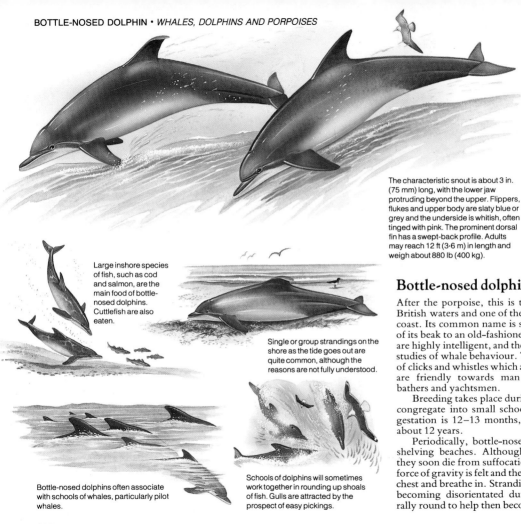

The characteristic snout is about 3 in. (75 mm) long, with the lower jaw protruding beyond the upper. Flippers, flukes and upper body are slaty blue or grey and the underside is whitish, often tinged with pink. The prominent dorsal fin has a swept-back profile. Adults may reach 12 ft (3·6 m) in length and weigh about 880 lb (400 kg).

Large inshore species of fish, such as cod and salmon, are the main food of bottle-nosed dolphins. Cuttlefish are also eaten.

Single or group strandings on the shore as the tide goes out are quite common, although the reasons are not fully understood.

Bottle-nosed dolphins often associate with schools of whales, particularly pilot whales.

Schools of dolphins will sometimes work together in rounding up shoals of fish. Gulls are attracted by the prospect of easy pickings.

Bottle-nosed dolphins are often seen of south and west coasts in summer. The may appear in small schools.

Bottle-nosed dolphin *Tursiops truncatus*

After the porpoise, this is the whale most commonly seen in British waters and one of the few that is often seen off the south coast. Its common name is said to derive from the resemblance of its beak to an old-fashioned gin bottle. Bottle-nosed dolphins are highly intelligent, and they have been the subject of intensive studies of whale behaviour. They are able to emit a wide variety of clicks and whistles which are used for echo location. Dolphins are friendly towards man and sometimes swim alongside bathers and yachtsmen.

Breeding takes place during the summer, when the dolphins congregate into small schools of mixed sexes. The period of gestation is 12–13 months, and both sexes reach maturity at about 12 years.

Periodically, bottle-nosed dolphins are stranded on gently shelving beaches. Although they are air-breathing creatures they soon die from suffocation, because out of the water the full force of gravity is felt and the animal lacks the strength to raise its chest and breathe in. Stranding may be caused by some dolphins becoming disorientated during migration, and others which rally round to help then become stranded themselves.

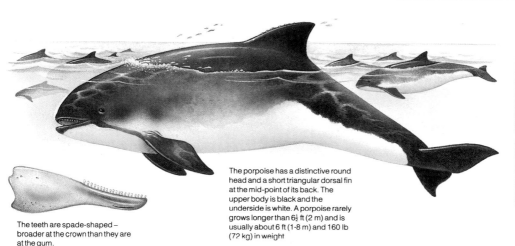

The teeth are spade-shaped – broader at the crown than they are at the gum.

The porpoise has a distinctive round head and a short triangular dorsal fin at the mid-point of its back. The upper body is black and the underside is white. A porpoise rarely grows longer than 6½ ft (2 m) and is usually about 6 ft (1·8 m) and 160 lb (72 kg) in weight

The porpoise, although air-breathing, cannot survive out of water for long as it cannot expand its chest on dry land.

Stranded porpoises are usually found singly. Most strandings occur during late summer.

The young are born in summer, but are rarely seen. This suggests they are born well away from coastal waters.

The porpoise dives to the sea-bed for sand eels, flatfish and crabs. It also feeds on herring and mackerel.

Common porpoise *Phocoena phocoena*

Smallest of all British whales, the common porpoise is the only one of the genus *Phocoena* found in British waters, and occurs on all parts of the coast. Like other whales in British waters porpoises migrate seasonally, sometimes in very large schools, and have been recorded as travelling more than 13 miles (20 km) in four hours. Like dolphins they have an extremely accurate echo-locating ability and acute hearing.

Female porpoises have only one calf each year, after an 11 month pregnancy, born after July. The breeding locality is not known. Maturity is reached in 3–4 years and life expectancy is 10–15 years, providing the porpoise does not fall foul of its chief predator, the killer whale.

Common porpoises have a voracious appetite and an adult probably consumes about 50 herring-size fish a day. Their diet may be their downfall, for porpoises have recently shown signs of pesticide poisoning. It is thought that the fish they eat contain minute amounts of insecticides brought down to the sea by rivers. Over the years these chemicals can build up in a porpoise's body until harmful concentrations are reached. The evidence is not conclusive, but porpoises are becoming scarcer.

147

Pilot whale
Globicephala melaena

A toothed whale, usually seen off south-western coasts in groups of about 20. Up to 20 ft (6 m) long.

Bottle-nosed whale
Hyperoodon ampullatus

Usually seen in small groups off northern coasts in autumn and winter. A toothed whale, up to 30 ft (9 m) long.

Risso's dolphin
Grampus griseus

A toothed whale, seen in small groups off southern and western coasts and off Scotland. Up to 13 ft (4 m) long.

Common dolphin
Delphinus delphis

Normally seen off southern and western coasts, leaping and diving in large groups. A toothed whale, up to 7½ ft (2·3 m) long.

Whales – marine-dwelling mammals

Among the largest mammals in the world, whales are mostly seen at close quarters around Britain only when they become stranded on a beach. The baleen whales feed on plankton and small creatures which they consume by taking in large mouthfuls of water. Instead of teeth the mouth contains flat plates fringed with bristles that filter the food from the water. The toothed whales feed on squids and fish, which they catch and devour in the same manner as other carnivorous mammals. Toothed whales generally have a bulbous, bulging forehead, often with a projecting narrow snout, whereas baleen whales have a flat-topped head. Most of the toothed whales are small and fast moving. Baleen whales are large and ponderous. All whales breathe air through a blowhole which is on top of the head. The position of a large whale is often betrayed by the distinctive 'blow' of air and water vapour as the animal surfaces.

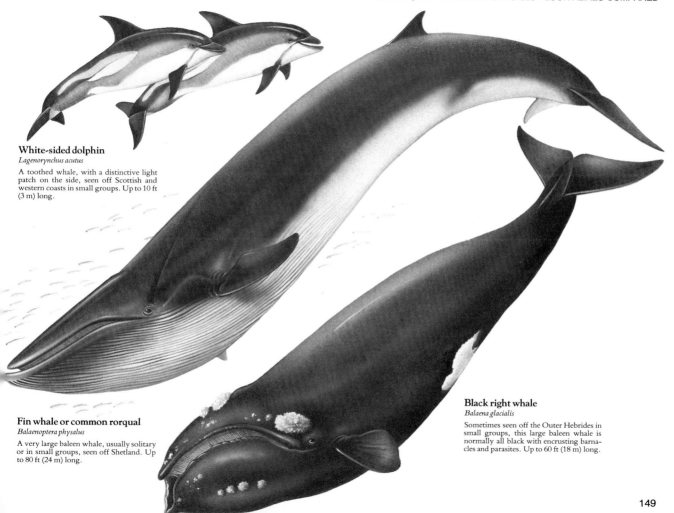

White-sided dolphin
Lagenorynchus acutus

A toothed whale, with a distinctive light patch on the side, seen off Scottish and western coasts in small groups. Up to 10 ft (3 m) long.

Fin whale or common rorqual
Balaenoptera physalus

A very large baleen whale, usually solitary or in small groups, seen off Shetland. Up to 80 ft (24 m) long.

Black right whale
Balaena glacialis

Sometimes seen off the Outer Hebrides in small groups, this large baleen whale is normally all black with encrusting barnacles and parasites. Up to 60 ft (18 m) long.

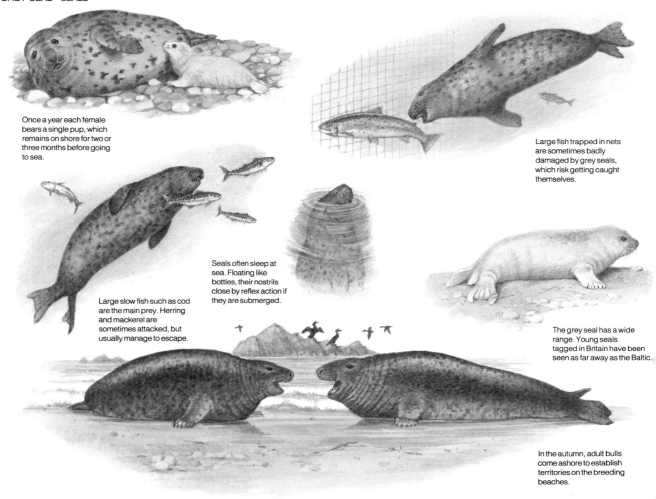

Once a year each female bears a single pup, which remains on shore for two or three months before going to sea.

Large fish trapped in nets are sometimes badly damaged by grey seals, which risk getting caught themselves.

Large slow fish such as cod are the main prey. Herring and mackerel are sometimes attacked, but usually manage to escape.

Seals often sleep at sea. Floating like bottles, their nostrils close by reflex action if they are submerged.

The grey seal has a wide range. Young seals tagged in Britain have been seen as far away as the Baltic.

In the autumn, adult bulls come ashore to establish territories on the breeding beaches.

Female grey seals have dark blotches on a light background. At birth the pup has a white coat, but this is moulted when the pup is about ten days old.

Grey seal *Halichoerus grypus*

The grey seal is one of the world's rare seals, and some 80–85,000 live around our coast – more than 60 per cent of the world population. It is also Britain's largest carnivorous mammal. Breeding colonies are confined mainly to the west coast and the Scottish islands, particularly the Shetlands and the Hebrides, but there is a large colony on the Farne Islands, off the coast of Northumberland, which has existed since the 12th century.

Breeding takes place in the autumn, when the males come ashore to establish territories. The females, already pregnant from the previous breeding season, then join them to give birth to their pups which are suckled by the mother for at least three weeks. They are then left to fend for themselves and the adults mate again. Mating is not confined to pairs; one female may mate several times with more than one male. Females may live for 30 years or more, but males rarely exceed 20 years.

Grey seals have been legally protected in Britain for half a century and their numbers have steadily increased. Culling, the legal killing of the surplus population, is sometimes necessary to protect fisheries and to prevent overcrowding of colonies, but has led to considerable public hostility.

Grey seals may be seen in the water or basking in the sun on rocks or sand-banks. The female has a flat profile, a slender muzzle and grows to about 6 ft (1·8 m); the male has a convex profile, a wide, heavy muzzle and is larger, growing to 7 ft (2·1 m).

151

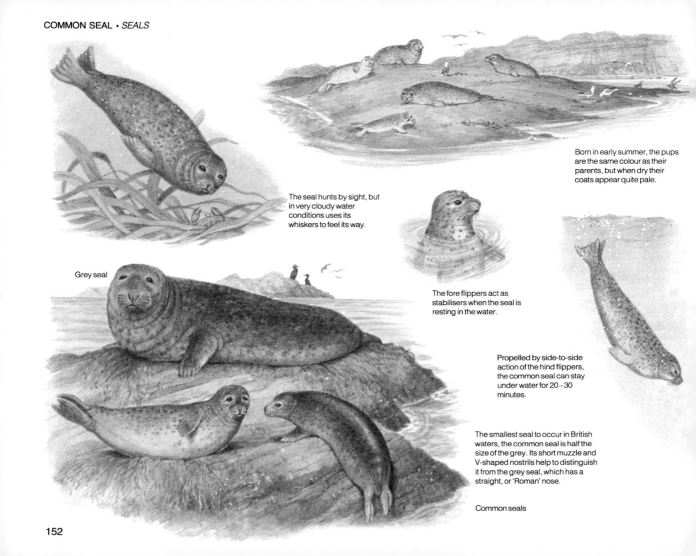

The seal hunts by sight, but in very cloudy water conditions uses its whiskers to feel its way.

Born in early summer, the pups are the same colour as their parents, but when dry their coats appear quite pale.

Grey seal

The fore flippers act as stabilisers when the seal is resting in the water.

Propelled by side-to-side action of the hind flippers, the common seal can stay under water for 20–30 minutes.

The smallest seal to occur in British waters, the common seal is half the size of the grey. Its short muzzle and V-shaped nostrils help to distinguish it from the grey seal, which has a straight, or 'Roman' nose.

Common seals

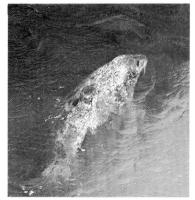

Although able to swim soon after birth, the pup often accepts a lift from its mother. It blends so well with its mother's fur that it can hardly be seen.

Common seal *Phoca vitulina*

Despite its name the common seal is less abundant in British waters than its cousin, the grey seal. About 20,000 live around our coasts, the largest groups gathering around The Wash and on the Norfolk coast to produce their young in June and July. The single seal pup is born on a sand-bank, rocks or the shore and is able to swim and dive from birth, often taking to the water on the next tide. They suckle on land or in water for a month or more.

Fish are the common seal's main diet, and they may attack fish trapped in nets, particularly salmon and herring. Consequently they are considered a pest by fishermen. They do not travel widely, though they may follow migrating fish shoals, and will remain in one area for long periods, often leaving the sea to bask on a rock or sand-bank as the tide recedes.

Common seals are hunted for their skins in Shetland, Orkney, the west coast of Scotland, and also in The Wash where an average of about 550 pups are taken yearly. Those that survive can expect to live for 20–30 years. They have no other natural predators in British waters, though oil pollution can sometimes smother seals.

The common seal has a rounded head and comparatively short snout; seen in the water from a distance it resembles a buoy or fishing float. Males 5½ ft (1·7 m) long; females smaller.

153

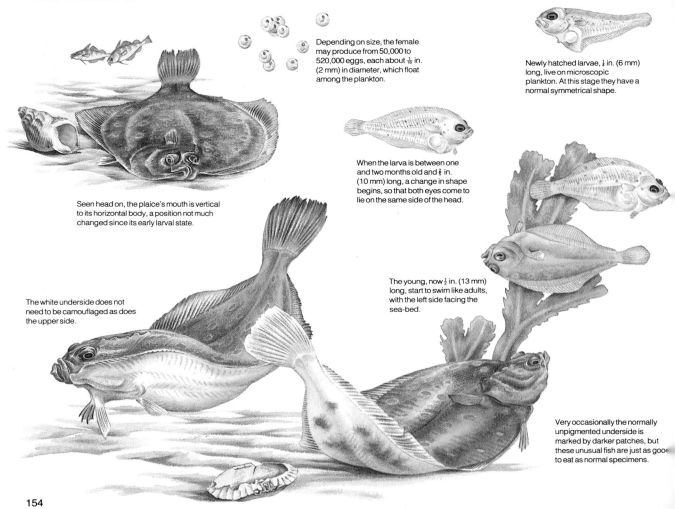

Depending on size, the female may produce from 50,000 to 520,000 eggs, each about $\frac{1}{16}$ in. (2 mm) in diameter, which float among the plankton.

Newly hatched larvae, $\frac{1}{4}$ in. (6 mm) long, live on microscopic plankton. At this stage they have a normal symmetrical shape.

When the larva is between one and two months old and $\frac{2}{5}$ in. (10 mm) long, a change in shape begins, so that both eyes come to lie on the same side of the head.

Seen head on, the plaice's mouth is vertical to its horizontal body, a position not much changed since its early larval state.

The young, now $\frac{1}{2}$ in. (13 mm) long, start to swim like adults, with the left side facing the sea-bed.

The white underside does not need to be camouflaged as does the upper side.

Very occasionally the normally unpigmented underside is marked by darker patches, but these unusual fish are just as good to eat as normal specimens.

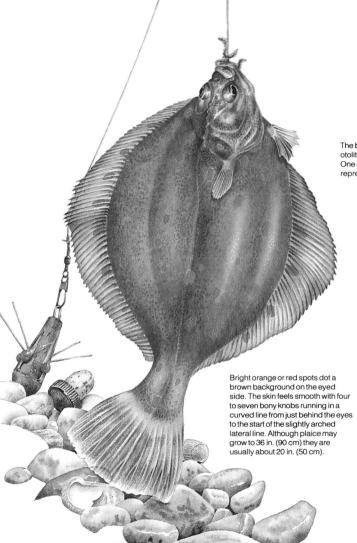

Bright orange or red spots dot a brown background on the eyed side. The skin feels smooth with four to seven bony knobs running in a curved line from just behind the eyes to the start of the slightly arched lateral line. Although plaice may grow to 36 in. (90 cm) they are usually about 20 in. (50 cm).

The bands on the ear bone, or otolith, indicate the plaice's age. One dark and one light band represent one year.

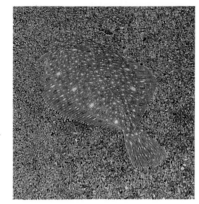

The plaice, which is widely distributed from the Mediterranean to Iceland and northern Norway, is well camouflaged on its upper surface.

Plaice *Pleuronectes platessa*

In the early 19th century, plaice were distributed free among London's poor at the direction of the Lord Mayor in order to dispose of a glut of these fish. Today commercial catches vary, however, due to migration of the fish, poor spawning success and local overfishing. Even so, over 150,000 tons (150 million kg) are caught annually on the continental shelf.

Prodigious numbers of eggs – over half a million – are produced by each female between December and April, though these are heavily preyed on by other marine animals. The exact time of spawning depends on the temperature of the water; it occurs earlier in warm, southern latitudes. The fish migrate to coastal waters prior to spawning. After about three weeks, the eggs hatch. Within two months, the larva takes on its characteristic flatfish shape and becomes a bottom-dweller.

The young plaice remains inshore for up to four years and then migrates to the sandy beds of deeper waters. There it feeds on bristle worms, sandhoppers and thin-shelled bivalves such as tellins and razor shells. The male plaice requires two to six years to reach maturity, the female three to seven. They may live up to 20 years and weigh almost 8 lb (3·6 kg).

Identifying flatfish

Many bottom-living fish have developed flattened, camouflaged bodies which merge with the sea-bed. Fish such as skates and rays are flattened from top to bottom, so that they lie belly-down, but the flatfishes are compressed from side to side. At rest on the sea-bed they lie on either the left or right side, and have become adapted to this way of life by having both eyes on one side – the upper side – of the head. Some species have the eyes on the right, some on the left, but reversed individuals have been recorded. When flatfish hatch they look like any other fish, and feed in the surface water. As they grow the head becomes asymmetrical so that both eyes are on the same side. As this change occurs the body becomes deepened and compressed, and the fish begins to swim on its side, blind side down, at a lower level. Eventually, the change complete, most species become exclusively bottom-dwelling.

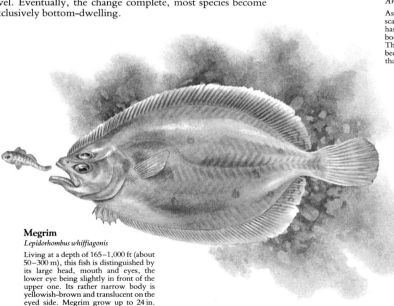

Scaldfish
Arnoglossus laterna

As they are fragile and easily rub off, the scales of this small fish are often missing. It has a small head and a slender, translucent body that is a brownish-grey in colour. This fish is of no importance commercially because of its small size. It is rarely more than 7½ in. (19 cm) long.

Megrim
Lepidorhombus whiffiagonis

Living at a depth of 165–1,000 ft (about 50–300 m), this fish is distinguished by its large head, mouth and eyes, the lower eye being slightly in front of the upper one. Its rather narrow body is yellowish-brown and translucent on the eyed side. Megrim grow up to 24 in. (60 cm) long.

Norwegian topknot
Phrynorhombus norvegicus

Though fairly common, this fish is rarely seen because of its small size and habit of concealing itself on rough grounds. The eyed side is a yellowish-brown with darker, irregular shaped markings and is covered in large toothed scales that make the fish feel rough. Both the dorsal and anal fins extend to form two distinct lobes under the tail. It is the smallest flatfish in British waters and reaches a maximum length of about 4¾ in. (12 cm).

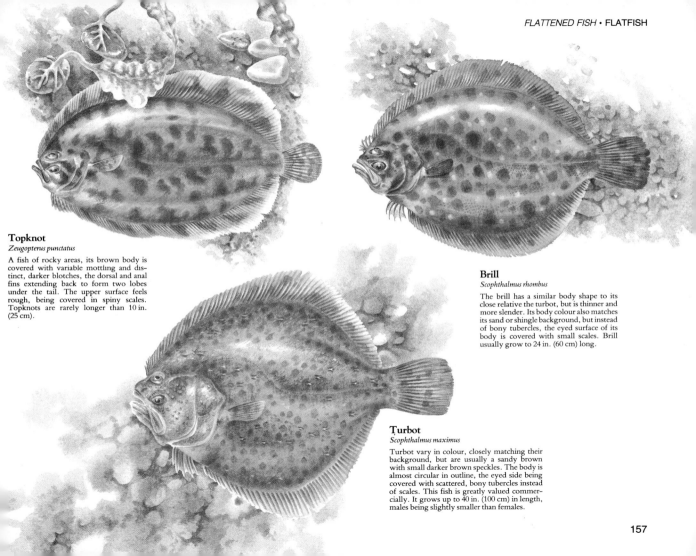

Topknot
Zeugopterus punctatus

A fish of rocky areas, its brown body is covered with variable mottling and distinct, darker blotches, the dorsal and anal fins extending back to form two lobes under the tail. The upper surface feels rough, being covered in spiny scales. Topknots are rarely longer than 10 in. (25 cm).

Brill
Scophthalmus rhombus

The brill has a similar body shape to its close relative the turbot, but is thinner and more slender. Its body colour also matches its sand or shingle background, but instead of bony tubercles, the eyed surface of its body is covered with small scales. Brill usually grow to 24 in. (60 cm) long.

Turbot
Scophthalmus maximus

Turbot vary in colour, closely matching their background, but are usually a sandy brown with small darker brown speckles. The body is almost circular in outline, the eyed side being covered with scattered, bony tubercles instead of scales. This fish is greatly valued commercially. It grows up to 40 in. (100 cm) in length, males being slightly smaller than females.

157

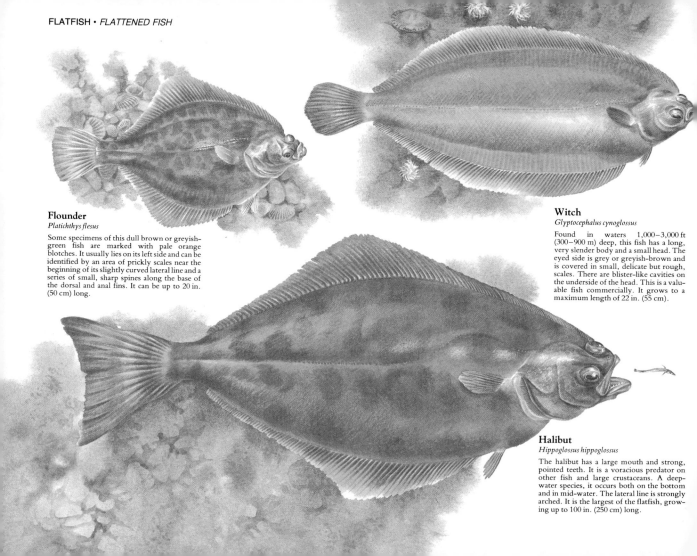

Flounder
Platichthys flesus

Some specimens of this dull brown or greyish-green fish are marked with pale orange blotches. It usually lies on its left side and can be identified by an area of prickly scales near the beginning of its slightly curved lateral line and a series of small, sharp spines along the base of the dorsal and anal fins. It can be up to 20 in. (50 cm) long.

Witch
Glyptocephalus cynoglossus

Found in waters 1,000–3,000 ft (300–900 m) deep, this fish has a long, very slender body and a small head. The eyed side is grey or greyish-brown and is covered in small, delicate but rough, scales. There are blister-like cavities on the underside of the head. This is a valuable fish commercially. It grows to a maximum length of 22 in. (55 cm).

Halibut
Hippoglossus hippoglossus

The halibut has a large mouth and strong, pointed teeth. It is a voracious predator on other fish and large crustaceans. A deep-water species, it occurs both on the bottom and in mid-water. The lateral line is strongly arched. It is the largest of the flatfish, growing up to 100 in. (250 cm) long.

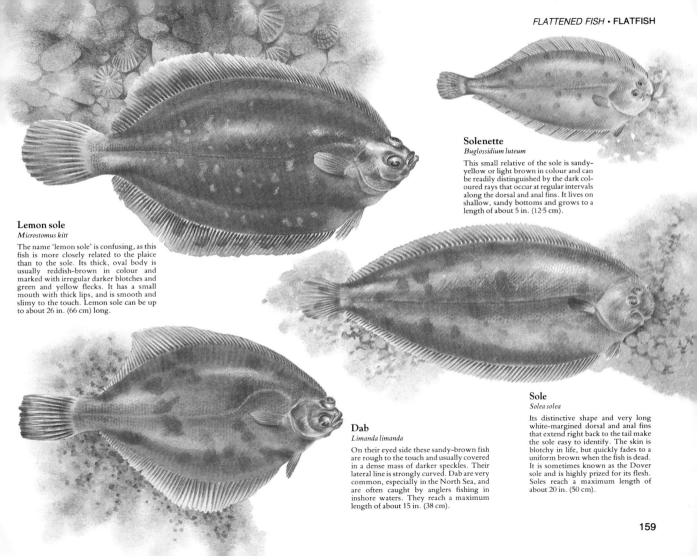

Solenette
Buglossidium luteum

This small relative of the sole is sandy-yellow or light brown in colour and can be readily distinguished by the dark coloured rays that occur at regular intervals along the dorsal and anal fins. It lives on shallow, sandy bottoms and grows to a length of about 5 in. (12·5 cm).

Lemon sole
Microstomus kitt

The name 'lemon sole' is confusing, as this fish is more closely related to the plaice than to the sole. Its thick, oval body is usually reddish-brown in colour and marked with irregular darker blotches and green and yellow flecks. It has a small mouth with thick lips, and is smooth and slimy to the touch. Lemon sole can be up to about 26 in. (66 cm) long.

Dab
Limanda limanda

On their eyed side these sandy-brown fish are rough to the touch and usually covered in a dense mass of darker speckles. Their lateral line is strongly curved. Dab are very common, especially in the North Sea, and are often caught by anglers fishing in inshore waters. They reach a maximum length of about 15 in. (38 cm).

Sole
Solea solea

Its distinctive shape and very long white-margined dorsal and anal fins that extend right back to the tail make the sole easy to identify. The skin is blotchy in life, but quickly fades to a uniform brown when the fish is dead. It is sometimes known as the Dover sole and is highly prized for its flesh. Soles reach a maximum length of about 20 in. (50 cm).

159

The large thorns which cover the upper body have a stout button-like base called a buckler.

When resting on the bottom, the ray draws water into the gill chamber through the spiracle, a hole behind the eye, instead of through the mouth.

The male has pelvic claspers by which spermatozoa are introduced into the genital opening of the female. Spines on the claspers probably help to grip the female during mating.

The egg capsule, commonly known as a mermaid's purse, measures 2⅜ in. (60 mm) long by 1½ in. (40 mm) wide. The corners are drawn out into short, hollow 'horns'.

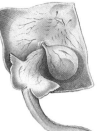

After four to five months' incubation, the young thornback emerges from the mermaid's purse with its yolk sac still attached.

The most common ray around British coasts, the thornback is usually found over sand and mud.

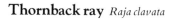

The upper body is covered with coarse prickles, even when the fish is young. On adults there are also large spines with swollen bases. The body is mottled grey or fawn, often with brown marbling. Thornbacks grow to about 34 in. (86 cm) long.

Thornback ray *Raja clavata*

Most of the 'skate' found in fishmongers is likely to be thornback ray, or roker – the commonest ray in British waters. It can be distinguished from other species by its colouring and by the arrangement of the spines. Males have four rows on either side of the wings, females are more spiny and have additional spines on the underside.

Although mainly a non-migratory species found all round the British Isles, the fish does move inshore in some places during the winter and early spring. It is normally found in depths of 30–200 ft (10–60 m), but during the breeding season mature fish keep to the shallower parts of this range.

Mature females – about nine years and over – move into shallow water during the spring, to be followed later by mature males – about seven years and over. Females lay egg capsules from March to August; these take 16–20 weeks to hatch. At first the young ray obtains nourishment from its yolk sac. Gradually it begins to hunt small crustaceans such as bottom-dwelling shrimps. It then adds fish, such as sand eels, herrings, sprats and small flatfish to its diet, with shore and swimming crabs and brown shrimps as its main food.

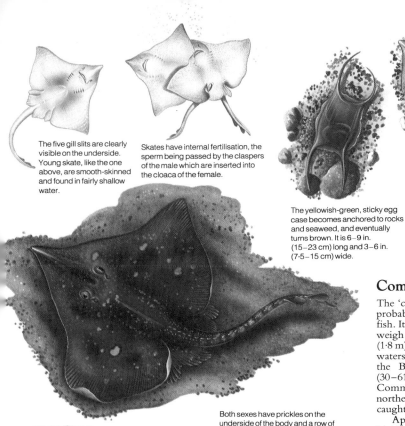

The five gill slits are clearly visible on the underside. Young skate, like the one above, are smooth-skinned and found in fairly shallow water.

Skates have internal fertilisation, the sperm being passed by the claspers of the male which are inserted into the cloaca of the female.

The yellowish-green, sticky egg case becomes anchored to rocks and seaweed, and eventually turns brown. It is 6–9 in. (15–23 cm) long and 3–6 in. (7·5–15 cm) wide.

The embryo develops inside the egg case for two to five months before it emerges.

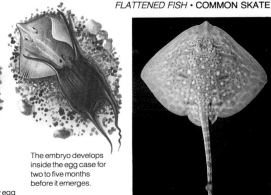

Adults, which mostly live at depths of more than 300 ft (90 m), may be found inshore in deeps off rocky coasts.

The arrangement of the teeth in rows is an adaptation for crushing and grinding prey.

Both sexes have prickles on the underside of the body and a row of large spines down the mid-line of the tail. The male is spiny over the entire upper surface but the female has spines only on the front part. The upper surface is grey or brown with lighter or darker spots; the underside is blue-grey. Females grow to about 9 ft (2·7 m) long; males are slightly smaller.

Common skate *Raja batis*

The 'common' skate is no longer as common as it used to be, probably due to over-exploitation of this important commercial fish. It is the largest and heaviest European ray, and females may weigh up to 250 lb (113 kg) and measure 9 ft (2·7 m) long and 6 ft (1·8 m) wide. Skate are widely distributed throughout European waters from the north of Russia to the Mediterranean and from the Baltic to Madeira, living in depths of 100–2,000 ft (30–610 m), though the young fish live in shallow water. Commercially they are caught by long-line or trawl in the northern North Sea and off Iceland. They are also occasionally caught by sport anglers in boats.

Apart from man the adult skate has few predators, but is a highly predatory fish itself, feeding on other species of ray, dogfish, flatfish, herrings and pilchards. It feeds in mid-water as well as close to the sea-bed where it preys on lobster and crabs. There is a slow migration during the summer months; this may be in search of food fishes.

Breeding takes place in spring and summer, and hatching of the young takes place after several months, otherwise little is known about the biology of this fish.

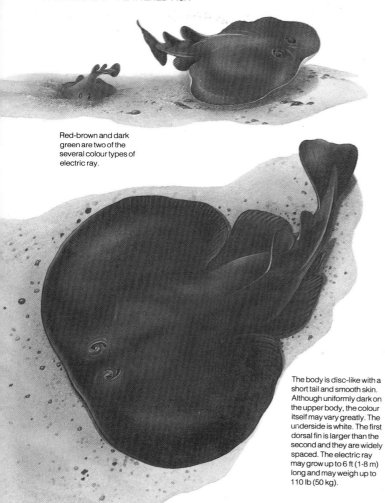

Red-brown and dark green are two of the several colour types of electric ray.

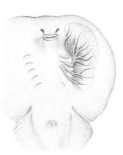

Electric organs on each side of the head are capable of producing pulses ranging from 45 to 220 volts at up to 8 amps – enough to disable a person.

Bottom-living on sand and mud in shallow waters, electric rays are found mainly along south and west coasts of Britain.

The body is disc-like with a short tail and smooth skin. Although uniformly dark on the upper body, the colour itself may vary greatly. The underside is white. The first dorsal fin is larger than the second and they are widely spaced. The electric ray may grow up to 6 ft (1·8 m) long and may weigh up to 110 lb (50 kg).

Electric ray *Torpedo nobiliana*

A shock from a large electric ray can be strong enough to disable a grown man. Like other species of electric fish, the ray uses its shock to stun prey – mainly bottom-living fish – and to defend itself. The electric organs, which constitute about one-sixth of the total body weight, are columns of prism-like structures forming a network of cells that resembles a honeycomb. The upper surface of the ray is positive, the lower negative. The electrical discharge is believed to be a reflex action stimulated by touch. Repeated electrical pulses become weaker and the fish must rest before it becomes recharged.

Electric rays are warm-water fish and only two species – the common and the much rarer marbled – extend northwards to British waters. Although electric rays are not abundant, they do occur in commercial catches, particularly off the south and south-west coasts, at depths of 16–165 ft (5–50 m).

Breeding is thought to take place in more southerly waters than those around Britain, so small electric rays are rarely found here. The eggs are retained in the female until they hatch, after which the fully formed young – less than 20 – are produced. They are believed to be fairly slow-growing, long-lived fish.

The fish is distinguished from its shark relatives by the flattened, ray-like body, broad pectoral and pelvic fins and the wide mouth. Each nostril is covered by a broad flap of skin. The upper body is sandy or greyish-brown and the underside is white. Monkfish grow to about 6 ft (1·8 m) long.

Although more common during summer, monkfish are found throughout the year around the coast of Britain.

Like some other sharks, but unlike rays, the monkfish gives birth to fully formed young.

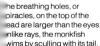

he breathing holes, or piracles, on the top of the ead are larger than the eyes. nlike rays, the monkfish wims by sculling with its tail.

Monkfish *Squatina squatina*

The fish takes one of its common names from the shape of its head, which resembles a monk's cowl. The other name for the predatory monkfish – angel fish – is derived from the shape of the broad pectoral fins, which look like the folded wings of an angel. Although these 'wings' give the fish a superficial resemblance to the rays, it is in fact a bottom-living member of the shark family. The two dorsal fins have no spines and there is no anal fin. The monkfish can swim rapidly but it is more often found partially buried in the sand of the sea-bed waiting for prey, such as flatfish, molluscs, crustaceans and worms.

Of the 14 species of monkfish, only *Squatina squatina* lives in European waters. It can be found throughout the year in shallow water, 16–23 ft (5–7 m) deep, although some have been caught at over 100 ft (30 m). During summer, the mature female gives birth to some 9–20 young, each about 9½ in. (24 cm) long.

Although monkfish are not now fished commercially, in the past their skins were used for polishing wood and also to cure some skin ailments. Anglers fish for large specimens, and there is reputed to be some trade in the flesh, which when cut in strips and deep fried is very similar to expensive scampi.

163

Six common rays

The members of this family have flattened, cartilaginous bodies with broad pectoral fins which give them their typical diamond shape. They are related to the sharks. Most of the larger species are known as skates and have long snouts, while the smaller, short-snouted ones are called rays. All species that are sold in fishmongers are classified as 'skate' and several thousand tons are sold in Britain every year. For this reason there is considerable concern for the 20 or so species found in European waters, as they are extremely vulnerable to overfishing. This is because they mature slowly and produce few eggs. For example, the spotted ray takes nine years to reach maturity and an adult female produces only about 70 eggs annually. Each egg, individually wrapped in its mermaid's purse, may take between 4 and 15 months to hatch.

Flat spine

Stingray
Dasyatis pastinaca

Seen mainly in southern waters, the species is greyish, olive or brown above. The tail is whip-like and bears one long sharp spine or, sometimes, two. Stingrays grow to about 40 in. (100 cm) from tip to tail.

The venom-carrying tail spine is flat and serrated on the sides. It measures about 5 in. (12·5 cm) long.

Each large dorsal spine is mounted on a massive, ridged base.

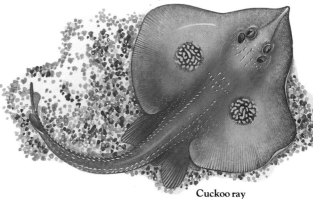

Starry ray
Raja radiata

Prickles cover the dorsal surface and large spines run down the centre of the back and tail. Starry rays grow to 30 in. (76 cm) long.

Cuckoo ray
Raja naevus

Distinctive black and yellow 'roundels' mark the wings. Cuckoo rays can grow to 28 in. (70 cm) long.

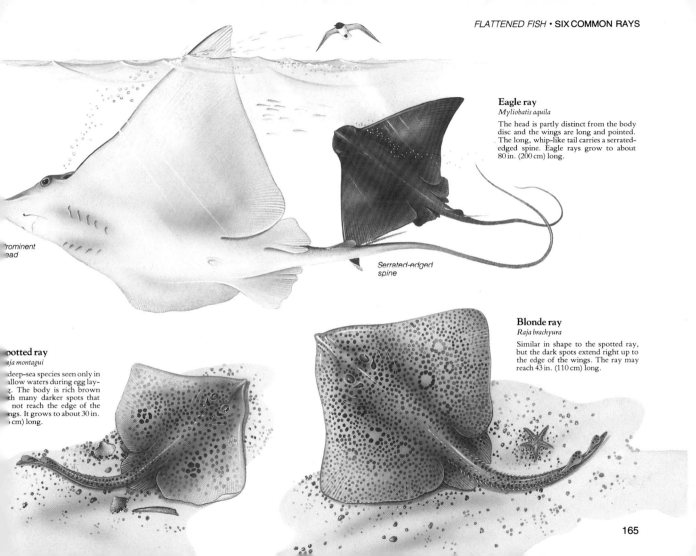

Eagle ray
Myliobatis aquila

The head is partly distinct from the body disc and the wings are long and pointed. The long, whip-like tail carries a serrated-edged spine. Eagle rays grow to about 80 in. (200 cm) long.

rominent
ead

Serrated-edged
spine

Blonde ray
Raja brachyura

Similar in shape to the spotted ray, but the dark spots extend right up to the edge of the wings. The ray may reach 43 in. (110 cm) long.

potted ray
aja montagui

deep-sea species seen only in allow waters during egg lay-g. The body is rich brown th many darker spots that not reach the edge of the gs. It grows to about 30 in. cm) long.

165

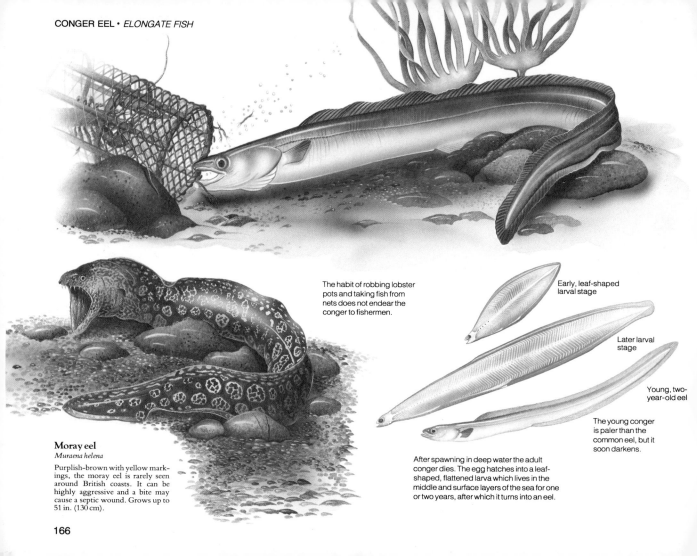

The habit of robbing lobster pots and taking fish from nets does not endear the conger to fishermen.

Early, leaf-shaped larval stage

Later larval stage

Young, two-year-old eel

The young conger is paler than the common eel, but it soon darkens.

Moray eel
Muraena helena

Purplish-brown with yellow markings, the moray eel is rarely seen around British coasts. It can be highly aggressive and a bite may cause a septic wound. Grows up to 51 in. (130 cm).

After spawning in deep water the adult conger dies. The egg hatches into a leaf-shaped, flattened larva which lives in the middle and surface layers of the sea for one or two years, after which it turns into an eel.

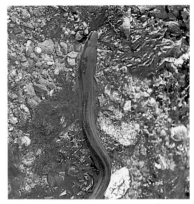

Common on rocky shores and offshore, congers also lurk in man-made habitats such as harbour walls, loose stone groynes and sunken wrecks.

Like the common eel, the conger has a smooth, apparently scaleless skin and no pelvic fins. It differs in having a slightly protruding lower jaw and a dorsal fin that starts near the head. Also, the gill openings are larger. It is usually dull brown on the upper body and creamy or golden brown on the underside. It can grow to a maximum length of 9 ft (2·7 m).

During the day congers hide in rock crevices and are particularly plentiful among submerged wrecks.

Conger eel *Conger conger*

Almost anything that moves on the sea-bed – squid, octopus, fish and crustaceans – is prey to the voracious conger eel. The conger is entirely marine and is found from Iceland to the Atlantic coast of Africa. Around Britain, congers are more common off rocky western coasts, with smaller specimens found in inshore waters and larger ones — up to 143 lb (65 kg) – in deeper water away from the coast. They are caught mainly on a long line by sport and commercial fishermen, and several thousand tons are landed each year in Continental ports.

As maturity approaches, conger eels cease feeding, their skeletons begin to decalcify and certain organs, apart from their sexual organs, start to regress. Like freshwater eels, congers undertake a considerable migration prior to summer spawning. Between Gibraltar and the Azores congers will spawn at enormous depths – 10,000 to 13,000 ft (3,000 to 4,000 m). The female is larger than the male and can carry up to 8 million eggs. After spawning, the adults die, having lived for up to 15 years.

The moray eel has a more southerly distribution and only about six records of capture in British waters exist. Its bite, which may contain a toxin, often causes blood poisoning.

167

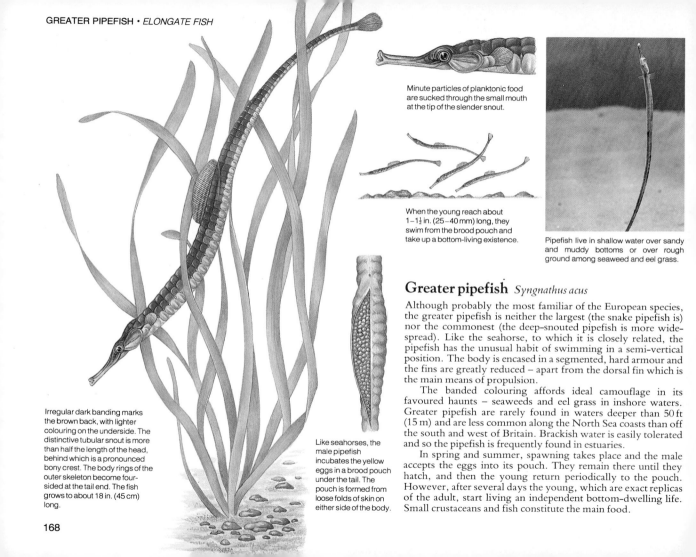

Minute particles of planktonic food are sucked through the small mouth at the tip of the slender snout.

When the young reach about 1–1½ in. (25–40 mm) long, they swim from the brood pouch and take up a bottom-living existence.

Pipefish live in shallow water over sandy and muddy bottoms or over rough ground among seaweed and eel grass.

Irregular dark banding marks the brown back, with lighter colouring on the underside. The distinctive tubular snout is more than half the length of the head, behind which is a pronounced bony crest. The body rings of the outer skeleton become four-sided at the tail end. The fish grows to about 18 in. (45 cm) long.

Like seahorses, the male pipefish incubates the yellow eggs in a brood pouch under the tail. The pouch is formed from loose folds of skin on either side of the body.

Greater pipefish *Syngnathus acus*

Although probably the most familiar of the European species, the greater pipefish is neither the largest (the snake pipefish is) nor the commonest (the deep-snouted pipefish is more widespread). Like the seahorse, to which it is closely related, the pipefish has the unusual habit of swimming in a semi-vertical position. The body is encased in a segmented, hard armour and the fins are greatly reduced – apart from the dorsal fin which is the main means of propulsion.

The banded colouring affords ideal camouflage in its favoured haunts – seaweeds and eel grass in inshore waters. Greater pipefish are rarely found in waters deeper than 50 ft (15 m) and are less common along the North Sea coasts than off the south and west of Britain. Brackish water is easily tolerated and so the pipefish is frequently found in estuaries.

In spring and summer, spawning takes place and the male accepts the eggs into its pouch. They remain there until they hatch, and then the young return periodically to the pouch. However, after several days the young, which are exact replicas of the adult, start living an independent bottom-dwelling life. Small crustaceans and fish constitute the main food.

Identifying other pipefish

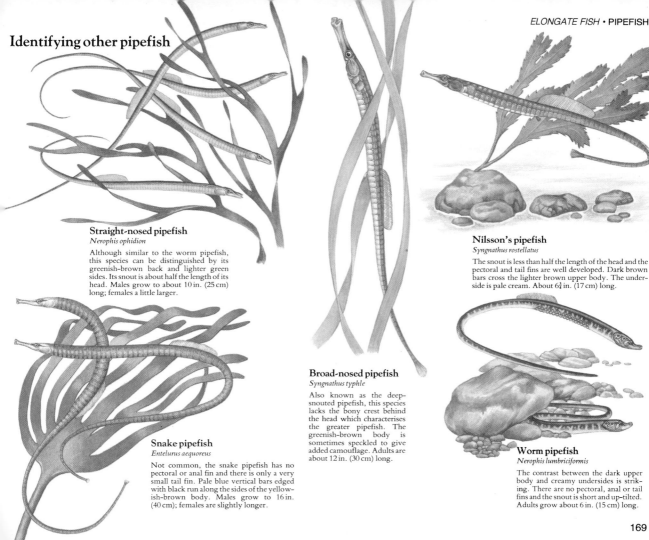

Straight-nosed pipefish
Nerophis ophidion

Although similar to the worm pipefish, this species can be distinguished by its greenish-brown back and lighter green sides. Its snout is about half the length of its head. Males grow to about 10 in. (25 cm) long; females a little larger.

Snake pipefish
Entelurus aequoreus

Not common, the snake pipefish has no pectoral or anal fin and there is only a very small tail fin. Pale blue vertical bars edged with black run along the sides of the yellowish-brown body. Males grow to 16 in. (40 cm); females are slightly longer.

Broad-nosed pipefish
Syngnathus typhle

Also known as the deep-snouted pipefish, this species lacks the bony crest behind the head which characterises the greater pipefish. The greenish-brown body is sometimes speckled to give added camouflage. Adults are about 12 in. (30 cm) long.

Nilsson's pipefish
Syngnathus rostellatus

The snout is less than half the length of the head and the pectoral and tail fins are well developed. Dark brown bars cross the lighter brown upper body. The underside is pale cream. About 6¾ in. (17 cm) long.

Worm pipefish
Nerophis lumbriciformis

The contrast between the dark upper body and creamy undersides is striking. There are no pectoral, anal or tail fins and the snout is short and up-tilted. Adults grow about 6 in. (15 cm) long.

169

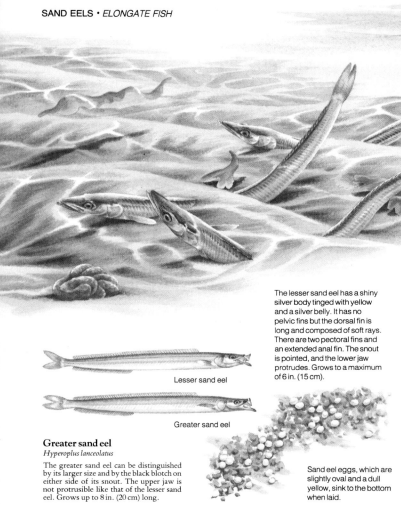

Shoals of shiny-scaled sand eels can sometimes be seen in the shallow waters off sandy beaches.

The lesser sand eel has a shiny silver body tinged with yellow and a silver belly. It has no pelvic fins but the dorsal fin is long and composed of soft rays. There are two pectoral fins and an extended anal fin. The snout is pointed, and the lower jaw protrudes. Grows to a maximum of 6 in. (15 cm).

Lesser sand eel

Greater sand eel

Greater sand eel
Hyperoplus lanceolatus

The greater sand eel can be distinguished by its larger size and by the black blotch on either side of its snout. The upper jaw is not protrusible like that of the lesser sand eel. Grows up to 8 in. (20 cm) long.

Sand eel eggs, which are slightly oval and a dull yellow, sink to the bottom when laid.

Lesser sand eel *Ammodytes tobianus*

Despite its name, the sand eel is not a true eel. The eel-like shape is a compromise designed for both swimming and burrowing in the sand. Of the five species of sand eel found around the coasts of the North Atlantic, the lesser sand eel is commoner than the greater. The other three species live in deeper waters and are encountered only by deep-sea trawlers.

Small shoals can be seen swimming below the surface in shallow water close to the shore's edge. They provide food for fish such as the herring and mackerel, as well as many seabirds. The sand eel itself feeds on smaller fish, marine worms and plankton. Spawning takes place over sand or gravel during the summer – the more southerly the latitude, the earlier the spawning. Up to 20,000 eggs, each no more than $\frac{1}{2}$ in. (1 mm) across, may be laid by one female during a spawning season.

Hundreds of thousands of tons of lesser and greater sand eel are caught each year by shallow-water trawlers. Most are turned into agricultural fishmeal although some are fried, salted or dried to be eaten locally around British fishing ports and in larger quantities on the Continent. Anglers catch them for bait by raking the sand immediately after the tide has gone out.

About six weeks after fertilisation, translucent, larvae, $\frac{1}{5}$ in. (5 mm) long, emerge from the eggs. The yolk sac is placed well forward.

The body is long and slippery to the touch. The anal fin is about half the length of the dorsal fin, which runs from the back of the head to the tail.

To protect the eggs, the female curls her body around the newly laid clump, compacting it into a ball about 1 in. (25 mm) in diameter.

The fish is common on rocky shores all round Britain's coast, especially in pools washed by the tide.

distinctive row of usually 12 black spots ringed with white, uns along the base of the orsal fin. The body is brown or reenish with lighter coloured ertical bars on the sides. A ark stripe runs through each ye to the corner of the mouth. he fish can reach up to 10 in. 25 cm) long.

Butterfish *Pholis gunnellus*

With its slippery, mucous-covered skin and long dorsal fin made up of short spines, the butterfish is very difficult to handle – hence its name. Also known as a gunnel, it is an inshore species often seen in the rock pools around the coasts of Britain, as well as those of Northern Europe, Iceland, Greenland and the Atlantic coast of North America. As butterfish are so abundant in the intertidal zone, they form an important link in the food chain of many of the larger, more commercially important food fish. They feed on worms, small crustaceans and molluscs, and in their turn are eaten by larger fish. They are of no commercial value although they make good bait for inshore anglers.

Spawning takes place over January and February in British waters. Several hundred eggs, each about $\frac{1}{20}$ in. (1·5 mm) across, are laid in sticky clumps between stones or inside empty shells. Both male and female stand guard over the eggs – although more usually the female alone – until the eggs hatch about one month after they were laid. Growth is slow – about 1$\frac{1}{4}$ in. (30 mm) in each of the first four years and $\frac{1}{2}$ in. (13 mm) each year thereafter. The life expectancy of the butterfish is thought to be about ten years.

171

Young shanny live mainly on small invertebrates but they also nip the limbs off feeding barnacles.

The fish is found in shallow, rocky waters and also in sandy areas where stones and seaweeds provide cover.

The colour varies with habitat but is usually a blotched dull brown or dull green. There are no tentacles on the head. The one long dorsal fin is notched half-way along and there is a black dot on the front edge. A full-grown shanny is about 6¼ in. (16 cm) long.

Eggs, about 1/20 in. (1·5 mm) in diameter, are laid under stones or in rock crevices. The male stands guard and fans the eggs with his pectoral fins.

Shanny *Lipophrys pholis*

Of all British inshore fish, the shanny is probably the most widespread. It is the rock-pool dweller that most children on seaside holidays will have spent hours trying to catch with a net or in their bare hands. But as they will have discovered, the shanny's slimy scaleless skin makes it very elusive, and by using its paired fins it can move between pools and crawl into crevices and under stones. It is the commonest of the four true blennies found in British waters and is most abundant on rocky coasts where there is plenty of weed for concealment. Shannies feed on barnacles, small crabs and other crustaceans.

After spawning, which takes place from April to August, the male attends the eggs constantly for up to two months before they hatch. In winter, shannies move from the rock pools to avoid the storms but they remain in relatively shallow water. The wide range of colour patterns – from green blotched with red to brown and yellow marbling – provides camouflage against wading birds at low tide and other fish at high tide.

Though resembling true blennies, Yarrell's blenny and the snake blenny are not of the same family. Nor is the viviparous blenny, which is an eelpout.

Blennies and similar fish

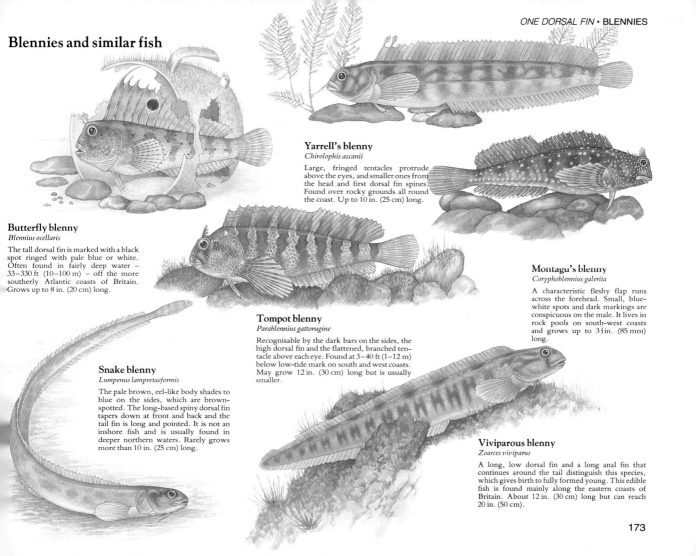

Yarrell's blenny
Chirolophis ascanii

Large, fringed tentacles protrude above the eyes, and smaller ones from the head and first dorsal fin spines. Found over rocky grounds all round the coast. Up to 10 in. (25 cm) long.

Butterfly blenny
Blennius ocellaris

The tall dorsal fin is marked with a black spot ringed with pale blue or white. Often found in fairly deep water – 33–330 ft (10–100 m) – off the more southerly Atlantic coasts of Britain. Grows up to 8 in. (20 cm) long.

Montagu's blenny
Coryphoblennius galerita

A characteristic fleshy flap runs across the forehead. Small, blue-white spots and dark markings are conspicuous on the male. It lives in rock pools on south-west coasts and grows up to 3⅜ in. (85 mm) long.

Tompot blenny
Parablennius gattorugine

Recognisable by the dark bars on the sides, the high dorsal fin and the flattened, branched tentacle above each eye. Found at 3–40 ft (1–12 m) below low-tide mark on south and west coasts. May grow 12 in. (30 cm) long but is usually smaller.

Snake blenny
Lumpenus lampretaeformis

The pale brown, eel-like body shades to blue on the sides, which are brown-spotted. The long-based spiny dorsal fin tapers down at front and back and the tail fin is long and pointed. It is not an inshore fish and is usually found in deeper northern waters. Rarely grows more than 10 in. (25 cm) long.

Viviparous blenny
Zoarces viviparus

A long, low dorsal fin and a long anal fin that continues around the tail distinguish this species, which gives birth to fully formed young. This edible fish is found mainly along the eastern coasts of Britain. About 12 in. (30 cm) long but can reach 20 in. (50 cm).

173

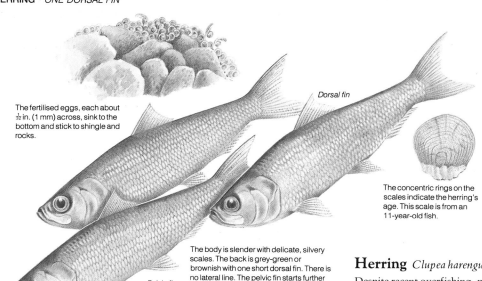

The fertilised eggs, each about ½ in. (1 mm) across, sink to the bottom and stick to shingle and rocks.

Dorsal fin

The concentric rings on the scales indicate the herring's age. This scale is from an 11-year-old fish.

Although overfished locally, such as in the North Sea, the herring remains abundant off northern Europe.

The body is slender with delicate, silvery scales. The back is grey-green or brownish with one short dorsal line. There is no lateral line. The pelvic fin starts further back than the front of the dorsal fin and, unlike the sprat, it has no keel scales. It may grow up to 16 in. (40 cm) long.

Pelvic fin

Herring *Clupea harengus*

Despite recent overfishing, more than 3 million tons of Atlantic herring are caught each year. The fish ranges from the eastern seaboard of North America across to the European continental shelf. Distinct races, such as the Baltic, White Sea and Norwegian herring, are found.

The Norwegian herring is the most important for the British fishing industry, although overfishing in the 1950s reduced stocks to almost non-commercial quantities. Between three and seven years old the female becomes mature and lays about 40,000 eggs. Spawning takes place off the Norwegian coast from February to April, after which the herring shoals migrate through the North Sea to spend summer off the coast of Iceland. During the course of a 24 hour period, the shoals may move several hundred feet vertically through the water in search of their planktonic food.

Two species related to the herring – the sprat and the sardine, or pilchard – are also commercially important. Sprats are found in the shallower waters around Britain and Norway; pilchards, which may grow up to 10 in. (25 cm), extend from the English Channel to the Mediterranean.

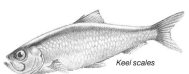

Keel scales

Sprat
Sprattus sprattus

The sprat is distinguished from the herring by a keel of sharply pointed scales and a pelvic fin that starts in front of, or directly below, the front of the dorsal fin. Sprats can grow up to 6¼ in. (16·5 cm) long.

Newly hatched larvae, about ¼ in. (6 mm) long, are attracted by light up to the plankton where they feed on algae and small animals.

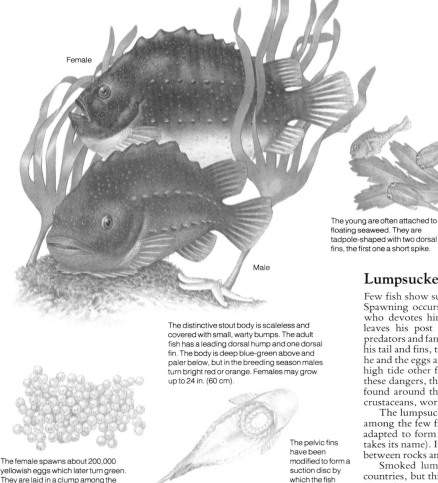

Female

Male

The distinctive stout body is scaleless and covered with small, warty bumps. The adult fish has a leading dorsal hump and one dorsal fin. The body is deep blue-green above and paler below, but in the breeding season males turn bright red or orange. Females may grow up to 24 in. (60 cm).

The female spawns about 200,000 yellowish eggs which later turn green. They are laid in a clump among the rocks at low-tide level, and are guarded by the male.

The pelvic fins have been modified to form a suction disc by which the fish attaches itself to rocks.

The young are often attached to floating seaweed. They are tadpole-shaped with two dorsal fins, the first one a short spike.

Common in seas to the north of the English Channel, lumpsuckers migrate to shallow coastal waters to spawn.

Lumpsucker *Cyclopterus lumpus*

Few fish show such parental devotion as the male lumpsucker. Spawning occurs between March and May and it is the male who devotes himself totally to guarding the eggs. He rarely leaves his post – even to eat – defending the eggs against predators and fanning a constant stream of water over them with his tail and fins, to make sure they are aerated. At low tide, both he and the eggs are exposed to attack by various seabirds, and at high tide other fish, such as halibut, are predators. In spite of these dangers, the lumpsucker remains relatively common. It is found around the entire British coastline, and feeds on small crustaceans, worms and fishes.

The lumpsucker and its relatives, the sea snails (*Liparis*), are among the few fish in British waters that have their pelvic fins adapted to form a powerful suction disc (from which the fish takes its name). It lives on the bottom, where it hides in crevices between rocks and under stones.

Smoked lumpsucker used to be eaten in most European countries, but this fish now has little commercial value. Sometimes the eggs are treated with salt and black dye and used as a substitute for caviar.

175

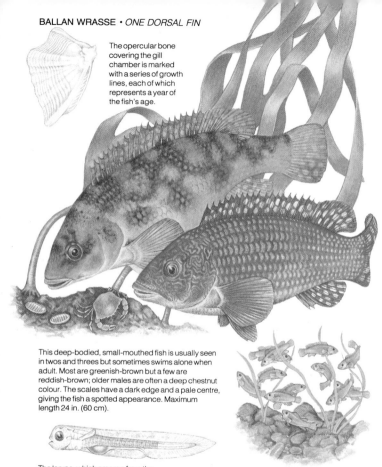

The opercular bone covering the gill chamber is marked with a series of growth lines, each of which represents a year of the fish's age.

This deep-bodied, small-mouthed fish is usually seen in twos and threes but sometimes swims alone when adult. Most are greenish-brown but a few are reddish-brown; older males are often a deep chestnut colour. The scales have a dark edge and a pale centre, giving the fish a spotted appearance. Maximum length 24 in. (60 cm).

The larvae, which emerge from the eggs after three weeks, pass through a long development stage, slowly drifting in the plankton into shallow inshore waters.

Young ballan wrasse are common in rock pools along the shore. Their bright green colour provides good camouflage where green seaweeds abound.

The wrasse has crushing teeth in its throat – pharyngeal teeth – used for breaking up molluscs and crustaceans. The lower set of teeth is known as the ballan cross because of its shape. It was sometimes carried by sailors as a charm against drowning.

The ballan wrasse is found all round Britain's coastline, especially where there are cliffs and rocky reefs.

Ballan wrasse *Labrus bergylta*

The ballan wrasse once presented a curious puzzle to naturalists, who could not understand why all small ones were female and all older, larger specimens were male. It is now known that all ballan wrasse are born female, taking six years to become sexually mature. After several spawning seasons, however, some of these females change sex. There are no external differences between the male and female. The changes that occur are internal and the mechanism that causes them is not yet fully understood.

Of the 300 species of wrasse, most are tropical or subtropical. Only seven are found in north European waters: they are all fairly long-bodied and have at least part of their dorsal fin composed of sharp spines. The ballan wrasse is the best known British species. It is a familiar sight in inshore waters and, because it prefers rocky areas, it is more abundant off the west coast than in the North Sea. The ballan wrasse feeds mainly on crustaceans and molluscs. Spawning occurs in June when the female builds a nest, usually in a crevice, from fine strands of seaweed. Young wrasse grow very slowly, but the fish is long-lived and some males reach 20 years old.

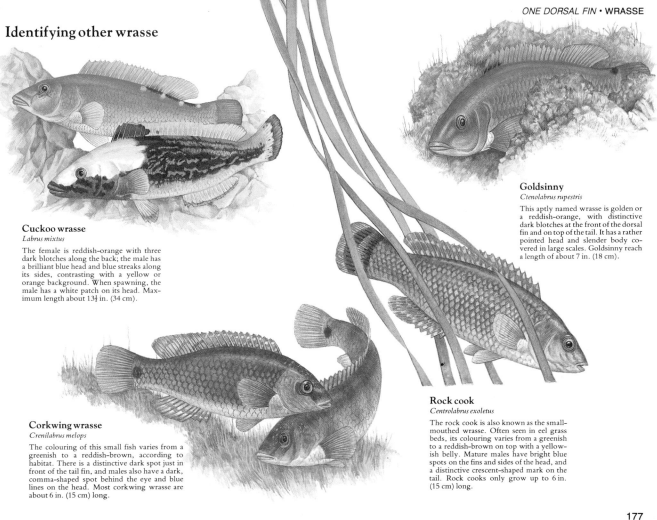

Identifying other wrasse

Goldsinny
Ctenolabrus rupestris

This aptly named wrasse is golden or a reddish-orange, with distinctive dark blotches at the front of the dorsal fin and on top of the tail. It has a rather pointed head and slender body covered in large scales. Goldsinny reach a length of about 7 in. (18 cm).

Cuckoo wrasse
Labrus mixtus

The female is reddish-orange with three dark blotches along the back; the male has a brilliant blue head and blue streaks along its sides, contrasting with a yellow or orange background. When spawning, the male has a white patch on its head. Maximum length about 13½ in. (34 cm).

Rock cook
Centrolabrus exoletus

The rock cook is also known as the small-mouthed wrasse. Often seen in eel grass beds, its colouring varies from a greenish to a reddish-brown on top with a yellow-ish belly. Mature males have bright blue spots on the fins and sides of the head, and a distinctive crescent-shaped mark on the tail. Rock cooks only grow up to 6 in. (15 cm) long.

Corkwing wrasse
Crenilabrus melops

The colouring of this small fish varies from a greenish to a reddish-brown, according to habitat. There is a distinctive dark spot just in front of the tail fin, and males also have a dark, comma-shaped spot behind the eye and blue lines on the head. Most corkwing wrasse are about 6 in. (15 cm) long.

Shallow pools high on the shore are often diluted by rain and freshwater run-off, and are subject to wide temperature changes. Only versatile animals such as shore crabs and sand hoppers can tolerate these conditions, together with a few seaweeds such as the green *Enteromorpha* which is often grazed on by limpets.

Many of the deep pools on the middle shore are lined with pink coral weed, which cannot survive out of water, and the rocks are often encrusted with pink *Lithothamnion* seaweed. These pools provide shelter for many small fish, prawns, worms, anemones and crabs, all of which can remain actively feeding after the tide has gone out.

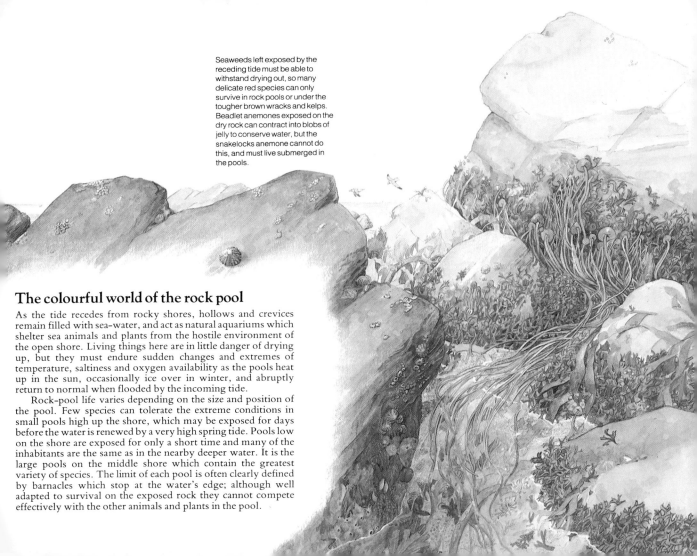

Seaweeds left exposed by the receding tide must be able to withstand drying out, so many delicate red species can only survive in rock pools or under the tougher brown wracks and kelps. Beadlet anemones exposed on the dry rock can contract into blobs of jelly to conserve water, but the snakelocks anemone cannot do this, and must live submerged in the pools.

The colourful world of the rock pool

As the tide recedes from rocky shores, hollows and crevices remain filled with sea-water, and act as natural aquariums which shelter sea animals and plants from the hostile environment of the open shore. Living things here are in little danger of drying up, but they must endure sudden changes and extremes of temperature, saltiness and oxygen availability as the pools heat up in the sun, occasionally ice over in winter, and abruptly return to normal when flooded by the incoming tide.

Rock-pool life varies depending on the size and position of the pool. Few species can tolerate the extreme conditions in small pools high up the shore, which may be exposed for days before the water is renewed by a very high spring tide. Pools low on the shore are exposed for only a short time and many of the inhabitants are the same as in the nearby deeper water. It is the large pools on the middle shore which contain the greatest variety of species. The limit of each pool is often clearly defined by barnacles which stop at the water's edge; although well adapted to survival on the exposed rock they cannot compete effectively with the other animals and plants in the pool.

Fish that live in shore pools

Pools on the shore are very harsh places for animals; they warm up, cool down, become diluted and evaporate. Yet they harbour a variety of fish. Some rely on good camouflage, like the shanny or the worm pipefish which hides among seaweeds that it resembles. Others hide in crevices and holes. Some, such as the 15-spined stickleback, are stranded marine species; others, such as gobies and blennies, are true shore fish.

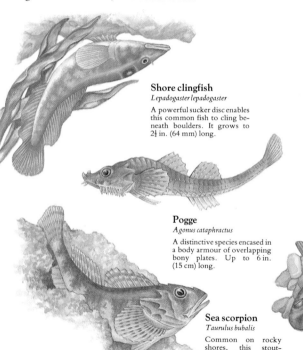

Shore clingfish
Lepadogaster lepadogaster

A powerful sucker disc enables this common fish to cling beneath boulders. It grows to 2¼ in. (64 mm) long.

Pogge
Agonus cataphractus

A distinctive species encased in a body armour of overlapping bony plates. Up to 6 in. (15 cm) long.

Sea scorpion
Taurulus bubalis

Common on rocky shores, this stout-bodied fish has a long, strong spine above the gill cover. Up to 7 in. (18 cm) long.

Butterfish
Pholis gunnellus

A long, slender fish with a row of dark spots along the back, common beneath stones in pools. Up to 10 in. (25 cm) long. (Page 171.)

Rock goby
Gobius paganellus

Fairly large and found under stones among seaweeds. Up to 4¼ in. (12 cm) long. (Page 187.)

Montagu's sea snail
Liparis montagui

This small crevice-dwelling fish has a sucker disc on the belly for gripping rocks. Up to 2½ in. (65 mm) long.

Bull rout
Myoxocephalus scorpius

A large, spiny fish related to the sea scorpion. Females may grow 12 in. (30 cm) long.

Sea snail
Liparis liparis

Rarely seen, this rounded, well-camouflaged fish hides in crevices between rocks. It has a sucker disc under the belly for gripping rocks. About 4 in. (10 cm) long.

Sand smelt
Atherina presbyter

Small shoals of young sand smelt are sometimes found in rock pools. The bright silver line along the side is distinctive.

Two-spotted goby
Gobiusculus flavescens

Slender-bodied, with (in males) two dark spots on each side, this is one of the many gobies which inhabit rock pools. Up to 2⅜ in. (60 mm) long. (Page 187.)

Fifteen-spined stickleback
Spinachia spinachia

Easily identified by the row of 14–16 short spines on the back, this slender-bodied fish usually grows 6 in. (15 cm) long.

Montagu's blenny
Coryphoblennius galerita

Recognisable by the fleshy flap on the forehead, this small blenny is common in rock pools lined with pink coral weed. Up to 3½ in. (90 mm) long. (Page 173.)

Corkwing wrasse
Crenilabrus melops

Young specimens are very common in pools among seaweeds such as sea lettuce. The corkwing is a deep-bodied fish with a distinctive dark spot near the tail. (Page 177.)

Worm pipe fish
Nerophis lumbriciformis

An elongated relative of the seahorse, normally found among seaweeds which it matches in colour. Up to 6 in. (15 cm) long. (Page 169.)

Shanny
Lipophrys pholis

A very common shore fish, which feeds on barnacles and other small crustaceans. Up to 6¼ in. (16 cm) long. (Page 172.)

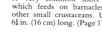

Tompot blenny
Parablennius gattorugine

Found in pools on the lower shore among seaweeds, this blenny has a long, regular dorsal fin. Normally less than 12 in. (30 cm) long. (Page 173.)

181

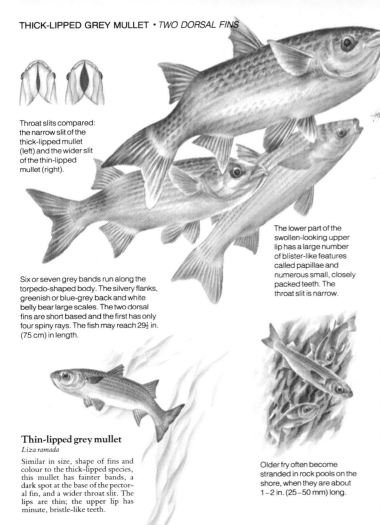

Throat slits compared: the narrow slit of the thick-lipped mullet (left) and the wider slit of the thin-lipped mullet (right).

Six or seven grey bands run along the torpedo-shaped body. The silvery flanks, greenish or blue-grey back and white belly bear large scales. The two dorsal fins are short based and the first has only four spiny rays. The fish may reach 29½ in. (75 cm) in length.

The lower part of the swollen-looking upper lip has a large number of blister-like features called papillae and numerous small, closely packed teeth. The throat slit is narrow.

Thin-lipped grey mullet
Liza ramada
Similar in size, shape of fins and colour to the thick-lipped species, this mullet has fainter bands, a dark spot at the base of the pectoral fin, and a wider throat slit. The lips are thin; the upper lip has minute, bristle-like teeth.

Older fry often become stranded in rock pools on the shore, when they are about 1–2 in. (25–50 mm) long.

Newly hatched fry are seen in summer in shallow inshore waters where the adults come to spawn.

The thick-lipped mullet is common in coastal waters and can tolerate the low salinity of estuaries and creeks.

Thick-lipped grey mullet *Chelon labrosus*

The commonest of the grey mullets, the thick-lipped grey mullet can be immediately identified by its wide and swollen upper lip. It is also the largest British species, growing to a length of 29½ in. (75 cm) and a weight of 10 lb (4·5 kg). During the summer, mullets may move up the estuaries of south coast rivers into fresh water, possibly in search of food. In the 19th century, fishermen reported catches as far as 20 miles from the open sea.

Grey mullet have an unusual feeding method – they swallow mud and digest the animal and plant life it holds. The stomach has a thick, muscular wall to grind the soil and food to a smooth consistency, making digestion more efficient. Spawning takes place in spring, when there is a marked increase in numbers in shallow waters as the fish move inshore. Eggs are shed in open water and hatch about a week later.

Apart from providing sport for anglers, the thick-lipped mullet is caught by seine and trammel nets during summer along the English Channel and off south-west Ireland. Like the less common thin-lipped mullet and the even rarer golden mullet, it will leap out of the water to escape nets or predatory fish.

During the autumn, British stocks of bass migrate to warmer wintering areas – for example, off western Cornwall.

The bass – streamlined for hunting – has a greyish-green back, silver sides and a dark patch on the gill covers. The scales are large and there are small pointed spines on the lower edge of the gill covers. Fully grown bass are about 24 in. (60 cm) but may grow to 40 in. (100 cm).

Bass up to four or five years old are often seen in estuaries up to the tidal limit. Young fish are sometimes dark spotted.

After four days floating in the sea, the larva, about ⅛ in. (4 mm) long, emerges from its egg.

Bass *Dicentrarchus labrax*

Because of its voracity and strength the bass is a good 'sporting' fish for anglers and is fished for mainly from the beaches on the southern and western coasts of England and Wales and around the Thames estuary. The largest caught by rod and line in British waters – on the Eddystone Reef in 1975 – weighed over 18½ lb (8·34 kg), but much larger specimens are found in the warmer waters of the Mediterranean and off the coast of Spain.

The bass – a very good fish for eating – is immediately recognisable by its two dorsal fins, the forward one being spiny, and by the distinctive dark patch on the gill covers. It is a voracious predatory fish, feeding on small gobies, sand eels and crustaceans when young and moving on to a diet of sprats, herrings, crabs and pilchards when older. Spawning takes place in inshore waters during early summer.

British waters are at the northern edge of the bass's range, and its growth rate here is slower than in warmer waters. There is concern about over-fishing by sport anglers – it is estimated that more than 250,000 anglers fish for bass in the UK – and a minimum legal size limit of 12½ in. (32 cm) has been introduced to protect the species around Britain.

183

The ling has a single, long barbel on the lower jaw. The greenish-brown body has light but distinct marbling. Both the second dorsal fin and the anal fin are elongated; the anal and dorsal fins are white tipped, and a dark spot marks the hind edge of the first dorsal fin. May grow to 80 in. (200 cm).

Hake
Merluccius merluccius

Closely related to the cod, the hake can be distinguished by its slim shape and the black lining to the mouth. It is brown or grey above and silver below. Hake move into shallow water in the summer, where they are often caught by anglers.

The head of the fry of the ling is turned downwards and the paired fins are very long.

The female produces millions of eggs, each about $\frac{1}{32}$ in. (1 mm) across. Each egg contains a pale green oil globule that provides buoyancy and helps it to float in the plankton until it hatches.

Large ling inhabit deep water but smaller ones are caught inshore, especially off rocky coasts or around wrecks.

Common ling *Molva molva*

A long-bodied member of the cod family, the ling is an important food species which was formerly held in high regard as a table fish. Today most of the 50,000 tons caught annually in northern waters around Britain, Scandinavia and Iceland are salted and dried for export to southern Europe. There are three species: the common ling, Spanish ling and blue ling, but the common ling accounts for 90 per cent of the commercial catch.

Young fish inhabit fairly shallow waters, and are often caught by sea anglers. After two years they move into deeper water, occurring in the greatest numbers at depths of 1,000–1,300 ft (300–400 m), where they lurk among wrecks, or on a rocky sea-bed. Voracious predators, they will attack fish such as eels, flatfish, mackerel and other fish of the cod family. Spawning takes place between March and July at a depth of about 650 ft (200 m); the eggs float to the surface layers of the sea and hatch in approximately ten days.

The hake is another commercially valuable relative of the cod, some 150,000 tons being caught in European waters each year. It is found at depths of 330–1,000 ft (100–300 m), from Iceland to the North African coast.

Eggs and the transparent, silvery-blue larvae are found among the plankton.

The slender-bodied fish is reddish-brown on the upper body and lighter below. It has one barbel on the chin, a pair on the upper lip and a pair on the front nostrils. Adults are about 10 in. (25 cm) long.

Rocklings, found all round Britain, usually live in rock pools or in depressions in the sea-bed at the base of pilings.

Shore rockling
Gaidropsarus mediterraneus

Three-bearded rockling
Gaidropsarus vulgaris

Five-bearded rockling

Shore rockling

Other rocklings

The shore rockling, similar in colour to the five-bearded, has only three barbels. The three-bearded (the largest of the rocklings) is pinkish to light brown with dark blotches and bars across the back and upper sides. Shore rocklings grow up to 10 in. (25 cm); three-bearded rocklings reach about 16 in. (40 cm).

Five-bearded rockling *Ciliata mustela*

At least seven species of rockling are found in European waters, and the five-bearded is the most common inshore species. It is easily differentiated by the number of its barbels – the beard of its name – from the three-bearded and four-bearded forms, the big-eyed (three), shore (three) and silvery (three). The northern rockling also has five barbels but has a much larger head. The five-bearded rockling ranges from northern Scandinavia and the southerly tip of Iceland to the Portuguese coast.

During winter and spring, floating eggs are shed into the sea at some distance from the shore. Before darkening or dulling in colour, young rockling are called 'mackerel midge' because large shoals of them are hunted by mackerel. The midge stage lasts for several months before the young move inshore in early summer to become bottom dwellers.

Intertidal pools in gravel, mud, sand or rocks are favoured habitats. At this later stage of its life, the rockling's main food consists of small shrimps, prawns and shore crabs, although gobies and blennies may be included. The rockling is of no interest to sport fishermen or as a commercial fish, but it is an important food for larger, commercially valuable fish.

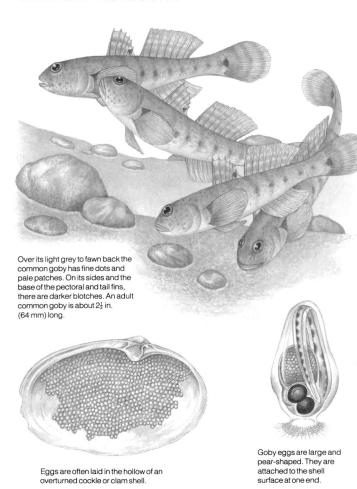

Over its light grey to fawn back the common goby has fine dots and pale patches. On its sides and the base of the pectoral and tail fins, there are darker blotches. An adult common goby is about 2½ in. (64 mm) long.

Eggs are often laid in the hollow of an overturned cockle or clam shell.

Goby eggs are large and pear-shaped. They are attached to the shell surface at one end.

The pelvic fins are united to form a fan-shaped sucker, which prevents the fish being swept away by waves.

Common gobies are sometimes found in shallow inshore waters, like these salt-marsh pools at Aberlady Bay in Scotland.

Common goby *Pomatoschistus microps*

Gobies are one of the most diverse families of fish – several hundred species have been recorded from places as varied as inshore marine waters, estuaries and fresh water. Of the numerous species that abound, 17 live in British waters. Among them is the common goby which is found all round the coast; tidal pools are its main habitat but is also occurs in estuaries and brackish drainage channels. The goby is a gregarious fish, swimming in shoals on or near the bottom where it finds its main food – small crustaceans. Because its colour provides such good camouflage it can be very difficult to spot.

Several spawnings may occur between April and September. The male is markedly territorial, fighting other males for a female and guarding her while she lays her eggs. She then leaves and he keeps watch over the eggs until they hatch. Both sexes pair with several partners during a season. Common gobies, like many small fish, are short-lived, usually dying after two years.

The serrated-edge scales and the absence of a lateral line distinguish gobies from blennies, with which they could be confused. Sensory pores arranged in characteristic lines across the head help in identifying different species.

How to identify other gobies

Rock goby
Gobius paganellus

This usually solitary species is brown with darker mottling. The adult males are deep purple-brown with an orange band on the dorsal fin. Adults are about 4¾ in. (12 cm) long.

Painted goby
Pomatoschistus pictus

The stout body is fawn or brown with dark spots on the edge of each scale and an orange band across the dorsal fins. The adult is about 3¾ in. (95 mm) long.

Leopard-spotted goby
Thorogobius ephippiatus

The head and body of this slender, fawn-coloured fish are marked with conspicuous orange blotches. There is also a black spot near the edge of the first dorsal fin. Adults are about 5 in. (12·5 cm) long.

Sand goby
Pomatoschistus minutus

The slender, sandy brown body is covered with fine spots and faint bars across the back. Males have a dark blue or black spot with a white rim on the first dorsal fin. The species grows to about 3¾ in. (95 mm) long.

Black goby
Gobius niger

Although the colour varies with habitat, this stout-bodied species is usually dark brown with darker blotches on the back and sides. It grows to about 6¾ in. (17 cm) long.

Two-spotted goby
Gobiusculus flavescens

The slender body is reddish-brown with pale and darker blue markings. There is a large black spot on the base of the tail fin and, on the male, a spot below the first dorsal fin. About 2⅜ in. (60 mm) long.

Transparent goby
Aphia minuta

Except for its pigmented eyes, silvery swim bladder and gut, the body is completely transparent. Grows to 2 in. (50 mm) long.

187

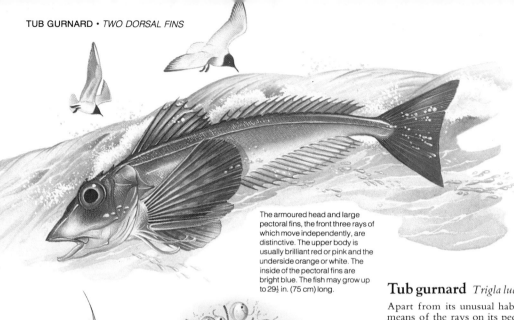

The tub gurnard is found on muddy or sandy sea-beds all round the coast, though less commonly off Scotland.

The armoured head and large pectoral fins, the front three rays of which move independently, are distinctive. The upper body is usually brilliant red or pink and the underside orange or white. The inside of the pectoral fins are bright blue. The fish may grow up to 29½ in. (75 cm) long.

Dragonet
Callionymus lyra

Like the tub gurnard, the dragonet is a bottom-feeding fish. The male has very long dorsal and tail fins, and is bright yellow with blue markings; the female is brownish. Male and female are engaged here in the start of their elaborate mating display. Males grow up to 12 in. (30 cm) long; females are smaller.

The female lays many thousands of small, floating eggs. After hatching, the larvae feed in the plankton for a period, then move to the sea-bed.

The fish 'walks' over the sea-bed, using its sensitive, finger-like pectoral fin rays to search for food.

Tub gurnard *Trigla lucerna*

Apart from its unusual habit of 'walking' on the sea-bed by means of the rays on its pectoral fins, the tub gurnard is also noted for the grunting noise it makes – in fact, its common name is said to come from the French *grogner*, 'to grunt'. The unusual sound, achieved by muscular contractions pushing air through the complex chambers of the swim bladder, may be a means by which individuals in a shoal keep in contact with each other.

British coasts represent the most northerly part of the tub gurnard's range. It therefore becomes less common off Scotland and is absent from the northern North Sea. It occurs most frequently between 160–480 ft (50–150 m) but also ranges widely outside these limits. Although the tub gurnard is the largest of the six species of gurnard found round Britain – exceptional specimens of 10 lb (4·5 kg) have been caught – the grey gurnard is the most common. As the name implies, the predominant colour is grey. All gurnards are bottom feeders, living off crustaceans, small fish and molluscs.

Though not related to the gurnards, the dragonet is somewhat similar in appearance, being a brightly coloured, spiny, small fish. They live in shallow water all round Britain's coast.

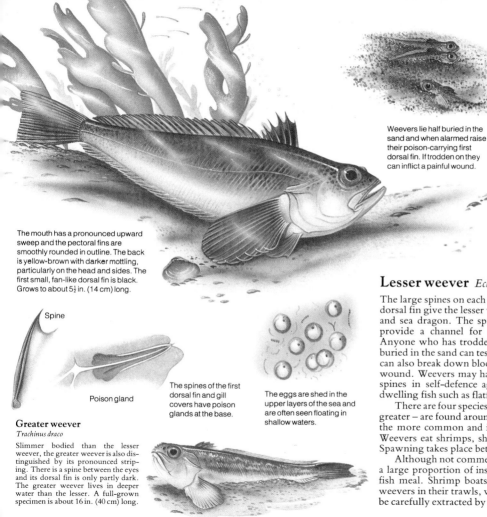

Weevers lie half buried in the sand and when alarmed raise their poison-carrying first dorsal fin. If trodden on they can inflict a painful wound.

Lesser weevers abound in shallow water over clean sand, particularly where shrimps are plentiful.

The mouth has a pronounced upward sweep and the pectoral fins are smoothly rounded in outline. The back is yellow-brown with darker mottling, particularly on the head and sides. The first small, fan-like dorsal fin is black. Grows to about 5½ in. (14 cm) long.

Spine

Poison gland

The spines of the first dorsal fin and gill covers have poison glands at the base.

The eggs are shed in the upper layers of the sea and are often seen floating in shallow waters.

Greater weever
Trachinus draco

Slimmer bodied than the lesser weever, the greater weever is also distinguished by its pronounced striping. There is a spine between the eyes and its dorsal fin is only partly dark. The greater weever lives in deeper water than the lesser. A full-grown specimen is about 16 in. (40 cm) long.

Lesser weever *Echiichthys vipera*

The large spines on each of the gill covers and at the front of the dorsal fin give the lesser weever its alternative names – sting fish and sea dragon. The spines are grooved along their length to provide a channel for poison from the glands at the base. Anyone who has trodden on the raised spines of a weever half buried in the sand can testify how painful this nerve poison is. It can also break down blood cells and thus hinder the healing of a wound. Weevers may have developed their poison-conducting spines in self-defence against being eaten by larger bottom-dwelling fish such as flatfish and rays.

There are four species worldwide and two – the lesser and the greater – are found around the British Isles. The lesser weever is the more common and inhabits the shallows of sandy shores. Weevers eat shrimps, shore crabs, small sand eels and flatfish. Spawning takes place between June and August.

Although not commercially fished for, weevers do make up a large proportion of inshore trawl catches and are turned into fish meal. Shrimp boats often pick up enormous numbers of weevers in their trawls, which wastes time because they have to be carefully extracted by hand.

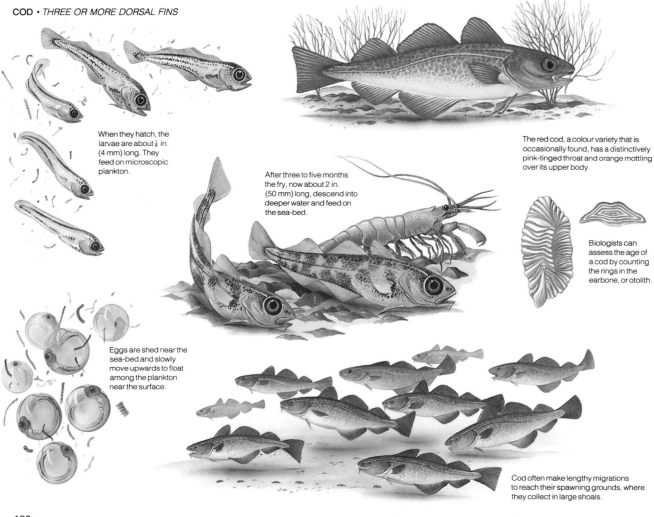

When they hatch, the larvae are about ⅛ in. (4 mm) long. They feed on microscopic plankton.

After three to five months the fry, now about 2 in. (50 mm) long, descend into deeper water and feed on the sea-bed.

The red cod, a colour variety that is occasionally found, has a distinctively pink-tinged throat and orange mottling over its upper body.

Biologists can assess the age of a cod by counting the rings in the earbone, or otolith.

Eggs are shed near the sea-bed and slowly move upwards to float among the plankton near the surface.

Cod often make lengthy migrations to reach their spawning grounds, where they collect in large shoals.

190

Cod have three dorsal and two ventral fins and a conspicuous single barbel under the lower jaw. The fish's colour varies according to its habitat, from brownish to pale grey, but it always has dark mottling over its upper body. It normally grows up to 47 in. (119 cm) long.

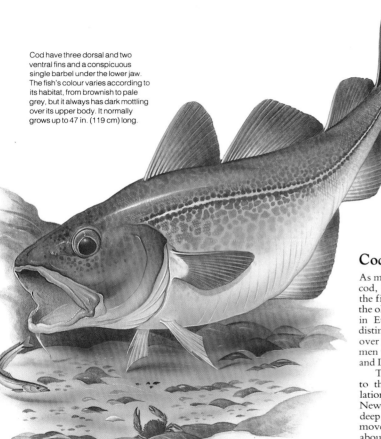

In the autumn, during their migration period, medium-sized cod can be caught with rod and line from beaches all round Britain.

Cod *Gadus morhua*

As many as nine million eggs have been found in an adult female cod, though two to three million is more normal. It is because the fish is so prolific that it has become the mainstay of one of the oldest, and still one of the most important, fishing industries in Europe. The migrations and fluctuations in numbers of distinct cod populations led to arguments in the 16th century over national fishing rights between French and Spanish fishermen which were similar to the quarrels between Britain, Iceland and Denmark during the 'cod war' of the 1960s and 1970s.

The cod is found in all the cool seas from the north of Russia to the eastern seaboard of North America. Distinct populations are centred on the North Sea, Iceland, Greenland and Newfoundland. Cod are found in waters up to 2,000 ft (600 m) deep, but during spawning between February and April, they move into water about 330 ft (100 m) deep. The fry hatch after about 12 days. As they grow larger they descend into deeper water and become bottom feeders, living first on crustaceans and then on fish. Cod weighing 30 lb (13 kg) are considered to be a good size by modern standards, but fish of 200 lb (90 kg) were recorded in the 17th century.

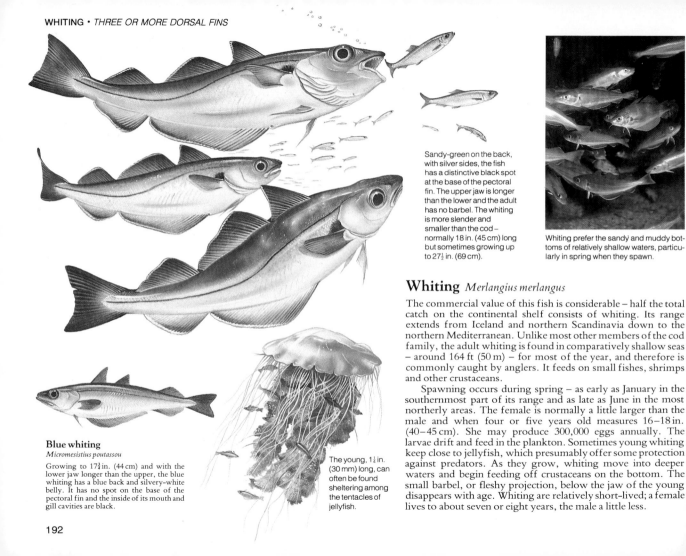

Sandy-green on the back, with silver sides, the fish has a distinctive black spot at the base of the pectoral fin. The upper jaw is longer than the lower and the adult has no barbel. The whiting is more slender and smaller than the cod – normally 18 in. (45 cm) long but sometimes growing up to 27½ in. (69 cm).

Whiting prefer the sandy and muddy bottoms of relatively shallow waters, particularly in spring when they spawn.

Blue whiting
Micromesistius poutassou

Growing to 17¼ in. (44 cm) and with the lower jaw longer than the upper, the blue whiting has a blue back and silvery-white belly. It has no spot on the base of the pectoral fin and the inside of its mouth and gill cavities are black.

The young, 1¼ in. (30 mm) long, can often be found sheltering among the tentacles of jellyfish.

Whiting *Merlangius merlangus*

The commercial value of this fish is considerable – half the total catch on the continental shelf consists of whiting. Its range extends from Iceland and northern Scandinavia down to the northern Mediterranean. Unlike most other members of the cod family, the adult whiting is found in comparatively shallow seas – around 164 ft (50 m) – for most of the year, and therefore is commonly caught by anglers. It feeds on small fishes, shrimps and other crustaceans.

Spawning occurs during spring – as early as January in the southernmost part of its range and as late as June in the most northerly areas. The female is normally a little larger than the male and when four or five years old measures 16–18 in. (40–45 cm). She may produce 300,000 eggs annually. The larvae drift and feed in the plankton. Sometimes young whiting keep close to jellyfish, which presumably offer some protection against predators. As they grow, whiting move into deeper waters and begin feeding off crustaceans on the bottom. The small barbel, or fleshy projection, below the jaw of the young disappears with age. Whiting are relatively short-lived; a female lives to about seven or eight years, the male a little less.

Bib are common inshore fish and are often found in rocky areas swimming in shoals about reefs and wrecks.

Deep bodied with three dorsal and two anal fins, the bib has a striking coppery-brown upper body across which run four or five darker bands. The sides are yellowish and the underside is white. There is a conspicuous black spot at the base of the pectoral fin and a long barbel on the lower jaw. It may reach 16¼ in. (41 cm) in length.

Poor cod

Trisopterus minutus

Like the bib, it has a long chin barbel but is distinguished by its overlapping upper jaw. The back is yellowish, the underside silvery-grey. A small black spot marks the upper part of the pectoral fin base. Grows up to 9 in. (23 cm) long.

Small bib are abundant over the sandy bottoms of shallow coastal waters.

Bib *Trisopterus luscus*

A very common shallow-water fish, the bib can be distinguished from all the other members of the cod family by its much deeper body, and by the distinctive coppery colour which has earned it the name 'brassy' in some areas. It is also known as the pout, pouting and whiting pout. It is found all round the coast of Britain, southern Scandinavia, France, Spain and the north-west Mediterranean, and is frequently caught in trawl nets. Because the fish is relatively small, however, usually weighing less than 5 lb (2 kg), and because its soft flesh is of low market value, such catches are normally sold for fish-meal processing.

The bib feeds in large shoals, on molluscs, small fish and crustaceans, particularly shrimps. Spawning occurs in March and April at depths of 165–230 ft (50–70 m); the eggs float in the surface waters and hatch in 10–12 days. Initially growth is rapid, but once the fish has achieved maturity at the end of the first year the growth rate slows down. It has a short life-span of only six to eight years.

The poor cod is similar to the bib in appearance and habits, but is more of an offshore fish, and is preyed on by commercially valuable species such as the hake and whiting.

Large pollack may be found feeding in small shoals over rocky grounds, weedy reefs or on wrecks.

The lower jaw extends further forwards than the upper, and has no barbel. Unlike whiting, the pollack has no black pectoral spot and the dark lateral line curves in an arch above the pectoral fin. The upper body is dark greenish-brown and the belly is white. It grows up to 51 in. (130 cm) long.

Saithe
Pollachius virens

Also known as coley or coalfish, saithe have jaws of equal length, a straight, lightcoloured lateral line and dark green fins. They grow up to 24 in. (60 cm) long.

Young pollack feed in shoals in estuaries or in water with a sandy bottom, eating worms and various small shellfish.

Pollack *Pollachius pollachius*

Anglers fishing with a bait or lure near the surface often catch pollack. It is a good sporting fish, as it is fairly active and can weigh up to 20 lb (9 kg). Pollack are also caught by trawlers, usually along with cod, although they are not an important commercial catch – only about 6,000 tons a year are caught in all European waters.

The pollack is a typical member of the cod family, having three dorsal fins, two anal fins and relatively large eyes. It is found inshore all round the coast of Britain, although larger pollack tend to spend the winter in deeper waters and move inshore only for the summer months. The pollack prefers rough, rocky waters and reefs and wrecks where it feeds on a variety of small crustaceans, smaller members of the cod and herring families, and sand eels. Spawning occurs between January and April in depths down to 330 ft (100 m). The eggs and newly hatched fry float freely, slowly drifting inshore to the shallower waters preferred by young pollack.

Saithe are smaller relatives of the pollack and have a similar way of life. They are more important commercially. About 250,000 tons are landed annually in European waters.

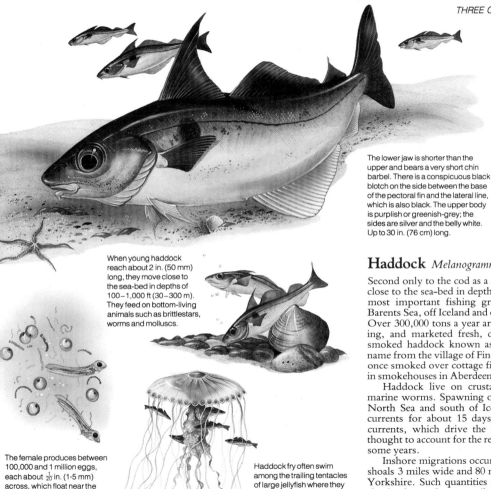

The lower jaw is shorter than the upper and bears a very short chin barbel. There is a conspicuous black blotch on the side between the base of the pectoral fin and the lateral line, which is also black. The upper body is purplish or greenish-grey; the sides are silver and the belly white. Up to 30 in. (76 cm) long.

The haddock lives close to the sea-bed in deep water, feeding on a diet of bottom-living animals.

When young haddock reach about 2 in. (50 mm) long, they move close to the sea-bed in depths of 100–1,000 ft (30–300 m). They feed on bottom-living animals such as brittlestars, worms and molluscs.

The female produces between 100,000 and 1 million eggs, each about $\frac{1}{25}$ in. (1·5 mm) across, which float near the surface in the plankton.

Haddock fry often swim among the trailing tentacles of large jellyfish where they are protected from predators.

Haddock *Melanogrammus aeglefinus*

Second only to the cod as a commercial fish, the haddock lives close to the sea-bed in depths of 100–1,000 ft (30–300 m). The most important fishing grounds are in the North Sea and Barents Sea, off Iceland and off the east coast of North America. Over 300,000 tons a year are caught, chiefly by bottom trawling, and marketed fresh, deep-frozen or smoked. A lightly smoked haddock known as finnan or finnan haddie takes its name from the village of Findon in Scotland where the fish were once smoked over cottage fires. The process is now carried out in smokehouses in Aberdeen.

Haddock live on crustaceans, shellfish, sea urchins and marine worms. Spawning occurs in the spring in the northern North Sea and south of Iceland, when the eggs drift in the currents for about 15 days before hatching. Changes in the currents, which drive the eggs ashore or into estuaries, are thought to account for the relative scarcity of young haddock in some years.

Inshore migrations occur in winter, and in the 19th century shoals 3 miles wide and 80 miles long gathered off the coast of Yorkshire. Such quantities are no longer seen, but occasional shoals of several square miles can be found off British coasts.

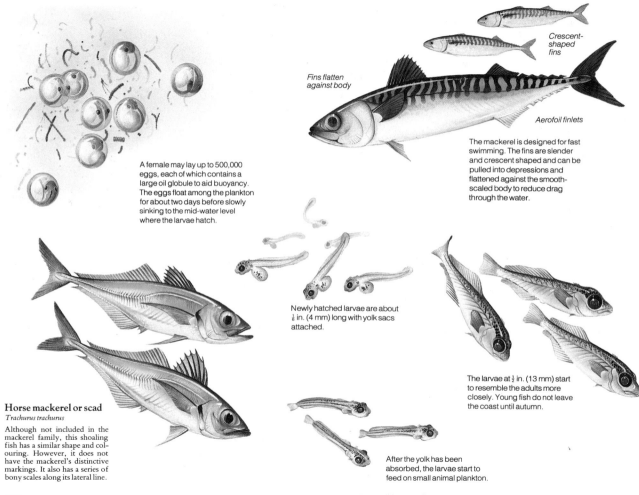

A female may lay up to 500,000 eggs, each of which contains a large oil globule to aid buoyancy. The eggs float among the plankton for about two days before slowly sinking to the mid-water level where the larvae hatch.

Fins flatten against body

Crescent-shaped fins

Aerofoil finlets

The mackerel is designed for fast swimming. The fins are slender and crescent shaped and can be pulled into depressions and flattened against the smooth-scaled body to reduce drag through the water.

Newly hatched larvae are about $\frac{1}{8}$ in. (4 mm) long with yolk sacs attached.

The larvae at $\frac{1}{2}$ in. (13 mm) start to resemble the adults more closely. Young fish do not leave the coast until autumn.

Horse mackerel or scad
Trachurus trachurus

Although not included in the mackerel family, this shoaling fish has a similar shape and colouring. However, it does not have the mackerel's distinctive markings. It also has a series of bony scales along its lateral line.

After the yolk has been absorbed, the larvae start to feed on small animal plankton.

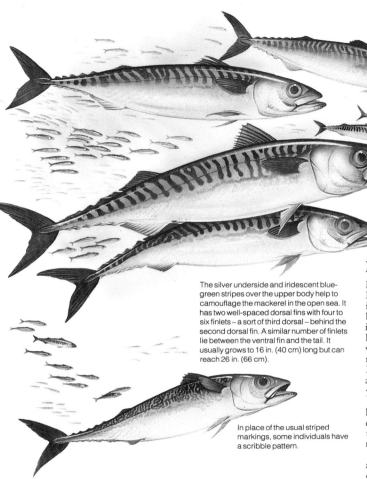

The silver underside and iridescent blue-green stripes over the upper body help to camouflage the mackerel in the open sea. It has two well-spaced dorsal fins with four to six finlets – a sort of third dorsal – behind the second dorsal fin. A similar number of finlets lie between the ventral fin and the tail. It usually grows to 16 in. (40 cm) long but can reach 26 in. (66 cm).

In place of the usual striped markings, some individuals have a scribble pattern.

During summer, large numbers of this North Atlantic deep-water fish migrate to shallower waters all round the coasts of Britain.

Mackerel *Scomber scombrus*

For centuries the mackerel has been an important food fish in Britain – even the laws governing Sunday trading were relaxed in 1698 to allow fishmongers to sell mackerel on that day because the flesh quickly goes bad due to the action of chemicals in the muscle tissue after death. Enormous catches of the fish have been recorded – in 1808 a Brighton boat lost all its gear when its nets became choked with mackerel. Today, mackerel no longer occur in the huge shoals once seen in the North Sea. British fishermen have to compete fiercely with east European and Russian trawlers, which take half the mackerel caught each year in European waters – largely around British coasts.

Like many other members of the tunny family, the mackerel has a body designed for sustained speed with minimum energy expenditure. Its aerofoil-like finlets, the depressions into which the fins can retract, and the jelly-like surrounds to the eyes, all reduce drag as it chases its prey – schools of smaller fish.

Spawning extends from March to September and peaks around May and June when huge shoals can be seen, particularly off Britain's western coasts. During winter, mackerel disperse to deeper waters, where feeding virtually stops.

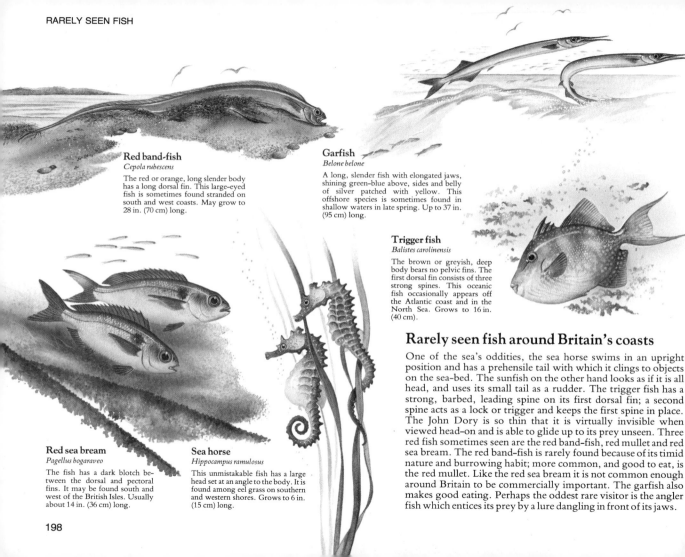

Red band-fish
Cepola rubescens

The red or orange, long slender body has a long dorsal fin. This large-eyed fish is sometimes found stranded on south and west coasts. May grow to 28 in. (70 cm) long.

Garfish
Belone belone

A long, slender fish with elongated jaws, shining green-blue above, sides and belly of silver patched with yellow. This offshore species is sometimes found in shallow waters in late spring. Up to 37 in. (95 cm) long.

Trigger fish
Balistes carolinensis

The brown or greyish, deep body bears no pelvic fins. The first dorsal fin consists of three strong spines. This oceanic fish occasionally appears off the Atlantic coast and in the North Sea. Grows to 16 in. (40 cm).

Red sea bream
Pagellus bogaraveo

The fish has a dark blotch between the dorsal and pectoral fins. It may be found south and west of the British Isles. Usually about 14 in. (36 cm) long.

Sea horse
Hippocampus ramulosus

This unmistakable fish has a large head set at an angle to the body. It is found among eel grass on southern and western shores. Grows to 6 in. (15 cm) long.

Rarely seen fish around Britain's coasts

One of the sea's oddities, the sea horse swims in an upright position and has a prehensile tail with which it clings to objects on the sea-bed. The sunfish on the other hand looks as if it is all head, and uses its small tail as a rudder. The trigger fish has a strong, barbed, leading spine on its first dorsal fin; a second spine acts as a lock or trigger and keeps the first spine in place. The John Dory is so thin that it is virtually invisible when viewed head-on and is able to glide up to its prey unseen. Three red fish sometimes seen are the red band-fish, red mullet and red sea bream. The red band-fish is rarely found because of its timid nature and burrowing habit; more common, and good to eat, is the red mullet. Like the red sea bream it is not common enough around Britain to be commercially important. The garfish also makes good eating. Perhaps the oddest rare visitor is the angler fish which entices its prey by a lure dangling in front of its jaws.

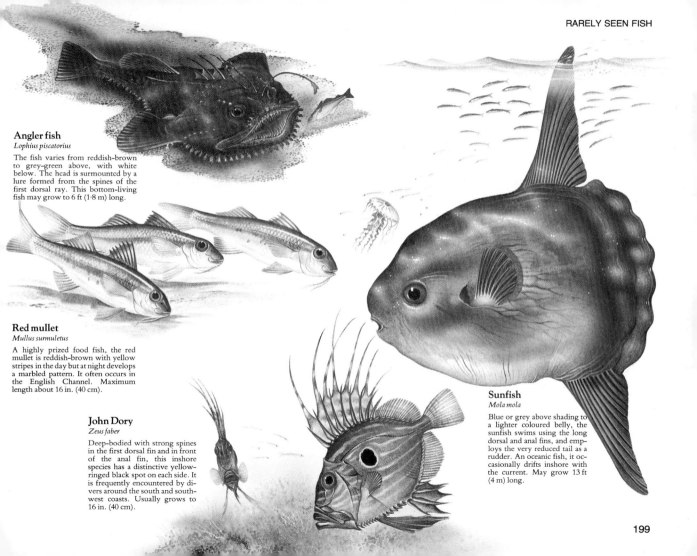

Angler fish
Lophius piscatorius

The fish varies from reddish-brown to grey-green above, with white below. The head is surmounted by a lure formed from the spines of the first dorsal ray. This bottom-living fish may grow to 6 ft (1·8 m) long.

Red mullet
Mullus surmuletus

A highly prized food fish, the red mullet is reddish-brown with yellow stripes in the day but at night develops a marbled pattern. It often occurs in the English Channel. Maximum length about 16 in. (40 cm).

John Dory
Zeus faber

Deep-bodied with strong spines in the first dorsal fin and in front of the anal fin, this inshore species has a distinctive yellow-ringed black spot on each side. It is frequently encountered by divers around the south and south-west coasts. Usually grows to 16 in. (40 cm).

Sunfish
Mola mola

Blue or grey above shading to a lighter coloured belly, the sunfish swims using the long dorsal and anal fins, and employs the very reduced tail as a rudder. An oceanic fish, it occasionally drifts inshore with the current. May grow 13 ft (4 m) long.

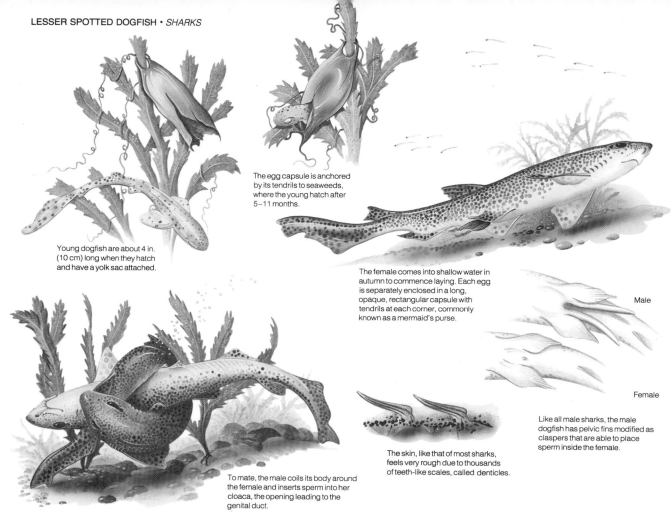

The egg capsule is anchored by its tendrils to seaweeds, where the young hatch after 5–11 months.

Young dogfish are about 4 in. (10 cm) long when they hatch and have a yolk sac attached.

The female comes into shallow water in autumn to commence laying. Each egg is separately enclosed in a long, opaque, rectangular capsule with tendrils at each corner, commonly known as a mermaid's purse.

Male

Female

Like all male sharks, the male dogfish has pelvic fins modified as claspers that are able to place sperm inside the female.

To mate, the male coils its body around the female and inserts sperm into her cloaca, the opening leading to the genital duct.

The skin, like that of most sharks, feels very rough due to thousands of teeth-like scales, called denticles.

Often seen close to the shore, the dogfish can be found all round the coast of Britain, but occurs most commonly in southern waters.

The upper body and sides are sandy brown dappled with small, darker spots. The underside is creamy-white. This small shark grows to nearly 40 in. (100 cm) in length.

Nursehound
Scyliorhinus stellaris

Sometimes known as the greater spotted dogfish, the nursehound usually lives over rocky sea-beds. The back and sides are sandy or greyish-brown with large dark blotches. The dorsal fin is further forward than that of the lesser spotted dogfish. An average sized adult is about 48 in. (120 cm) long.

Lesser spotted dogfish   Nursehound

Unlike the lesser spotted dogfish, the nasal flaps of the nursehound are distinctly separate and do not reach the upper lip.

Lesser spotted dogfish *Scyliorhinus canicula*

Known also as the rough hound or, simply, the dogfish, this harmless species is the commonest European shark. Of the 60 species found in tropical or temperate seas, only three live in European waters. They are most common in the southern waters of the British Isles but they also reach north to the Shetlands and south to the Mediterranean. Dogfish can be found at depths from 6 ft (1·8 m) to 700 ft (210 m) but the majority are caught at about 165 ft (50 m). They are mainly bottom-living and feed on molluscs, crustaceans and bottom-dwelling fish. Sometimes faster swimming fish are taken, but as the dogfish hunts by scent rather than by sight, these are few.

Mating occurs in autumn and, as fertilisation is internal, there is a delay of several weeks before egg laying. This starts in late autumn in shallow water and may continue through winter and spring into the next summer. Egg capsules, commonly called mermaid's purses, are often found washed ashore.

As the familiar rock salmon, or flake, of fish-and-chip shops, the dogfish has considerable commercial value, and several thousand tons are caught annually by British trawlers and long-liners.

201

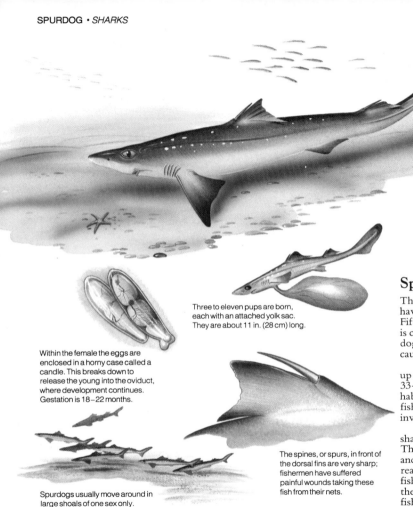

The spurdog is easily distinguished from other small sharks by the slender spine in front of each dorsal fin. Dark grey, with indistinct white spots running down the back and sides, it grows to about 48 in. (120 cm) in length.

The sea off Morte Point, north Devon, is one of the many places around Britain where spurdogs are found.

Three to eleven pups are born, each with an attached yolk sac. They are about 11 in. (28 cm) long.

Within the female the eggs are enclosed in a horny case called a candle. This breaks down to release the young into the oviduct, where development continues. Gestation is 18–22 months.

The spines, or spurs, in front of the dorsal fins are very sharp; fishermen have suffered painful wounds taking these fish from their nets.

Spurdogs usually move around in large shoals of one sex only.

Spurdog *Squalus acanthias*

The spiny sharks are a widely distributed group, all of which have strong spines in front of the dorsal fins, and lack anal fins. Fifteen species occur in European waters, but only the spurdog is common round the coasts of Britain. Also known as the spiny dogfish, it is commercially very valuable; over 50,000 tons are caught annually and sold as rock eel, rock salmon or flake.

Spurdogs gather in large shoals which move unpredictably up and down the coast, inshore and offshore at depths of 33–3,300 ft (10–1,000 m). They are opportunist in their eating habits and will take advantage of any abundant prey. Schooling fish such as herrings and sand eels, bottom-living flatfish, and invertebrates such as squids, worms and crabs are all eaten.

An average of four or five young are born in the spring in shallow water, nearly two years after fertilisation of the eggs. This is one of the longest gestation periods in the animal world, and since the spurdog, which has a life-span of 20 years, does not reach sexual maturity until the age of five to eight years many fish do not survive long enough to reproduce. For these reasons the species is vulnerable to uncontrolled exploitation by the fishing industry.

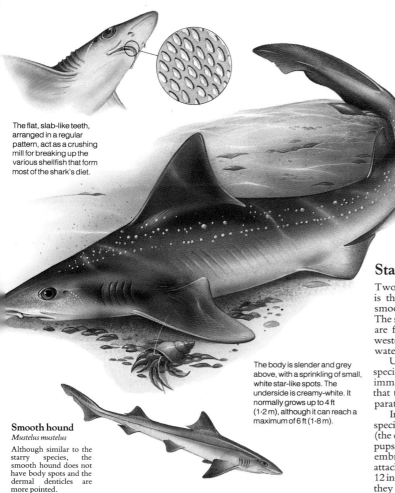

The flat, slab-like teeth, arranged in a regular pattern, act as a crushing mill for breaking up the various shellfish that form most of the shark's diet.

The dermal denticles, or close-packed tooth-like projections on the skin, are weakly ridged.

The shark, a bottom feeder, moves inshore during summer where it is often caught by beach fishermen.

The body is slender and grey above, with a sprinkling of small, white star-like spots. The underside is creamy-white. It normally grows up to 4 ft (1·2 m), although it can reach a maximum of 6 ft (1·8 m).

Smooth hound
Mustelus mustelus

Although similar to the starry species, the smooth hound does not have body spots and the dermal denticles are more pointed.

Starry smooth hound *Mustelus asterias*

Two characteristics give the starry smooth hound its name. One is that, compared with other sharks, it has a comparatively smooth skin; the other, that it is usually seen travelling in packs. The starry smooth hound and the closely related smooth hound are found all around Britain, but more frequently off rocky western coasts. Both species live close to the bottom in inshore waters, feeding on lobsters, crawfish, crabs and molluscs.

Until recently, it was thought that there was only one species, the spotted, or 'starry', form being considered an immature stage of the smooth. However, it is now established that the starry smooth hound is more common than the comparatively rare smooth hound.

In addition, the development of the embryo in the two species is different. The starry smooth hound is ovoviviparous (the eggs are retained in the mother until they hatch) and up to 30 pups are produced. The smooth hound is viviparous (each embryo is nourished in the uterus from a yolk-sac placenta attached to the mother), and produces up to 15 pups, each about 12 in. (30 cm) long. The young are found in the shallows, where they remain long after the parents return to deeper water.

203

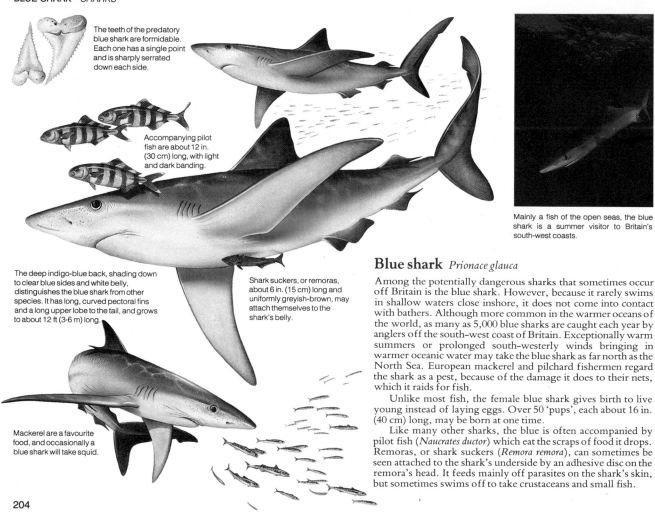

The teeth of the predatory blue shark are formidable. Each one has a single point and is sharply serrated down each side.

Accompanying pilot fish are about 12 in. (30 cm) long, with light and dark banding.

Mainly a fish of the open seas, the blue shark is a summer visitor to Britain's south-west coasts.

The deep indigo-blue back, shading down to clear blue sides and white belly, distinguishes the blue shark from other species. It has long, curved pectoral fins and a long upper lobe to the tail, and grows to about 12 ft (3·6 m) long.

Shark suckers, or remoras, about 6 in. (15 cm) long and uniformly greyish-brown, may attach themselves to the shark's belly.

Mackerel are a favourite food, and occasionally a blue shark will take squid.

Blue shark *Prionace glauca*

Among the potentially dangerous sharks that sometimes occur off Britain is the blue shark. However, because it rarely swims in shallow waters close inshore, it does not come into contact with bathers. Although more common in the warmer oceans of the world, as many as 5,000 blue sharks are caught each year by anglers off the south-west coast of Britain. Exceptionally warm summers or prolonged south-westerly winds bringing in warmer oceanic water may take the blue shark as far north as the North Sea. European mackerel and pilchard fishermen regard the shark as a pest, because of the damage it does to their nets, which it raids for fish.

Unlike most fish, the female blue shark gives birth to live young instead of laying eggs. Over 50 'pups', each about 16 in. (40 cm) long, may be born at one time.

Like many other sharks, the blue is often accompanied by pilot fish (*Naucrates ductor*) which eat the scraps of food it drops. Remoras, or shark suckers (*Remora remora*), can sometimes be seen attached to the shark's underside by an adhesive disc on the remora's head. It feeds mainly off parasites on the shark's skin, but sometimes swims off to take crustaceans and small fish.

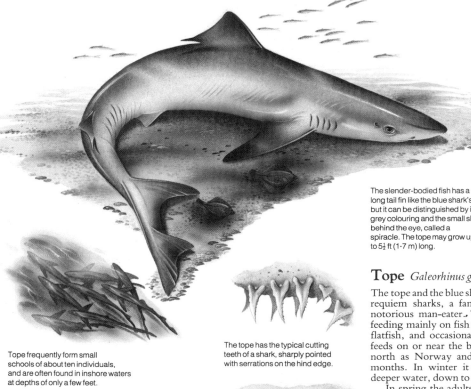

The slender-bodied fish has a long tail fin like the blue shark's, but it can be distinguished by its grey colouring and the small slit behind the eye, called a spiracle. The tope may grow up to 5½ ft (1·7 m) long.

Tope normally inhabit coastal waters, feeding on bottom-living fish at depths of 130–330 ft (40–100 m).

Tope frequently form small schools of about ten individuals, and are often found in inshore waters at depths of only a few feet.

The tope has the typical cutting teeth of a shark, sharply pointed with serrations on the hind edge.

The 20 or so young are born during the summer after a gestation period of ten months. Each measures about 15 in. (38 cm) in length.

The sharks of this family have a special cover to protect the eye, called the nictitating membrane.

Tope *Galeorhinus galeus*

The tope and the blue shark are members of the Carcharinidae or requiem sharks, a family which includes the tiger shark, a notorious man-eater. The tope has a less ambitious appetite, feeding mainly on fish such as whiting, cod and bottom-living flatfish, and occasionally taking molluscs and crustaceans. It feeds on or near the bottom in coastal waters, ranging as far north as Norway and southern Iceland during the summer months. In winter it migrates southwards and moves into deeper water, down to 650 ft (200 m).

In spring the adults move inshore, and during the summer each female gives birth to a litter of around 20 pups, depending on age. The young tope are gregarious, feeding in shoals in extremely shallow inshore water, but with age they develop a solitary way of life.

Although of no commercial value, the tope is an important sporting fish in European waters. Mature fish are not considered edible, for their flesh is unpalatable and tough, but in other parts of the world young specimens are eaten. It is one of the species employed in the manufacture of shark's fin soup, and the liver is used as a source of vitamin A.

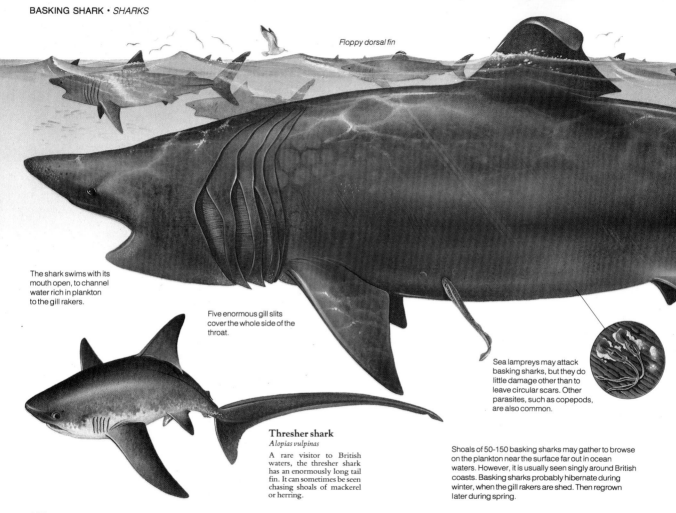

Floppy dorsal fin

The shark swims with its mouth open, to channel water rich in plankton to the gill rakers.

Five enormous gill slits cover the whole side of the throat.

Sea lampreys may attack basking sharks, but they do little damage other than to leave circular scars. Other parasites, such as copepods, are also common.

Thresher shark
Alopias vulpinas

A rare visitor to British waters, the thresher shark has an enormously long tail fin. It can sometimes be seen chasing shoals of mackerel or herring.

Shoals of 50-150 basking sharks may gather to browse on the plankton near the surface far out in ocean waters. However, it is usually seen singly around British coasts. Basking sharks probably hibernate during winter, when the gill rakers are shed. Then regrown later during spring.

The distinctive, floppy dorsal fin of the basking shark is sometimes seen breaking the surface of deep waters near rocky coasts around Britain.

The gill rakers are slender, horny bristles that filter plankton from the water before it passes out of the gills.

Basking shark *Cetorhinus maximus*

Despite its awesome appearance, the basking shark is quite harmless to man, as it feeds exclusively on plankton. It is one of the largest of the true fishes, occasionally reaching a formidable 36 ft (11 m) in length and a weight of over 3 tons (3,000 kg). Five very long gill slits covering the sides of the throat distinguish it from other sharks. Some 1,500 tons (1.5 million kg) of water an hour can pass through the slits as the shark feeds.

In summer, the basking shark may be sighted browsing off ocean coasts around the British Isles. Little is known of basking shark biology, but the female is thought to incubate eggs in its body until they hatch. At birth, the 'pups' are about 5 ft (1·5 m) long and grow to almost 12 ft (3·6 m) in two and a half years. The fish was once hunted by harpooners from Ireland, Norway and Scotland. Its gigantic liver – nearly a quarter of the total body weight – can yield as much as 500 gallons (2,270 litres) of oil, which was used in oil lamps and in candlemaking.

The thresher, or fox shark is as active as the basking shark is slow-moving, and is considered a fine sporting fish. During the summer, threshers weighing up to 300 lb (135 kg) are occasionally seen in European waters.

207

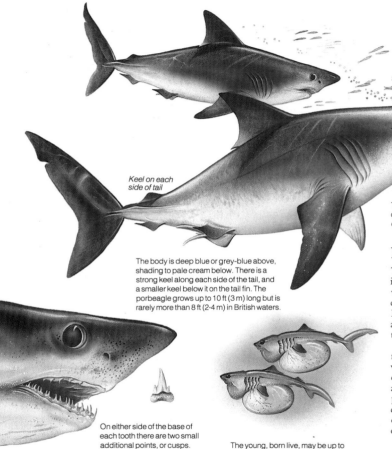

Keel on each side of tail

The body is deep blue or grey-blue above, shading to pale cream below. There is a strong keel along each side of the tail, and a smaller keel below it on the tail fin. The porbeagle grows up to 10 ft (3 m) long but is rarely more than 8 ft (2·4 m) in British waters.

On either side of the base of each tooth there are two small additional points, or cusps. These make the teeth more efficient for stabbing and tearing the flesh of prey.

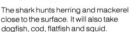

The young, born live, may be up to 20 in. (50 cm) long at birth. The pups' bellies are swollen from eating unfertilised eggs while still in the uterus.

The shark hunts herring and mackerel close to the surface. It will also take dogfish, cod, flatfish and squid.

Rare inshore specimens, caught off Anglesey about 200 years ago, were mistakenly called Beaumaris sharks.

Porbeagle *Lamna nasus*

The mackerel sharks, of which the porbeagle is the commonest, include the mako and the most feared of all sharks – the great white. However, although the porbeagle is theoretically capable of attacking people, there are no records of this happening in British waters. Though a small porbeagle may occasionally approach the shore, an adult rarely comes within 10 miles, and therefore has no contact with bathers.

The porbeagle can be seen off British coasts all the year round, especially during the summer when it migrates northwards. It normally lives near the surface, hunting among shoals of mackerel or herring, but it is sometimes caught at depths of about 420 ft (130 m). The flesh is edible and the carcass can be processed for oil, but apart from sport fishing it is of little commercial interest in Britain. It is eaten much more commonly on the Continent. Fishermen usually regard it as a pest because of the damage it causes to lines and nets which it raids for fish.

Some evidence suggests that the porbeagle, unlike other sharks and rays, may have colour vision as there are colour-receptive cells in the retina of its eyes. As with many sharks, its keen sight is its main means of locating its prey.

The mako is a surface-living, open ocean species that is considered a fine sporting fish by sea anglers.

The large, slender body is deep blue or blue-grey above, with a sharp transition to white underneath. The snout is pointed. Makos grow to 13 ft (4 m) in length and 1,000 lb (454 kg) in weight.

The teeth are long, triangular, narrow and have only one cusp, or point.

The mako gives birth to fully formed young which have been nourished on the yolk and, later, have fed on unfertilised eggs in the uterus of the mother.

Mako shark *Isurus oxyrinchus*

Widely distributed throughout the warmer waters of the Atlantic, Pacific and Indian oceans, the mako shark is relatively uncommon in British waters. It was first reported in 1955, but specimens which occurred before this date were probably mistaken for the very similar porbeagle. It appears to be a summer migrant from further south, and in warm years it may follow the Gulf Stream right up to northern Norway.

Like the porbeagle, the mako is an open-sea predator which feeds on shoaling fish such as herring, mackerel, pilchards and anchovies. It has a large, powerful tail which enables it to rapidly pursue anything that moves. This feature is characteristic of the mackerel shark family which also includes the man-eating great white shark. In European waters the mako does not grow big enough to be a serious threat, but large tropical specimens have the reputation of being dangerous to man. It is of no commercial value, but is considered fine sport by sea anglers.

In common with the majority of shark species, the mako is ovoviviparous – it produces eggs which are retained within the mother while they hatch and develop from embryos into miniatures of the adult. It is not known to breed around Britain.

209

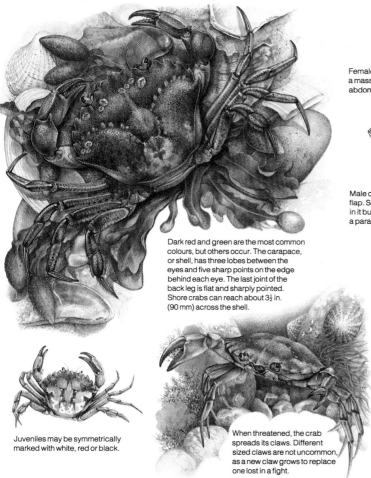

Dark red and green are the most common colours, but others occur. The carapace, or shell, has three lobes between the eyes and five sharp points on the edge behind each eye. The last joint of the back leg is flat and sharply pointed. Shore crabs can reach about 3½ in. (90 mm) across the shell.

Juveniles may be symmetrically marked with white, red or black.

When threatened, the crab spreads its claws. Different sized claws are not uncommon, as a new claw grows to replace one lost in a fight.

Females are often found carrying a mass of eggs beneath the broad abdomen flap.

Male crabs have only a narrow abdomen flap. Sometimes it appears to have eggs in it but these smooth, round forms are a parasite, *Sacculina*.

One of Britain's commonest forms of crab, the shore crab lives on all types of beach, and even in some estuaries.

Shore crab *Carcinus maenas*

Widespread around the coasts of Britain, the shore crab can be found hiding under stones and among seaweed. It is hardy and able to survive for long periods out of water. The discarded shells are a common sight on the shore, but it also lives below the low-tide mark.

A shore crab adjusts to growth by shedding, or moulting, its old hard shell to expose the soft replacement that has been developing underneath. By taking in water, the crab can expand the new shell before it hardens, but while it is still soft it is particularly vulnerable and hides away in crevices.

Crabs mate only when the female is soft and the male carries her with him for several days waiting for her to moult. The female carries the orange-coloured mass of eggs beneath her abdominal flap for 12–18 weeks before they hatch. The larvae float in the plankton for several weeks before descending to the sea-bed in their adult form.

Shore crabs are unfussy scavengers, eating almost anything live or dead. They are themselves eaten by fish such as ballan wrasse and by seabirds. They are not fished commercially in British waters, although they are eaten on the Continent.

The male's abdomen (top), is narrow, the female's broad.

Hairy legs and massive black-tipped claws project from beneath a slightly granular, brownish-pink oval carapace, or shell. It has about nine 'pie-crust' lobes at each side. The shell may reach 8 in. (20 cm) across.

Edible crabs live in rough, rocky areas but also live on sediment, where they dig for bivalve molluscs.

Edible crab *Cancer pagurus*

The edible crab is the largest of the common crab species in Britain. A large specimen can weigh up to several pounds and live to eight years or more. Juveniles and small adults are found under the rocks and among the kelp of the shore. Larger animals move into deeper waters.

Females first mate when they are about five or six years old and 5 in. (12·5 cm) broad. As with other crustaceans, mating occurs when the female is newly soft after moulting, or shedding her shell. The male will often attend her for several weeks beforehand and help with the moult. A large female may lay over 3 million eggs, but few survive. Those that do, float in the plankton and change into tiny crabs through a series of moults and sink from mid-water to the sea-bed after about one month. Young crabs moult several times each year but older ones only once a year or less.

Edible crabs feed mainly on other shellfish but are also great scavengers. They are caught in baited pots all around the coast of Britain, but particularly the south coast. Only specimens with shells at least 4½ in. (11·5 cm) broad are allowed to be taken, and it is also illegal to land females with eggs or with soft shells.

Hairy crab
Pilumnus hirtellus
Long hairs cover the legs and brownish-red shell, up to ⅘ in. (20 mm) across.

Favourite hiding places are rock crevices where the crab can easily defend itself.

211

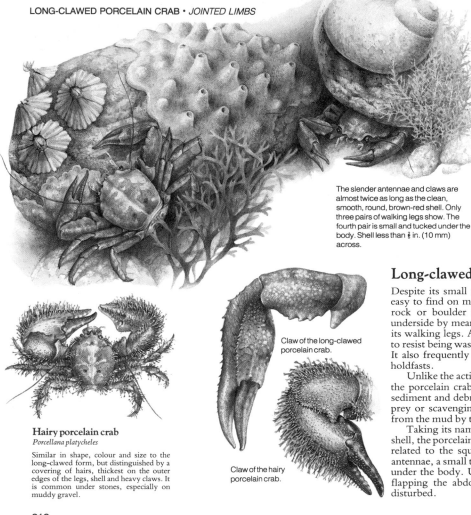

The slender antennae and claws are almost twice as long as the clean, smooth, round, brown-red shell. Only three pairs of walking legs show. The fourth pair is small and tucked under the body. Shell less than ⅜ in. (10 mm) across.

Common on shores all round Britain, this crab lives under stones and in seaweeds, often among kelp holdfasts.

Hairy porcelain crab
Porcellana platycheles

Similar in shape, colour and size to the long-clawed form, but distinguished by a covering of hairs, thickest on the outer edges of the legs, shell and heavy claws. It is common under stones, especially on muddy gravel.

Claw of the long-clawed porcelain crab.

Claw of the hairy porcelain crab.

Long-clawed porcelain crab *Pisidia longicornis*

Despite its small size, the long-clawed porcelain crab is quite easy to find on most rocky shores round Britain. Turn over a rock or boulder and it will be found clinging close to the underside by means of the sharply pointed spines at the ends of its walking legs. Anchoring its flattened body this way helps it to resist being washed away during the ebb and flow of the tides. It also frequently shelters among the tangled rootlets of kelp holdfasts.

Unlike the active and aggressive shore and swimming crabs, the porcelain crab is a filter feeder. It uses its claws to scoop sediment and debris from under rocks, rather than for hunting prey or scavenging. Minute particles of food are then filtered from the mud by the hairs fringing the mouthparts.

Taking its name from the delicate sheen on the surface of its shell, the porcelain crab is not a true crab at all but is more closely related to the squat lobsters. Like them, it has long, slender antennae, a small tail and a fourth pair of legs tucked out of sight under the body. Unlike the squat lobsters, it cannot swim by flapping the abdomen and simply scuttles away sideways if disturbed.

The crab's soft body is hidden in the empty shell of some other creature, often a whelk (above). The head, large claws and two pairs of walking legs stick out, but if danger threatens they are drawn into the shell and the large right claw blocks the opening.

The claws are large and coarsely granulated.

Pagurus prideauxi

This smaller species of hermit crab lives below low-tide level, often in a shell covered by a cloak anemone. The claws are less grainy but have bristles. The eye-stalk has a red mark.

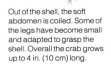

Out of the shell, the soft abdomen is coiled. Some of the legs have become small and adapted to grasp the shell. Overall the crab grows up to 4 in. (10 cm) long.

Hermit crabs are found on all types of shore – small ones in various snail shells, larger ones in whelk shells.

Hermit crab *Pagurus bernhardus*

The sharp-eyed searcher will find hermit crabs on almost any beach, including sand-flats. They live inside discarded shells, often of the whelk. These offer such effective protection from attack that the crab can be active during the day. In the search for food it walks on its two front pairs of legs, dragging its shell behind. The hermit crab is a great scavenger and will eat almost anything organic. It is preyed upon by gulls and other seabirds – provided they can break the shell.

In order to grow, the hermit crab, like other crabs, has to shed the shell covering of its claws and front legs. Periodically it also changes its borrowed shell. A hermit crab will often share its shell with other animals. The *Nereis fucata* worm feeds on scraps at the shell entrance; the parasitic anemone, the sea fir *Hydractinia*, and the orange sponge *Suberites* may also live on the shell. Yellowish lumps under the crab's tail are almost unrecognisable parasitic barnacles.

After pairing with a male, the female will carry the eggs on her abdomen until they hatch into minute larvae. After floating with the plankton for several weeks and moulting several times, they take on adult form.

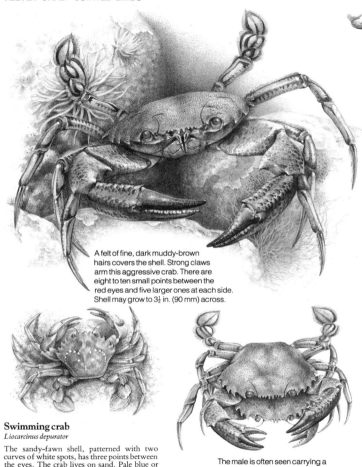

A felt of fine, dark muddy-brown hairs covers the shell. Strong claws arm this aggressive crab. There are eight to ten small points between the red eyes and five larger ones at each side. Shell may grow to 3½ in. (90 mm) across.

The last joint of the rear leg is flattened and rounded to aid swimming. Blue-black lines pattern all the legs and there are red marks at the joints.

Velvet crabs live in rocky habitats, especially where there is kelp, both on and offshore along most coasts.

Swimming crab
Liocarinus depurator

The sandy-fawn shell, patterned with two curves of white spots, has three points between the eyes. The crab lives on sand. Pale blue or violet tinges the flattened last joint of its back leg. Shell 2 in. (50 mm) across.

The male is often seen carrying a female ready to moult, or cast her shell, after which the pair can mate.

Velvet crab *Liocarcinus puber* (syn. *Macropipus puber*)

The name of this beautiful crab is derived from the dense, velvety covering of fine hairs on its shell. It also has bold blue-black and red markings on its legs, and bright red eyes which help to intimidate attackers. The flattened paddle-like hind legs are used to swim rapidly sideways to avoid predators such as cuttlefish. If cornered, the fiddler crab, as this aggressive crab is sometimes known, will try to frighten off an enemy by sitting back and holding its claws wide apart to make itself look as large as possible. The velvet crab will eat almost anything it can catch or find as it hunts through the kelp forest and under cover of seaweed on the shore.

In a fight or an accident, the crab may lose a claw or even a leg. However, as with most crustaceans, the limb breaks along predetermined lines to minimise damage and may grow again – often larger than the original.

Velvet-crab males carry their smaller female partners around for several days before mating and egg laying. The females carry their eggs for some weeks before they hatch into larvae. For the first few weeks, the larvae float among the mid-water plankton until, in tiny adult form, they descend to the sea-bed.

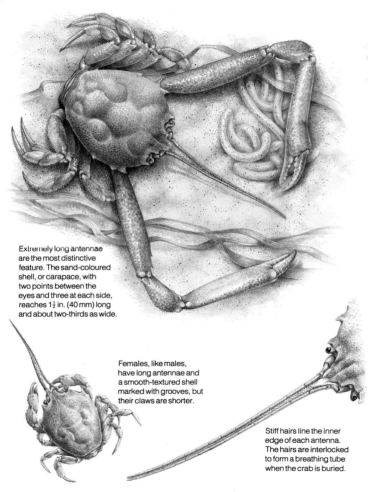

Extremely long antennae are the most distinctive feature. The sand-coloured shell, or carapace, with two points between the eyes and three at each side, reaches 1½ in. (40 mm) long and about two-thirds as wide.

Females, like males, have long antennae and a smooth-textured shell marked with grooves, but their claws are shorter.

Stiff hairs line the inner edge of each antenna. The hairs are interlocked to form a breathing tube when the crab is buried.

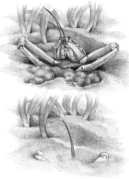

After the crab has buried itself during the daytime, only the antennae and parts of the claws show above the sand. The crab emerges at night.

Masked crabs live buried in clean sand on the lower shore, and also in shallow water all round Britain.

Masked crab *Corystes cassivelaunus*

The furrows on the upper shell of the masked crab resemble a human face, and it is from this characteristic that the crab takes its name. During daylight it lies buried several inches in the sand, with only the tips of its two antennae in the water above. It breathes by taking in water between its antennae and passing it over its gills. The antennae are joined by an interlocking network of fine hairs designed to prevent sand grains from blocking the gills.

Under cover of darkness the crab emerges to forage for shrimps, worms and other small animals. Seabirds and fish, such as skate, eat masked crabs. If threatened, the crab quickly digs itself into the sand by sitting on its rear while pulling down with its hind legs and pushing sand away with its large claws.

In early summer, large numbers of masked crabs come ashore to mate. Females attract males by releasing chemicals called pheromones and, unlike other crabs, the female does not need to moult before mating. She carries her eggs on her abdomen for several weeks before they hatch. After floating among the plankton for a few weeks, the larvae sink to the sea-bed to begin adult life.

215

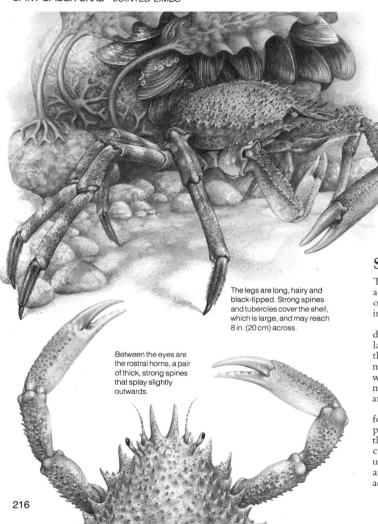

The legs are long, hairy and black-tipped. Strong spines and tubercles cover the shell, which is large, and may reach 8 in. (20 cm) across.

Between the eyes are the rostral horns, a pair of thick, strong spines that splay slightly outwards.

Immature crabs with small claws are sometimes found in low-tide pools. They camouflage themselves by draping seaweed or debris over the shell.

Living among rocks and sand on south and south-west coasts, the crabs mass into big heaps for summer breeding.

Spiny spider crab *Maja squinado*

The long, spindly, hairy legs and squat appearance give the crab a spider-like look. It is also known as the thornback crab because of the sharp spines on the top of its body shell. When half-buried in the sand, the spines can be painful if trodden upon.

During the summer, sexually mature adults migrate from deeper to shallower inshore waters to mate. They congregate in large mating mounds, sometimes with as many as 100 crabs. In the centre of the mound are those females that have recently moulted. The hard-shelled males remain on the outside, except when copulating, and protect the soft-shelled females. The mating season ends in the autumn when the mounds disperse and the crabs move back into deeper waters.

After fertilisation, the female crab carries up to 150,000 eggs for nine months. The larvae, when hatched, float among the plankton and drift with the currents until, after a few weeks, those that survive settle on the sea-bed as tiny crabs. Young crabs are rarely seen close inshore and the female will not breed until she is 4–6 in. (10–15 cm) long. Spider crabs feed on small animals and on seaweeds. Using their thin pincers, they are adept at extracting food from narrow crevices and small holes.

Identifying other spider crabs

Long, spidery legs and a triangular body characterise spider crabs. They all drape themselves with seaweeds, sponges and other debris as camouflage, which may hide identification details. The species illustrated are widespread on the shore. These and other similar species can also be found in shallow water.

A broad, U-shaped cleft divides the rostral horns.

Scorpion spider crab
Inachus dorsettensis

A reddish–brown crab which often holds its short, stout claws bent. The shell front bears a row of four small tubercles with one larger tubercle rising behind them. The shell may reach 1 in. (25 mm) across its base.

The tips of the pair of rostral horns projecting between the eyes converge.

The rostral horns are separated by a very narrow slit.

Leach's spider crab
Inachus phalangium

Similar to the scorpion spider crab but with only two tubercles near the front of the shell and a larger one behind them. The shell grows to ¾ in. (20 mm) across its base.

Long-legged spider crab
Macropodia rostrata

The colour may be greyish to yellowish or reddish–brown. The claws are short but the legs are very long, thin and hairy with curved tips. The prominent eyes have no sockets. The shell is about ⅝ in. (15 mm) wide across its base.

There is no gap between the straight rostral horns.

Sea toad
Hyas araneus

Small tubercles and bristles cover this pear-shaped, dirty brown or dull purple crab. The eyes can be withdrawn into prominent sockets. The shell is about 4¼ in. (11 cm) long.

217

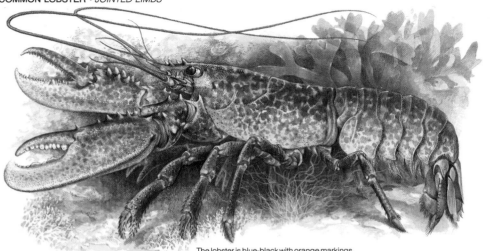

Lobsters inhabit rocky areas all round Britain, often making shelters under rocks by digging out sediment.

The lobster is blue-black with orange markings. The massive claws are unequal, one used for crushing and the smaller one for cutting. Lobsters may reach 20 in. (50 cm) in length.

Norway lobster
Nephrops norvegicus

Shaped like a lobster, but orange, slender and only about 6 in. (15 cm) long, it lives in mud burrows below low-tide mark. It is eaten as scampi or Dublin Bay prawn.

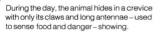

During the day, the animal hides in a crevice with only its claws and long antennae – used to sense food and danger – showing.

Common lobster *Homarus gammarus*

The summer holidaymaker may find a lobster in a deep pool on the shore, but the animal normally prefers rocky areas below low tide. Like the crab, it grows by periodically shedding its shell, pulling its soft body out through a split across the back; the new shell hardens in a few days. It may eat some discarded shell to replace lost calcium. The lobster normally walks over the sea-bed searching for food such as shellfish, but when frightened it can swim away backwards by flexing and flapping its tail and abdomen. Its powerful claws are used to crush food and to protect it from predators such as seals.

Females mate at about seven years old when they are some 10 in. (25 cm) long. An orange mass of some 150,000 eggs is laid in late summer. For nine to ten months this is carried under the abdomen attached to legs called swimmerets. Eventually, the eggs hatch into shrimp-like larvae about ⅜ in. (10 mm) long.

Britain's main commercial lobster fisheries lie off Scotland, Devon and Cornwall, where they are caught in baited traps. To protect this slow-growing species, it is illegal to take one with a carapace, or main body shell, less than 3⅓ in. (85 mm) long. Lobsters may live 30 years but diet, not age, determines size.

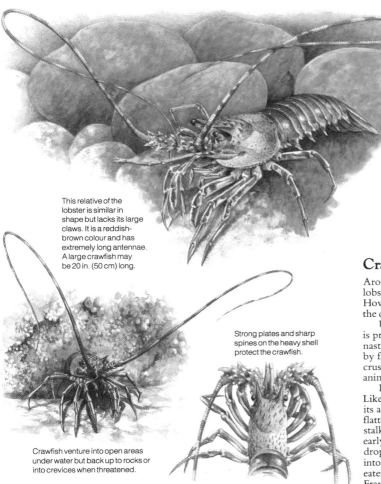

This relative of the lobster is similar in shape but lacks its large claws. It is a reddish-brown colour and has extremely long antennae. A large crawfish may be 20 in. (50 cm) long.

Crawfish venture into open areas under water but back up to rocks or into crevices when threatened.

Strong plates and sharp spines on the heavy shell protect the crawfish.

The fine, delicate claws are used to pick up food, and not for defence or for crushing prey.

Only south-western and western waters are warm enough for crawfish, which live mostly below 65 ft (20 metres).

Crawfish or spiny lobster *Palinurus elephas*

Around Britain the crawfish is much less common than the lobster, preferring the warmer waters of the Mediterranean. However, it is attracted to the warm Gulf Stream currents off the coasts of western Scotland, Cornwall and Devon.

Unlike the lobster the crawfish does not have large claws. It is protected by a heavy and very spiny shell which can inflict a nasty wound if mishandled. It can swim awkwardly backwards by flapping its abdomen and tail. Without front claws it cannot crush hard-shelled creatures and so lives on small, soft-bodied animals such as worms, and on dead fish.

During the winter, adults migrate to deeper water to mate. Like the lobster, the female crawfish carries its eggs attached to its abdomen for several months until they hatch. The larvae are flattened, transparent discs about ¼ in. (6 mm) across with stalked eyes and long feathery legs. During the weeks of their early growth they may be carried for miles by currents before dropping to the sea-bed. These young crawfish will not come into shallow water until they are fully mature. Although rarely eaten in Britain, crawfish are considered a great delicacy in France, where they are sold as *langouste*.

219

Red and blue markings make this the most vivid of British squat lobsters, especially when it is seen under water. Spines cover the large claws, the edge of the shell, and the legs. The tail and last rudimentary pair of legs are always held under the body, which may reach 4¾ in. (12 cm) long.

Montagu's plated lobster
Galathea squamifera

Sometimes there are red flecks on the greeny-brown shell of this species. The nipper joints of the warty claws have spines only on the outer edge, but the other joints have them mostly on the inner edge. May reach 2¾ in. (70 mm) in overall length.

The rostrum, or tip of the shell between the eyes, has four spines on each side of the central point, which are smallest next to the eye.

Between the eyes the shell terminates in a sharp point, with three spines on either side.

This squat lobster is common on all coasts in rocky areas. It hides in crannies by day and emerges at night to feed.

Spiny squat lobster *Galathea strigosa*

In spite of its bright colours, a careful search must be made to find a spiny squat lobster on the shore. Its flattened body enables it to hide under stones, in crevices, or under rock overhangs, where it clings upside down. A smaller species, *Galathea squamifera*, is more common on the shore, but its greenish colour makes it even more difficult to see.

Unlike true lobsters, which carry their tails extended, the squat lobster curls its tail and abdomen under its thorax. By suddenly straightening out its tail and flexing it, the lobster can swim in short bursts. Distinct flapping noises heard when rocks are turned over on the shore indicate a squat lobster trying to escape in this way.

The squat lobster defends itself vigorously and can give a painful nip. It feeds at night, mainly on small particles collected from the sea bottom with its mouthparts. Its front claws are used to tear pieces from larger dead animals.

Squat lobsters mate in pairs and the eggs are carried under the female's abdominal flap. Eggs hatch into larvae in spring and summer, to float among the plankton. After a few weeks they sink to the bottom and develop into adults.

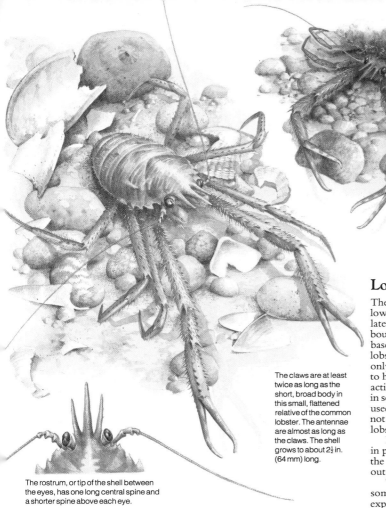

Often only the claws and their long, slender pincers are seen because this lobster habitually hides under rocks and in cracks.

Common around Scotland, the long-clawed squat lobster also occurs down Britain's west coast.

The claws are at least twice as long as the short, broad body in this small, flattened relative of the common lobster. The antennae are almost as long as the claws. The shell grows to about 2½ in. (64 mm) long.

The rostrum, or tip of the shell between the eyes, has one long central spine and a shorter spine above each eye.

Long-clawed squat lobster *Munida bamffica*

The aptly named long-clawed squat lobster is found below the low-water mark, particularly in areas where sediment accumulates between the rocks; in many Scottish sea lochs almost every boulder may have one of these creatures in residence beneath its base. By digging out a suitable hole under a boulder the squat lobster can remain hidden and protected during the day with only its long antennae and claws protruding. At night it emerges to hunt for food, though in deep, dark waters squat lobsters are active by day, crawling around over muddy and sandy bottoms in search of small crustaceans and worms. The delicate claws are used to probe crevices and for sifting through sediment, but are not as effective a defence as the stout claws of the spiny squat lobster.

Like the spiny squat lobster, long-clawed squat lobsters mate in pairs, the females carrying the eggs on their abdomen under the protection of the tail flap during the winter until they hatch out in the spring and summer.

The long-clawed squat lobster and the spiny squat lobster are sometimes eaten locally, but are too small for commercial exploitation.

221

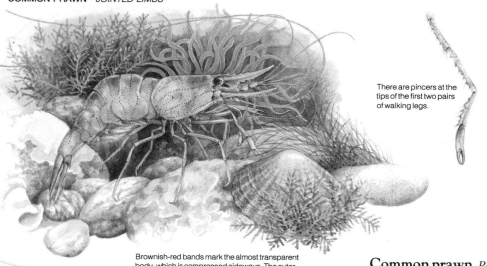

There are pincers at the tips of the first two pairs of walking legs.

Common in rock pools round south and west Britain, common prawns also live in eel grass on sandy shores.

Brownish-red bands mark the almost transparent body, which is compressed sideways. The outer pair of antennae are longer than the body and trail backwards. The inner pair, which often point forwards and up, are each divided into one short and two long branches. The second pair of legs are the biggest. Body is up to 4 in. (10 cm) long.

Palaemon elegans

Seven to nine teeth are spaced along the whole length of the top of the rostrum, with three underneath. This similar but smaller prawn species is common in rock pools all round the coast. Up to 2 in. (50 mm) long.

Between the eyes, the shell ends in a sharp-pointed tip, or rostrum, curving upwards. It has six or seven teeth on the top and four or five underneath, but none on the rostrum's front third.

Common prawn *Palaemon serratus*

Although common in rock pools in summer, the prawn is difficult to see because of its almost transparent body. It can also change the lines and spots of colour marking its body so that their pattern matches the background.

When feeding, the prawn walks delicately on the last three pairs of legs. The first two have pincers for picking up small worms, bits of seaweed and other edible debris. The long antennae and eyes on movable stalks can sense food and danger from any direction – a necessary protection as the prawn is a favourite food of fish, anemones and cuttlefish. When attacked, a prawn can escape rapidly, propelling itself by bending and flexing its abdomen. The animal can also swim slowly by beating its swimmerets – five pairs of short, flattened limbs under the abdomen.

In summer, males mate with soft, recently moulted females. Each female produces up to 2,500 eggs which are attached to the long hairs on her swimmerets. After hatching, the larvae float among the plankton until they develop into adults. Young prawns grow rapidly and will moult as often as every fortnight in the summer, each moult taking as little as 20 seconds.

The heavy claws on the first pair of legs are not of the typical nipper pattern. A spine across the claw end can be moved to close on a sharp point.

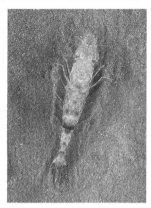

Shrimps are locally very abundant in sandy bays, beaches and estuaries. They are often found in beds of eel grass.

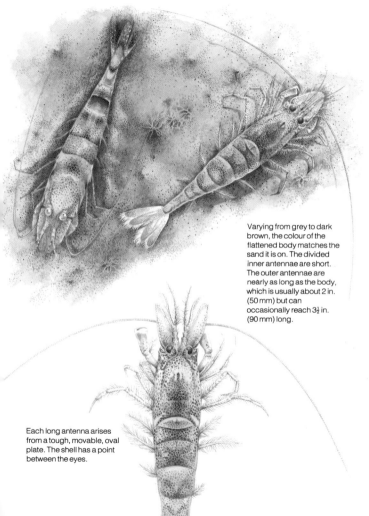

Varying from grey to dark brown, the colour of the flattened body matches the sand it is on. The divided inner antennae are short. The outer antennae are nearly as long as the body, which is usually about 2 in. (50 mm) but can occasionally reach 3½ in. (90 mm) long.

Each long antenna arises from a tough, movable, oval plate. The shell has a point between the eyes.

Common shrimp *Crangon crangon*

Well camouflaged, cautious and nocturnal by habit, the common shrimp can be a difficult animal to spot. During the day it conceals itself in the sand, and a captured shrimp placed in a sandy pool will quickly bury itself, shuffling its legs and beating the swimmerets under its abdomen to create a depression into which the animal gradually sinks. The rich, muddy sands of large bays and estuaries may contain very large populations of shrimps, and in areas such as Morecambe Bay and The Wash they are gathered commercially by dragging nets over the sand.

At night the shrimp emerges to feed. It rarely swims, but walks slowly over the surface of the sand on the last two pairs of legs, searching for food with its antennae and using the long third pair of legs to investigate any hard object it encounters. The stout claws enable the shrimp to tackle almost anything edible, including large worms, young fish, small crustaceans, plants and organic remains.

Shrimps are hardy animals, tolerant of a wide temperature range, and they may spawn in summer or winter. The eggs are attached to the walking legs and the swimmerets, and depending on the season they hatch after four to twelve weeks.

223

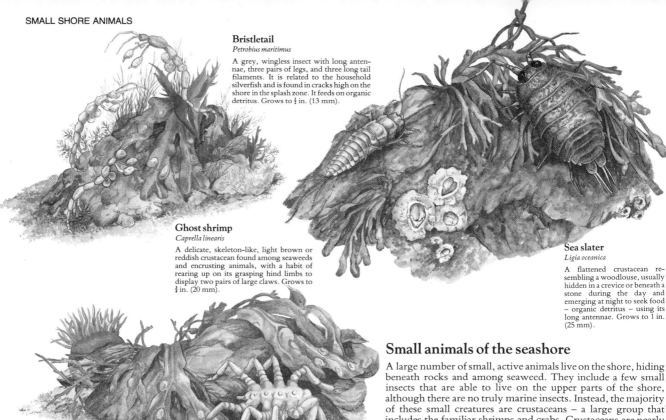

Bristletail
Petrobius maritimus

A grey, wingless insect with long antennae, three pairs of legs, and three long tail filaments. It is related to the household silverfish and is found in cracks high on the shore in the splash zone. It feeds on organic detritus. Grows to ½ in. (13 mm).

Ghost shrimp
Caprella linearis

A delicate, skeleton-like, light brown or reddish crustacean found among seaweeds and encrusting animals, with a habit of rearing up on its grasping hind limbs to display two pairs of large claws. Grows to ¾ in. (20 mm).

Sea slater
Ligia oceanica

A flattened crustacean resembling a woodlouse, usually hidden in a crevice or beneath a stone during the day and emerging at night to seek food – organic detritus – using its long antennae. Grows to 1 in. (25 mm).

Sea spider
Pycnogonum littorale

A squat, knobbly creature with eight thick segmented legs which terminate in strong claws. At the front is a conical snout bearing the mouth. Dirty yellow or pale brown, it lives under stones among debris on the lower shore, and feeds on sea anemones and hydroids. Up to 1 in. (25 mm).

Small animals of the seashore

A large number of small, active animals live on the shore, hiding beneath rocks and among seaweed. They include a few small insects that are able to live on the upper parts of the shore, although there are no truly marine insects. Instead, the majority of these small creatures are crustaceans – a large group that includes the familiar shrimps and crabs. Crustaceans are nearly as numerous in the sea as insects are on land. Insects, crustaceans and spiders (including sea spiders), all have segmented bodies and a hard outer shell which must be shed periodically so that the animal can grow. Some of the most common and widespread species are shown here. Searching under rocks and turning over debris left on the strandline is usually rewarding. When disturbed, many can escape rapidly by running or jumping, using their powerful back legs or tails as 'springs'.

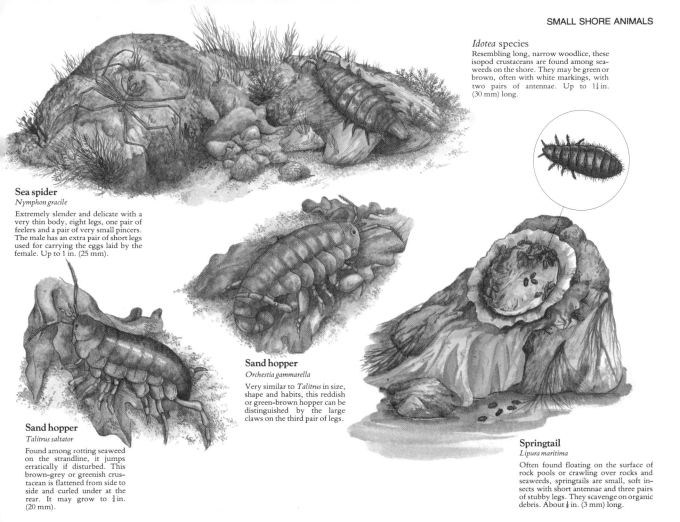

Sea spider
Nymphon gracile

Extremely slender and delicate with a very thin body, eight legs, one pair of feelers and a pair of very small pincers. The male has an extra pair of short legs used for carrying the eggs laid by the female. Up to 1 in. (25 mm).

Idotea species

Resembling long, narrow woodlice, these isopod crustaceans are found among seaweeds on the shore. They may be green or brown, often with white markings, with two pairs of antennae. Up to 1¼ in. (30 mm) long.

Sand hopper
Orchestia gammarella

Very similar to *Talitrus* in size, shape and habits, this reddish or green-brown hopper can be distinguished by the large claws on the third pair of legs.

Sand hopper
Talitrus saltator

Found among rotting seaweed on the strandline, it jumps erratically if disturbed. This brown-grey or greenish crustacean is flattened from side to side and curled under at the rear. It may grow to ¾ in. (20 mm).

Springtail
Lipura maritima

Often found floating on the surface of rock pools or crawling over rocks and seaweeds, springtails are small, soft insects with short antennae and three pairs of stubby legs. They scavenge on organic debris. About ⅛ in. (3 mm) long.

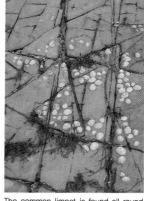

The common limpet is found all round Britain, in large numbers at every level of the rocky shore.

Algae and young seaweeds are rasped off the rock surface by the file-like tongue, or radula, leaving a pattern of small scratches.

China limpet
Patella aspera

Found on west, south-west and north-east coasts. The interior is shiny white with a pale orange centre. Up to 2¼ in. (60 mm) across.

Patella intermedia
Restricted to exposed south-western shores, this small, flat limpet has a ribbed shell marked with dark and light bands inside and out. Up to 1¼ in. (40 mm) across.

The limpet moves, like a land snail, on a muscular foot. When grazing, it extends the head and raises the shell to reveal a fringe of transparent tentacles. After feeding, it returns to the same spot each time, clamping itself down tightly and eroding a characteristic circular depression in the rock. The tough, ribbed, conical shell is usually tall with a blunt tip, but permanently submerged specimens may develop low, broad shells up to 2½ in. (60 mm) across.

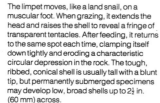

Dislodged, the limpet reveals its foot, head and transparent tentacles. The empty shell is pale yellow inside with a brown centre.

Common limpet *Patella vulgata*

It is almost impossible to dislodge a limpet except by a quick, surprise blow. At any sign of danger – the approach of a possible predator or pounding waves at high tide – it clings tightly to the rocks, its strong foot acting as a suction pad. Each limpet has a home base to which it returns and where, on soft rock, its shell edge has either ground down a depression or, on hard rock, has itself been worn down to fit the rock surface perfectly. Because of this perfect fit, water for breathing can be retained inside the shell at low tide.

To avoid desiccation, limpets feed when the tide is in or at night when the shore is cool and damp, browsing over the rocks using a file-like tongue or radula. They graze within about 3 ft (1 m) of their base, returning there probably by following their own trail with their tentacles. Grazing limpets feed mainly on young plants and often prevent seaweed from growing. If some disaster, such as an oil spill, kills off the limpets, the rocks soon become covered in young, bright green seaweeds.

Most limpets start life as males and change later into females. Spawning occurs between October and December. At first, the young settle below tide level, but gradually move up the shore.

Other limpet-like animals

Several limpet species are widely distributed along rocky stretches of the coast. The tortoiseshell lives on the lower shore and below, as do the keyhole limpet, slit limpet, white tortoiseshell limpet and ormer (which in Britain is restricted to the Channel Islands). The slipper limpet is found mainly up the east coast and along the Channel, and also round to Wales.

Slit limpet
Emarginula reticulata

Identifiable by the slot for the siphon at the front of the white, grey or green ribbed shell. Up to ¾ in. (20 mm) long.

Tortoiseshell limpet
Acmaea tessulata

Reddish-brown tortoiseshell markings pattern the shell; the apex is near the front edge. Up to 1 in. (25 mm) long.

White tortoiseshell limpet
Acmaea virginea

Usually found among kelp, it is similar to *A. tessulata* but smaller. The off-white shell has pink-brown rays. Up to ½ in. (13 mm) long.

Blue-rayed limpet
Patina pellucida

Patterned with vivid blue spots which fade with age, this small, smooth-shelled limpet feeds on kelp, excavating holes in the plant which can badly weaken it. Usually less than 1 in. (25 mm) long.

Keyhole limpet
Diodora apertura

The greyish, ribbed shell has an aperture for a breathing siphon at the apex. A fold of soft tissue, which can be withdrawn, covers the edge. Up to 1½ in. (40 mm) long.

Ormer
Haliotis tuberculata

A Mediterranean species that is common in the Channel Islands; the holes in the green-brown or reddish, ear-shaped shell allow water to escape after respiration. Up to 4 in. (10 cm) long.

Slipper limpet
Crepidula fornicata

The red-tinged, brownish, oval shells form chains, attaching themselves to oysters, mussels or stones, as they compete for food and space. Up to 2 in. (50 mm) long.

227

The sheltered world of boulders and crevices

All seashore animals must avoid drying out when the tide recedes. Barnacles retain moisture until the next high tide by closing their shells, and limpets trap water by clamping down firmly on to the rock. Many other animals solve the problem by sheltering in crevices or under boulders where surfaces remain damp.

Active crabs, snails, worms and starfish may hide temporarily, emerging when the tide returns, but squat lobsters and many other crevice dwellers emerge only at night, which helps to conceal them from predators. The tiny porcelain crabs have flattened bodies and strong claws, enabling them to cling beneath rocks and forage for food in safety. Other animals hide in kelp holdfasts, or among the convoluted plates of the ross coral; over 300 different animal species have been found living in holdfasts.

Encrusting and slow-moving animals such as sponges, sea squirts, saddle oysters and chitons live permanently attached to the undersides of rocks. Protected against drying out, they are also safe from attack, and do not have to compete for space with light-dependent seaweeds. Other animals such as daisy anemones and sea cucumbers are adapted to permanent life in crevices, and will only extend their feeding tentacles from a position of secure concealment.

The multi-branched holdfast of a large kelp plant provides shelter for a variety of small creatures, including sand hoppers, worms, sponges and brittlestars. Kelp plants are also used by man as a foodstuff and as a source of chemicals.

As the water recedes many animals such as the cowrie and the cushion star take refuge under rocks, emerging to browse on sea squirts and other organisms when the tide comes back in.

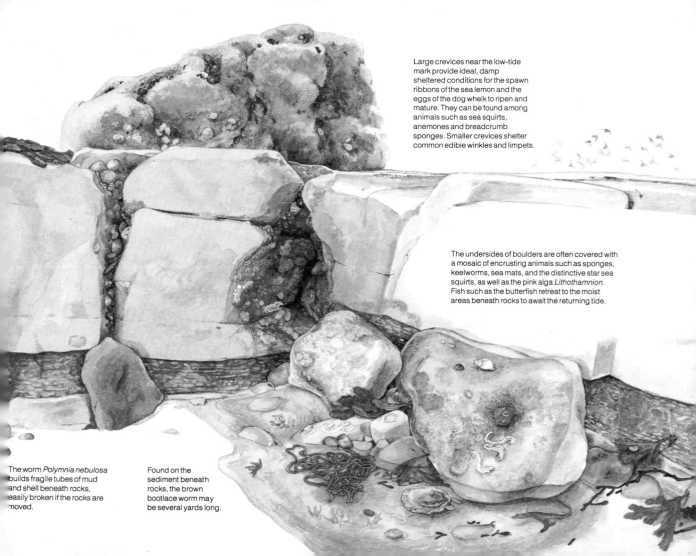

Large crevices near the low-tide mark provide ideal, damp sheltered conditions for the spawn ribbons of the sea lemon and the eggs of the dog whelk to ripen and mature. They can be found among animals such as sea squirts, anemones and breadcrumb sponges. Smaller crevices shelter common edible winkles and limpets.

The undersides of boulders are often covered with a mosaic of encrusting animals such as sponges, keelworms, sea mats, and the distinctive star sea squirts, as well as the pink alga *Lithothamnion*. Fish such as the butterfish retreat to the moist areas beneath rocks to await the returning tide.

The worm *Polymnia nebulosa* builds fragile tubes of mud and shell beneath rocks, easily broken if the rocks are moved.

Found on the sediment beneath rocks, the brown bootlace worm may be several yards long.

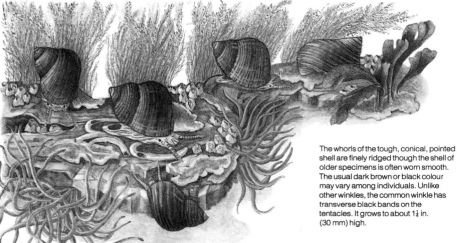

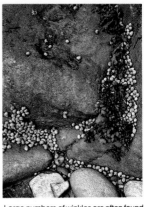

The whorls of the tough, conical, pointed shell are finely ridged though the shell of older specimens is often worn smooth. The usual dark brown or black colour may vary among individuals. Unlike other winkles, the common winkle has transverse black bands on the tentacles. It grows to about 1¼ in. (30 mm) high.

Large numbers of winkles are often found gathered in gullies and pools, browsing on seaweeds and detritus.

Common winkle *Littorina littorea*

'Triggering', or winkle gathering, has been both a pastime and a commercial activity in the British Isles for centuries. Despite extensive harvesting, the common winkle – the largest of the winkles – remains abundant on many rocky shores. It survives crashing waves by withdrawing into its tough shell, which can be rolled about by the roughest seas without damage. Water with low salinity presents no problems to this versatile species which is common in estuaries and in polluted waters such as marinas and the sea near sewage outfalls.

Winkles feed by scraping encrusting seaweeds from rocks, by rasping the surface of larger seaweeds, and by browsing on dead vegetation. They are sometimes deliberately introduced to oyster beds to keep down unwanted growths.

After mating, the female releases lens-shaped egg capsules, each about $\frac{1}{25}$ in. (1 mm) across, into the sea. Each floating capsule contains up to nine eggs which hatch into minute larvae after about six days. The larvae remain in the plankton for about two weeks before settling on the sea-bed and then crawling to the shore – usually during May and June. By their third year, the young are fully grown.

At low tide, the exposed winkle is held lip uppermost to the sides of boulders by a film of mucus. The dark shell may become bleached by the sun.

The outer lip of the opening runs almost parallel to the axis of the spire before its junction with the shell. The opening itself can be closed with a horny plate – the operculum – that protects the animal when the tide is out.

The spiral ridges on the shell of the young winkle are more pronounced than on that of the fully grown adult.

Winkles and snail-like animals

Winkles are common all round the coast of Britain and live among rocks, in crevices or on seaweed. Each species inhabits a preferred level on the seashore. In contrast to the rock dwellers, necklace and acteon shells burrow in the sand.

Flat winkle
Littorina littoralis

The distinctive, flat-topped shell is yellow, brown, red, orange, olive-green or banded. It is commonly seen browsing on seaweed such as bladder wrack on the middle and lower levels of sheltered, rocky shores. It is about ⅜ in. (10 mm) high.

Rough winkle
Littorina saxatilis

Distinct spiral ridges give the shell its rough appearance. This species lives on the upper shore in rock crevices and among channelled wrack. Its colour varies from red, to buff, to almost black. Unlike the edible winkle, the outer lip of the opening curves in to meet the spire almost at right-angles. The shell is less than ⅜ in. (10 mm) high.

Large necklace shell
Natica catena

Found near the low-water line and below, this species has only one row of red-brown streaks at the top of each whorl. Its umbilicus is fully open. The spawn collar forms only one incomplete coil. The shell may reach 1½ in. (40 mm) high.

Small or common necklace shell
Natica alderi

The broad foot ploughs through the sand to grip the bivalves on which it feeds. The shell is globular, shiny, and has several rows of brown-red streaks. On the underside is a large umbilicus, half closed by the shell lip. Eggs are spawned in spirals of jelly which have been strengthened with grains of sand. The shell grows to about ⅝ in. (15 mm) high.

Small winkle
Littorina neritoides

This tiny winkle lives in bare rock crevices at and above the high-water line. It is commonest on exposed shores where wave splash is greatest. The blue-black shell is smooth and sharply pointed, and only 1/16 in. (5 mm) high.

Acteon
Acteon tornatilis

The barrel-shaped shell is pink-brown with white bands and seven whorls. The lobed foot projects from an elongated opening. The acteon lives in the same habitat as necklace shells and may grow to ¾ in. (20 mm) long.

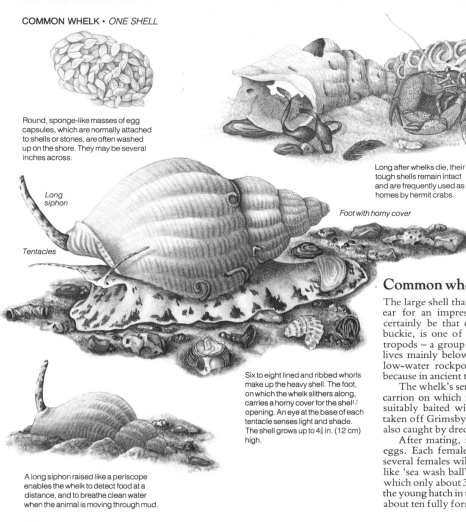

Round, sponge-like masses of egg capsules, which are normally attached to shells or stones, are often washed up on the shore. They may be several inches across.

Long after whelks die, their tough shells remain intact and are frequently used as homes by hermit crabs.

Long siphon

Tentacles

Foot with horny cover

Fully formed young emerge from the egg capsules. Common whelks have no floating planktonic larval stage.

Six to eight lined and ribbed whorls make up the heavy shell. The foot, on which the whelk slithers along, carries a horny cover for the shell opening. An eye at the base of each tentacle senses light and shade. The shell grows up to 4¾ in. (12 cm) high.

A long siphon raised like a periscope enables the whelk to detect food at a distance, and to breathe clean water when the animal is moving through mud.

Common whelk *Buccinum undatum*

The large shell that children find on the beach and hold to their ear for an impression of the sound of the sea will almost certainly be that of a large whelk. The common whelk, or buckie, is one of Britain's largest and most widespread gastropods – a group which includes all the snail-like shells – and lives mainly below low tide though it is sometimes found in low-water rockpools. It is also called the purple dogwhelk, because in ancient times it was a source of purple dye.

The whelk's sensitive siphon is able to detect the smell of the carrion on which it feeds and this makes it easy to catch in a suitably baited wicker or mesh pot. The largest number are taken off Grimsby, East Anglia and Whitstable, where they are also caught by dredging.

After mating, female whelks collect in a group to lay their eggs. Each female lays up to 2,000 fibrous egg capsules and several females will stick their eggs together to form a sponge-like 'sea wash ball'. Each capsule contains up to 1,000 eggs, of which only about 30 develop, the rest acting as a food source. As the young hatch in the capsule, they feed on each other until only about ten fully formed young finally emerge.

Identifying other whelks

Whelks are easily recognised by their thick, spiralling shells. The upper rim of the oval opening in the broad end of the shell is cut by the siphonal canal, a smooth channel from the shell interior through which the whelk can extend a long siphon, or tube. The whelk breathes through the siphon and senses food with it.

Thick-lipped dog whelk
Nassarius incrassatus

The shell, often banded with colour, has distinct whorls with clearly outlined ribs. Found on rocks or mud, it grows ⅝ in. (15 mm) long.

Netted dog whelk
Nassarius reticulatus

This carrion-eater lives on sand and mud low on the shore and beyond. The shell has a pattern of small squares. The transparent egg capsules are urn-shaped, and laid in rows. Up to 1½ in. (40 mm) long.

Spindle shell
Neptunea antiqua

The smooth-surfaced, prominently whorled shell is usually seen only when it is washed up empty on shore. Lives mainly below low-tide level and is commonest in Scotland. Eggs are similar to those of the common whelk but larger. Up to 6 in. (15 cm) long.

Sting winkle or oyster drill
Ocenebra erinacea

The rugged, lumpy shell has about five ribbed whorls. Lives on rocky shores and is a pest in oyster beds. It lays flask-shaped yellow egg capsules in crevices. Up to 2⅜ in. (60 mm) high.

Dog whelk
Nucella lapillus

Common on the shore among barnacles and mussels, which it eats. The shell is white, yellow or brown-banded. Egg capsules are yellow. Grows up to 1¼ in. (30 mm) long.

233

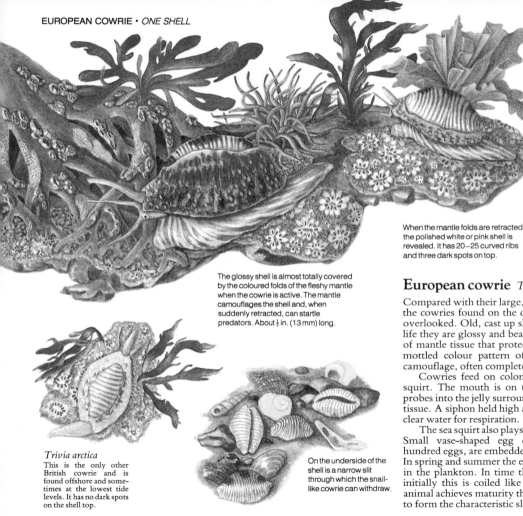

The glossy shell is almost totally covered by the coloured folds of the fleshy mantle when the cowrie is active. The mantle camouflages the shell and, when suddenly retracted, can startle predators. About ½ in. (13 mm) long.

When the mantle folds are retracted the polished white or pink shell is revealed. It has 20–25 curved ribs and three dark spots on top.

Widespread all round the coast of Britain, cowries are found under rocks on the lower shore and below.

Trivia arctica
This is the only other British cowrie and is found offshore and sometimes at the lowest tide levels. It has no dark spots on the shell top.

On the underside of the shell is a narrow slit through which the snail-like cowrie can withdraw.

European cowrie *Trivia monacha*

Compared with their large, brightly coloured tropical relatives, the cowries found on the coasts of Britain are small and easily overlooked. Old, cast up shells are often dull and worn, but in life they are glossy and beautifully shaped, maintained by folds of mantle tissue that protect and renew the shiny surface. The mottled colour pattern of the mantle provides an effective camouflage, often completely hiding the shell from predators.

Cowries feed on colonial sea squirts such as the star sea squirt. The mouth is on the end of a long proboscis which probes into the jelly surrounding the sea squirt and rasps out the tissue. A siphon held high above the bottom sediment draws in clear water for respiration.

The sea squirt also plays a part in the cowrie's breeding cycle. Small vase-shaped egg capsules, each containing several hundred eggs, are embedded in holes excavated in the sea squirt. In spring and summer the eggs hatch into larvae which float free in the plankton. In time they settle down to develop a shell; initially this is coiled like that of other sea snails, but as the animal achieves maturity the outer whorl grows over the others to form the characteristic slotted cowrie shell.

Identifying top shells

These shells are common on rocky shores all round the coast of Britain. They can be distinguished from winkles by their conical shape and broad base. Mother-of-pearl shows through where the shell is worn, and the foot has tentacles along its edges.

Purple or flat top shell
Gibbula umbilicalis

Although similar in size to the grey top shell, this species is flatter with a large umbilicus (cavity) underneath. The shell is green-tinged with broad purple stripes. It lives mainly on shore and is found well up on the middle levels.

Grey top shell or silver tommy
Gibbula cineraria

This species is commonly found on the rocks of the lower shore and on kelp fronds in deeper water. The shell is marked with very narrow, dark reddish-grey stripes. There is a small umbilicus, or cavity near the centre, on the underside. About ½ in. (13 mm) high.

Large top shell
Gibbula magus

The large top shell is found on sediment and gravel on the shore and below on the south and west coasts of Britain only. The broad shell has eight distinct angular whorls with irregular knobs, ridges and oblique purple streaks. It usually reaches about ¾ in. (20 mm) high and is broader than it is tall.

Painted top shell
Calliostoma zizyphinum

The sharply pointed conical shape and dramatic pink stripes make this species very easy to recognise. It is only found on the lower shore and below. It grows to about 1 in. (25 mm) high and measures the same across its base.

Thick top shell or toothed winkle
Monodonta lineata

Only found on the coasts of south-west England, the species lives on the middle shore, often on bare rock. The grey or greenish conical shell has six whorls with purple zigzag markings. If the shell pattern is worn this species may be mistaken for the edible winkle. However, it can be distinguished by the mother-of-pearl in the opening and the blunt 'tooth' on the lip. Up to 1 in. (25 mm) high.

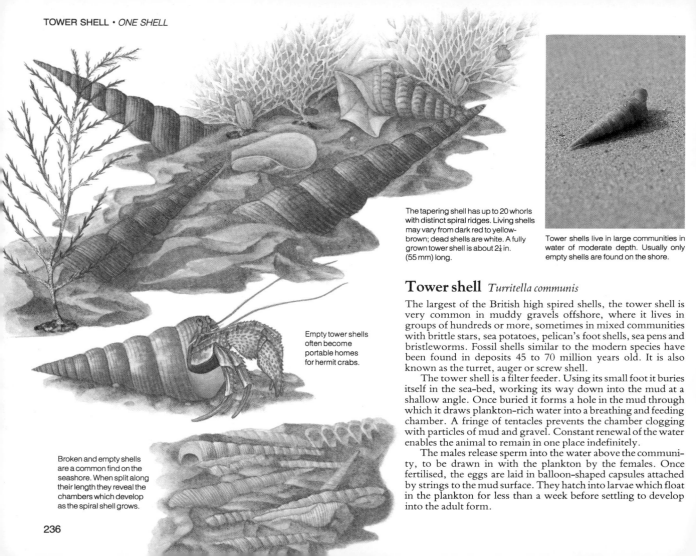

The tapering shell has up to 20 whorls with distinct spiral ridges. Living shells may vary from dark red to yellow-brown; dead shells are white. A fully grown tower shell is about 2¼ in. (55 mm) long.

Tower shells live in large communities in water of moderate depth. Usually only empty shells are found on the shore.

Empty tower shells often become portable homes for hermit crabs.

Broken and empty shells are a common find on the seashore. When split along their length they reveal the chambers which develop as the spiral shell grows.

Tower shell *Turritella communis*

The largest of the British high spired shells, the tower shell is very common in muddy gravels offshore, where it lives in groups of hundreds or more, sometimes in mixed communities with brittle stars, sea potatoes, pelican's foot shells, sea pens and bristleworms. Fossil shells similar to the modern species have been found in deposits 45 to 70 million years old. It is also known as the turret, auger or screw shell.

The tower shell is a filter feeder. Using its small foot it buries itself in the sea-bed, working its way down into the mud at a shallow angle. Once buried it forms a hole in the mud through which it draws plankton-rich water into a breathing and feeding chamber. A fringe of tentacles prevents the chamber clogging with particles of mud and gravel. Constant renewal of the water enables the animal to remain in one place indefinitely.

The males release sperm into the water above the community, to be drawn in with the plankton by the females. Once fertilised, the eggs are laid in balloon-shaped capsules attached by strings to the mud surface. They hatch into larvae which float in the plankton for less than a week before settling to develop into the adult form.

Identifying other tall-shelled animals

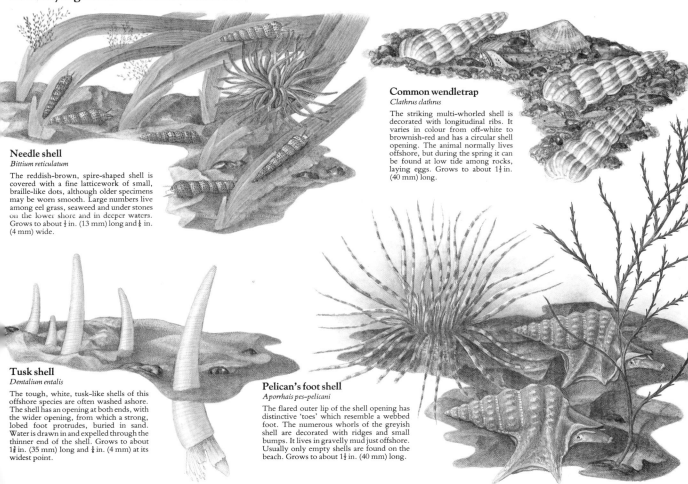

Needle shell
Bittium reticulatum

The reddish-brown, spire-shaped shell is covered with a fine latticework of small, braille-like dots, although older specimens may be worn smooth. Large numbers live among eel grass, seaweed and under stones on the lower shore and in deeper waters. Grows to about ½ in. (13 mm) long and ⅛ in. (4 mm) wide.

Common wendletrap
Clathrus clathrus

The striking multi-whorled shell is decorated with longitudinal ribs. It varies in colour from off-white to brownish-red and has a circular shell opening. The animal normally lives offshore, but during the spring it can be found at low tide among rocks, laying eggs. Grows to about 1½ in. (40 mm) long.

Tusk shell
Dentalium entalis

The tough, white, tusk-like shells of this offshore species are often washed ashore. The shell has an opening at both ends, with the wider opening, from which a strong, lobed foot protrudes, buried in sand. Water is drawn in and expelled through the thinner end of the shell. Grows to about 1⅜ in. (35 mm) long and ⅛ in. (4 mm) at its widest point.

Pelican's foot shell
Aporrhais pes-pelicani

The flared outer lip of the shell opening has distinctive 'toes' which resemble a webbed foot. The numerous whorls of the greyish shell are decorated with ridges and small bumps. It lives in gravelly mud just offshore. Usually only empty shells are found on the beach. Grows to about 1½ in. (40 mm) long.

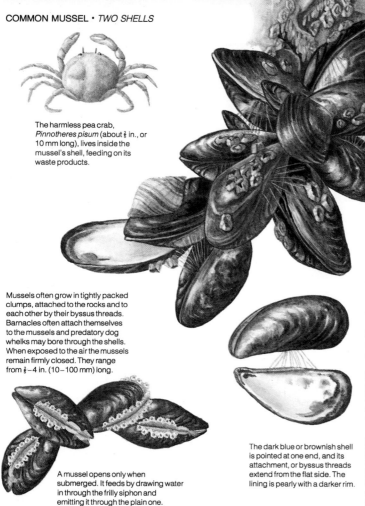

The harmless pea crab, *Pinnotheres pisum* (about ⅜ in., or 10 mm long), lives inside the mussel's shell, feeding on its waste products.

Mussels often grow in tightly packed clumps, attached to the rocks and to each other by their byssus threads. Barnacles often attach themselves to the mussels and predatory dog whelks may bore through the shells. When exposed to the air the mussels remain firmly closed. They range from ⅜–4 in. (10–100 mm) long.

A mussel opens only when submerged. It feeds by drawing water in through the frilly siphon and emitting it through the plain one.

The dark blue or brownish shell is pointed at one end, and its attachment, or byssus threads extend from the flat side. The lining is pearly with a darker rim.

Mussels, often in extensive beds, fix themselves to stones and rocks mainly on exposed shores and in estuaries.

Common mussel *Mytilus edulis*

Mussels are one of the commonest bivalves on Britain's rocky coasts. They have been eaten by man since early times, and the first mussel farms in Europe are reputed to have been established in France in the 13th century. Farmed mussels are grown on stakes, hanging ropes and in beds. Many natural mussel beds are also commercially exploited.

When collecting wild mussels check that the water they are in is not polluted. Mussels extract not only food particles from the surrounding water, but also any toxins present. These become concentrated in the mussel's body, making it unfit to eat. Some mussels contain many minute 'pearls' produced by wrapping up irritating particles in mother-of-pearl.

Spawning occurs in early spring, the larvae floating in the plankton before settling down to colonise new areas. The attachment, or byssus, threads are secreted as a thick fluid which hardens on contact with water, firmly anchoring the mussel in position. Mussels can reabsorb their threads and move about until they find an ideal place to settle; however, older mussels rarely move. In addition to man, mussels have several predators, including oystercatchers, starfish and dog whelks.

Mussels and mussel-like animals

By carefully searching rocky shores, three other species (apart from the common mussel) can be found. The fan mussel is not a true mussel though it resembles the group. It is found mainly below the low-tide line. Mussels grow singly or in clumps, although the horse mussel forms extensive offshore beds.

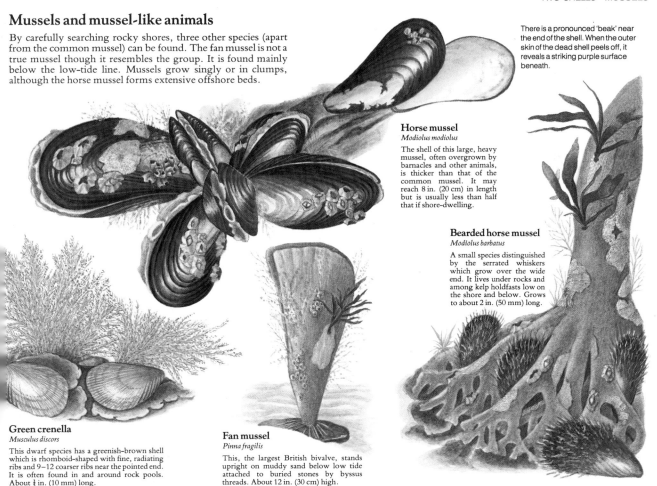

There is a pronounced 'beak' near the end of the shell. When the outer skin of the dead shell peels off, it reveals a striking purple surface beneath.

Horse mussel
Modiolus modiolus

The shell of this large, heavy mussel, often overgrown by barnacles and other animals, is thicker than that of the common mussel. It may reach 8 in. (20 cm) in length but is usually less than half that if shore-dwelling.

Bearded horse mussel
Modiolus barbatus

A small species distinguished by the serrated whiskers which grow over the wide end. It lives under rocks and among kelp holdfasts low on the shore and below. Grows to about 2 in. (50 mm) long.

Green crenella
Musculus discors

This dwarf species has a greenish-brown shell which is rhomboid-shaped with fine, radiating ribs and 9–12 coarser ribs near the pointed end. It is often found in and around rock pools. About ⅜ in. (10 mm) long.

Fan mussel
Pinna fragilis

This, the largest British bivalve, stands upright on muddy sand below low tide attached to buried stones by byssus threads. About 12 in. (30 cm) high.

239

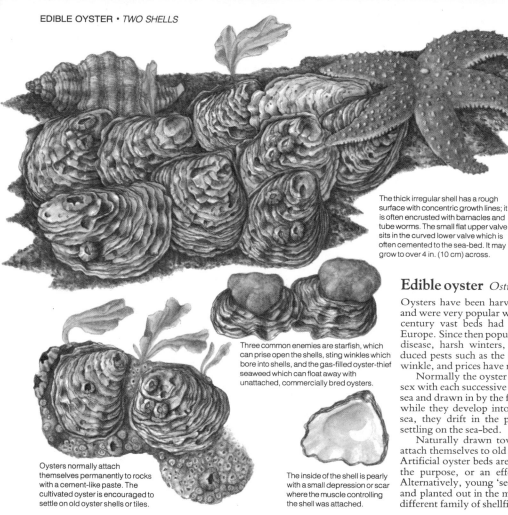

The thick irregular shell has a rough surface with concentric growth lines; it is often encrusted with barnacles and tube worms. The small flat upper valve sits in the curved lower valve which is often cemented to the sea-bed. It may grow to over 4 in. (10 cm) across.

Oysters prefer the shallow tidal waters of estuaries, bays or sea lochs where they live in natural or artificial beds.

Three common enemies are starfish, which can prise open the shells, sting winkles which bore into shells, and the gas-filled oyster-thief seaweed which can float away with unattached, commercially bred oysters.

Oysters normally attach themselves permanently to rocks with a cement-like paste. The cultivated oyster is encouraged to settle on old oyster shells or tiles.

The inside of the shell is pearly with a small depression or scar where the muscle controlling the shell was attached.

Edible oyster *Ostrea edulis*

Oysters have been harvested for food since prehistoric times, and were very popular with the Romans. By the end of the 18th century vast beds had been established on all the coasts of Europe. Since then populations have been drastically reduced by disease, harsh winters, overcropping and accidentally introduced pests such as the slipper limpet and the American sting winkle, and prices have risen accordingly.

Normally the oyster starts adult life as a male, then changes sex with each successive spawning. Sperm is discharged into the sea and drawn in by the female, which retains the fertilised eggs while they develop into free-floating larvae; released into the sea, they drift in the plankton for about two weeks before settling on the sea-bed.

Naturally drawn towards other oysters, the young often attach themselves to old shells, so increasing the size of the bed. Artificial oyster beds are often enlarged by providing shells for the purpose, or an effective substitute such as limed tiles. Alternatively, young 'seed' oysters may be reared in hatcheries and planted out in the main bed. The pearl oyster belongs to a different family of shellfish which lives in tropical seas.

The American oyster drill (*Urosalpinx cinerea*) bores through the shell and eats the contents. The pest has been accidently introduced to oyster beds.

The solid shell tends to be oblong in outline but it is often distorted and irregular. The almost flat upper shell sits within the deeply cupped lower one. Fully grown specimens may be 7 in. (18 cm) from the hinge to the opposite edge.

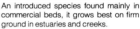

An introduced species found mainly in commercial beds, it grows best on firm ground in estuaries and creeks.

Slipper limpets (*Crepidula fornicata*) may attach themselves in chain formation to the oyster and compete with it for food and space.

Saddle oyster
Anomia ephippium

Although similar, this smaller and more delicate species is not a true oyster. Circular in outline, the shell may be distorted by close attachment to rocks. The thin, lower valve has a pear-shaped opening, through which the oyster attaches itself to rocks.

Pearly white inside and with a red-brown or purple muscle scar in each valve.

Portuguese oyster *Crassostrea angulata*

Introduced in 1890 for establishment in commercial oyster beds in southern England, the Portuguese oyster is considered inferior eating compared to the native species, but is easier to rear and grows to a greater size.

The breeding pattern differs from that of the native edible oyster, *Ostrea edulis*. Both sperm and eggs are released into the sea for fertilisation, and the larvae develop independently of the adults. The process requires a water temperature of at least 68°F (20°C), and so spawning rarely occurs in British waters. Commercial stocks are maintained in Britain by importing two-year-old 'seed' oysters from the Bay of Biscay, which are then fattened in beds situated mainly on the Essex coast and in the Helford river in Cornwall.

Unfortunately the Portuguese oyster is prone to disease, and over the past ten years European stocks have suffered badly from various infections; the French beds were nearly wiped out by a disease that destroys the gills. To combat this, seed oysters have been imported from America and Japan; incapable of even sporadic breeding in north European waters, these species are not regarded as part of the British marine fauna.

241

Held together by a conspicuous dark brown hinge, or ligament, the two halves of the tough oval shell have 22–28 radiating ribs, and concentric growth lines which indicate the animal's age. The two short, fringed siphons used for feeding and breathing are equipped with small primitive eyes. The cockle may grow to 2 in. (50 mm) across.

Buried near the surface of the sand, cockles are vulnerable to predators such as the oystercatcher, which may eat up to 300 in a day.

The shell of the edible cockle, unlike that of other cockle species, has internal grooves which run only a short distance inside the rim.

Edible cockles live in clean sand from mid to low-tide levels, in bays, coves and estuaries all round Britain.

Edible cockle *Cerastoderma edule*

One of the most familiar of all shellfish, the edible, or common, cockle occurs in vast numbers on sandy beaches where plenty of food is brought in with each tide. Some beds contain up to 10,000 cockles per sq. yd, and it was estimated that one bed off South Wales contained 460 million cockles.

Buried in the sand, the cockle feeds by drawing water in through one tube, known as a siphon, passing it through the gills to strain out plankton, and expelling the processed water through the second siphon. The siphons are relatively short, so the cockle has to remain close to the surface of the sand while it is feeding. This makes it easily accessible to predators such as gulls, oystercatchers, starfish, flatfish and human cockle-gatherers. It may escape by hauling itself further down with its large muscular foot, but the rounded shell is not adapted for rapid movement through the sand and many are caught.

The cockle is a hermaphrodite; it sheds both eggs and sperm into the water to allow cross-fertilisation, and the developing larvae or 'spat' float freely in the plankton before finally settling on the bed. Huge 'spat falls' occur every few years when conditions are ideal.

Four other types of cockle

Many cockles live buried in sediment at and below the low-tide mark. Each species can be distinguished by the ribs on the outside of the shell, by the grooves or absence of grooves inside the shell, and by their relative sizes. None are normally eaten.

Little cockle
Parvicardium exiguum

The outside of the shell has 20–22 flat ribs, along which are blunt bumps called tubercles. In adults, the tubercles are worn down except near the shell edge. The inside of the shell is smooth. Grows about ½ in. (13 mm) long.

Prickly cockle
Acanthocardia echinata

The shell has 18–22 prominent ribs, each decorated with backward-curving spines. The powerful digging foot is pink, in contrast to the white of most other species. The inside of the shell is grooved throughout its length. Grows up to 3 in. (75 mm) long.

Lagoon cockle
Cerastoderma glaucum

Similar to the edible cockle, this species is more triangular and has full-length grooves inside the shell. It lives in the brackish waters of lochs, estuaries and lagoons. Up to 2 in. (50 mm) long.

Dog cockle
Glycymeris glycymeris

Unrelated to the true cockles, this species lives just below the surface of sandy or muddy gravel into which it burrows with its powerful foot. The shell has zigzag brown markings and no ribs. A row of blunt 'teeth' runs along the inside of the hinge line. Grows about 2½ in. (64 mm) across.

Carpet shells occur on the shore and in shallow water all round Britain, buried in gravel, sand or mud.

Using its broad hatchet-shaped foot to burrow into gravel or coarse sand, the carpet shell feeds and breathes by circulating water through two short siphons, joined except at the ends which project above the sand surface. The thick, finely ridged shell may be cream, yellow or pink-brown, patterned with red-brown markings, and grows to 2½ in. (60 mm).

Banded carpet shell *Venerupis rhomboides*

Carpet shells and Venus shells belong to a large group of seashells found worldwide. There are 19 British species, whose tough, colourful shells are common along the shore. It is not always easy to identify different species, but the broad rhomboidal shape of the carpet shells does distinguish them from the roughly triangular Venus shells. To add to the confusion, several have recently been given new Latin names. The banded carpet shell is called *Tapes rhomboides* by some authorities.

As the shell is heavy, the animal only burrows a short way into the sediment and once buried does not move around much. The banded carpet shell lives in gravel or coarse sand but other species, like the striped Venus, prefer finer sand and some even prefer mud. Like most bivalves, carpet shells spawn through their outgoing siphon into the sea. The minute larvae drift in the plankton to new areas, where they eventually settle down.

Carpet and Venus shells are collected and eaten in some places. The American hard-shelled clam, *Mercenaria mercenaria*, belongs to this group and is a much prized food. Accidentally introduced into Britain around 1960, it is now cultivated on a small scale and flourishes, especially in Southampton Water.

Striped Venus
Venus striatula

Venus shells are similar to carpet shells but more triangular in shape, with the inside margin of the shell serrated. The striped Venus is a pale shell with three red-brown rays and a pattern of concentric ridges. Up to 1¾ in. (45 mm) long.

The smooth white interior of the shell is marked by faint scars where it was attached to the live animal within. This pattern helps identify the different species.

When the tide is in, the tellin extends two long, mobile siphon tubes. The longer one sucks in organic material from the sand and the shorter ejects waste water. The smooth, flattened shell may be ¾ in. (20 mm) long.

When the tide recedes, the siphons are retracted and the foot is used to haul the animal deeper into the sand.

The thin tellin lives in dense groups buried in fine sand, on shores all round Britain's coast.

Baltic tellin
Macoma balthica

Found in estuarine mud and muddy sands, the Baltic tellin is similar in colour to the thin tellin. However, it has a stouter and more rounded shell than the thin species and is slightly larger – up to 1 in. (25 mm) long.

The ligament joining the two glossy white, pink, yellow or orange shells is very strong, and holds the halves together long after the animal is dead.

Thin tellin *Tellina tenuis*

The beautifully coloured shells of the thin tellin can be found on most clean, sandy beaches. The animal spends its life buried in the sand, collecting food from the surface and ejecting waste by means of two long flexible tubes called siphons. It rarely lives more than 5 in. (12·5 cm) below the surface, and occurs on the shore and in shallow water no deeper than about 20 ft (6 m). Tellins often live in dense groups of 1,000 individuals per square yard – and as many as 8,000 per square yard in some places. Every incoming tide brings a fresh supply of food, although the smooth, flattened shell of the tellin allows it to move easily through the sand to new feeding grounds if necessary.

Tellins spawn in the spring. The larvae drift in the plankton before they eventually settle on the shore and develop into new adults. Every few years conditions are just right for a large number of larvae to settle successfully, thus maintaining dense populations.

A close relation, *Tellina fabula*, can be confused with the thin tellin. They are sometimes found together, but *T. fabula* has a more angular shell and usually lives very low on the shore and in deeper water down to 180 ft (55 m).

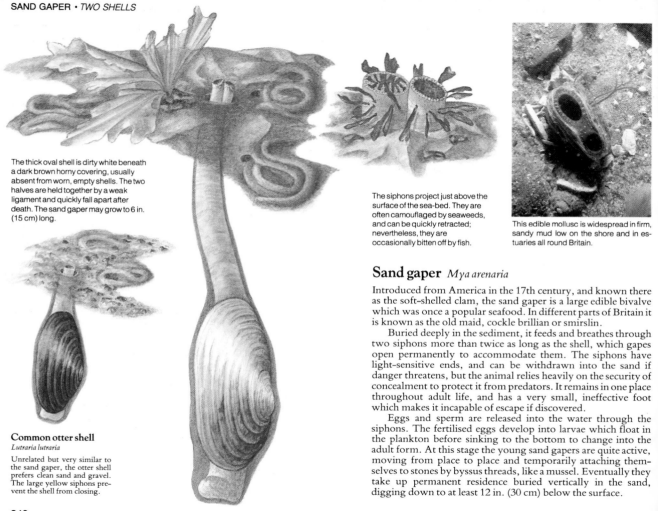

The thick oval shell is dirty white beneath a dark brown horny covering, usually absent from worn, empty shells. The two halves are held together by a weak ligament and quickly fall apart after death. The sand gaper may grow to 6 in. (15 cm) long.

The siphons project just above the surface of the sea-bed. They are often camouflaged by seaweeds, and can be quickly retracted; nevertheless, they are occasionally bitten off by fish.

This edible mollusc is widespread in firm, sandy mud low on the shore and in estuaries all round Britain.

Common otter shell
Lutraria lutraria

Unrelated but very similar to the sand gaper, the otter shell prefers clean sand and gravel. The large yellow siphons prevent the shell from closing.

Sand gaper *Mya arenaria*

Introduced from America in the 17th century, and known there as the soft-shelled clam, the sand gaper is a large edible bivalve which was once a popular seafood. In different parts of Britain it is known as the old maid, cockle brillian or smirslin.

Buried deeply in the sediment, it feeds and breathes through two siphons more than twice as long as the shell, which gapes open permanently to accommodate them. The siphons have light-sensitive ends, and can be withdrawn into the sand if danger threatens, but the animal relies heavily on the security of concealment to protect it from predators. It remains in one place throughout adult life, and has a very small, ineffective foot which makes it incapable of escape if discovered.

Eggs and sperm are released into the water through the siphons. The fertilised eggs develop into larvae which float in the plankton before sinking to the bottom to change into the adult form. At this stage the young sand gapers are quite active, moving from place to place and temporarily attaching themselves to stones by byssus threads, like a mussel. Eventually they take up permanent residence buried vertically in the sand, digging down to at least 12 in. (30 cm) below the surface.

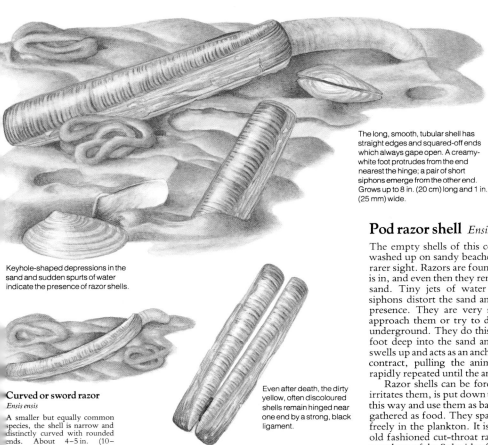

The long, smooth, tubular shell has straight edges and squared-off ends which always gape open. A creamy-white foot protrudes from the end nearest the hinge; a pair of short siphons emerge from the other end. Grows up to 8 in. (20 cm) long and 1 in. (25 mm) wide.

The shells are usually found buried vertically near the low-water mark and beyond on most clean, sandy shores.

Keyhole-shaped depressions in the sand and sudden spurts of water indicate the presence of razor shells.

Curved or sword razor
Ensis ensis

A smaller but equally common species, the shell is narrow and distinctly curved with rounded ends. About 4–5 in. (10–12·5 cm) long.

Even after death, the dirty yellow, often discoloured shells remain hinged near one end by a strong, black ligament.

Pod razor shell *Ensis siliqua*

The empty shells of this common bivalve are frequently seen washed up on sandy beaches at low tide. The living animal is a rarer sight. Razors are found near the surface only when the tide is in, and even then they remain almost completely buried in the sand. Tiny jets of water from their breathing and feeding siphons distort the sand and are usually the only sign of their presence. They are very sensitive to vibrations, and if you approach them or try to dig them up, they rapidly disappear underground. They do this by thrusting the tip of the massive foot deep into the sand and pumping blood into it so that it swells up and acts as an anchor, while the rest of the foot muscles contract, pulling the animal downwards. These actions are rapidly repeated until the animal is safely buried.

Razor shells can be forced to the surface when salt, which irritates them, is put down their burrows. Anglers catch them in this way and use them as bait and in the Orkney Islands they are gathered as food. They spawn in the spring, the larvae floating freely in the plankton. It is their similarity in appearance to an old fashioned cut-throat razor that has earned them and other members of the Solenidae family their common name.

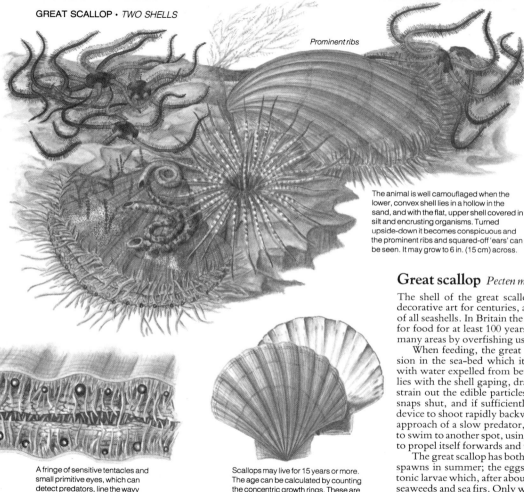

Prominent ribs

The animal is well camouflaged when the lower, convex shell lies in a hollow in the sand, and with the flat, upper shell covered in silt and encrusting organisms. Turned upside-down it becomes conspicuous and the prominent ribs and squared-off 'ears' can be seen. It may grow to 6 in. (15 cm) across.

The scallop occurs in beds in sands, firm muds, gravels and stony areas, mostly between 33–131 ft (10–40 m).

A fringe of sensitive tentacles and small primitive eyes, which can detect predators, line the wavy edges of the shell.

Scallops may live for 15 years or more. The age can be calculated by counting the concentric growth rings. These are closer together near the shell edge.

Great scallop *Pecten maximus*

The shell of the great scallop has been used as a symbol in decorative art for centuries, and it is probably the most familiar of all seashells. In Britain the scallop has been regularly gathered for food for at least 100 years, but stocks have been depleted in many areas by overfishing using the dredging technique.

When feeding, the great scallop occupies a shallow depression in the sea-bed which it excavates by blowing sand aside with water expelled from between the valves. Once installed it lies with the shell gaping, drawing water in through the gills to strain out the edible particles. At the threat of danger the shell snaps shut, and if sufficiently alarmed the animal can use this device to shoot rapidly backwards out of the way. Normally the approach of a slow predator, such as a starfish, merely causes it to swim to another spot, using water ejected from near the hinge to propel itself forwards and upwards in a series of jerks.

The great scallop has both male and female sexual organs and spawns in summer; the eggs hatch into free-swimming planktonic larvae which, after about three weeks, attach themselves to seaweeds and sea firs. Only when the shell is properly developed do they become independent adults.

Scallop look-alikes

Variegated scallop
Chlamys varia

Sometimes found attached by threads to rocks low on the shoreline it also lives unattached. The shell, with its distinctive colour variations of yellow, red, dark green and purple, has 23–35 narrow ribs and flattened spines. One 'ear' is two or three times longer than the other. Rarely longer than 2¼ in. (60 mm).

Tiger scallop
Chlamys tigerina

The shell is almost smooth, with fine radiating and concentric lines, although older specimens may develop up to 30 thicker ribs. The colour varies, but there are always distinctive streaks, spots and zigzags. It is found on gravel bottoms low on the shoreline and below, where it lives unattached. Rarely grows longer than 1 in. (25 mm).

Hunchback scallop
Chlamys distorta

Although the young are very similar to the variegated scallop, adults are smaller and are firmly cemented to rocks by their lower shell. The upper shell has 60–70 spiny ribs and is often distorted in outline – hence the scallop's name. About 1½ in. (40 mm) long.

Queen scallop
Chlamys opercularis

Although otherwise similar to the great scallop, both the shells of the 'queenie' are convex. It has about 20 broad radiating ribs and equal-sized 'ears'. It is a very active swimmer and lives in large colonies in muddy offshore waters where it is fished commercially in some areas. About 3½ in. (90 mm) long.

Gaping file shell
Lima hians

A free-swimming species, it builds its nest of threads among stones in water low on the shoreline and below. The spectacular fringe of red-and-orange tentacles cannot be retracted. When closed there is a gap between both shells. The shell is off-white with 50 or more ribs. Up to 1 in. (25 mm) long.

249

The shell is elongated, thin, white and criss-crossed by concentric ridges and radiating ribs. There are sharp spines at these intersections. A pair of tubes, or siphons, joined along their length, reach out into the water from within the burrow. Up to 6 in. (15 cm) long.

It is found in shallow waters and on the lower shores of south and south-west England and around Wales.

A sucker foot protrudes from a permanent gape between the two shell halves. The foot grips the sides of the hole and rocks the shell in a rasping see-saw motion.

Broken shells are often found on the beach. They show the strong ribs and spines at the front end which are used to bore into the rock.

Common piddock *Pholas dactylus*

Many shellfish are adapted for burrowing into sand or mud, where they are hidden from predators, but the common piddock is one of a small group of molluscs which are capable of boring into solid rock. It achieves this using its strong rasp-like shell, which is twisted in the burrow to enlarge and deepen it as the animal grows. The piddock may lie as deep as 6 in. (15 cm) in the rock.

It feeds and breathes through a pair of long joined siphons which extend to the mouth of the burrow. Plankton-rich water is drawn in for respiration and filtered to extract the food; as the water is pumped out it carries away the excavated material which is passed through the shellfish from the end of the tunnel.

Male and female release sperm and eggs directly into the water via the outgoing siphon. The fertilised eggs develop into larvae which float in the plankton, finally changing into the adult form which settles in a crevice and begins to grind its way down. It will tunnel into soft rock, stiff clay or occasionally wood, and if it encounters another piddock it is quite likely to bore straight through it. At night the animals are strongly luminescent, and glow with a greenish-blue light.

Molluscs that bore into rocks and wood

Timbers and rocks riddled with round holes will almost certainly be the work of piddocks and shipworms. These molluscs have sharp-edged shells which are specially adapted for excavating the bore-holes in which the animal lives. The four species shown here are commonly found on shores all round Britain.

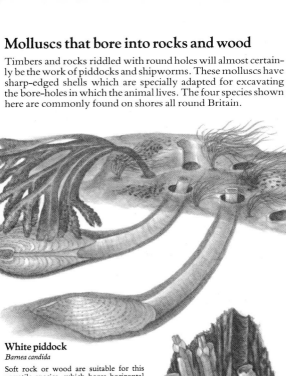

White piddock
Barnea candida

Soft rock or wood are suitable for this versatile species, which bores horizontal burrows. The oval-shaped shell, with rounded ends, can be almost completely closed. It rarely exceeds 2¼ in. (60 mm) long.

Shipworms bore only into wood. Driftwood is often riddled with their burrows.

Red nose or wrinkled rock borer
Hiatella arctica

This common species bores flask-shaped burrows in the soft rock of the sea-bed, or nestles in holes and crannies anchored by its bysuss threads. The tips of the breathing and feeding tubes, or siphons, are pink. The almost rectangular, often misshapen, off-white shell rarely exceeds 1½ in. (40 mm) long.

Shipworm
Teredo species

The worm-like body has only a small, three-lobed shell at the boring end. A chalky tube, which grows up to 8 in. (20 cm) long, lines the hole and protects the body.

The oval piddock can never completely close its shell.

Oval piddock
Zirfaea crispata

Found mainly in clay and shale, at and below the low-water mark, it is stubbier than the common piddock. The dirty white shell has a distinct furrow across both valves. Up to about 3¼ in. (90 mm) long.

251

Lepidochitona cinereus
The commonest shore species, with dull green, grey or red shell plates and a granular fleshy edge. It is found clinging tightly to rocks anywhere between tide marks. The shell width ratio is 3:4. Up to ¾ in. (20 mm) long.

Acanthochitona species
Easily recognised by 18 to 20 conspicuous tufts of bristles along the broad, thick, fleshy edge. Shell width ratio 2:5. Less than ¾ in. (20 mm) long.

Callochiton achatinus
The smooth, shiny, red-brown shell plates are patterned with light patches. The fleshy edge has numerous spines that lie flat in rows. The shell width ratio is 3:5. Up to ½ in. (13 mm) long.

Lepidopleurus asellus
Uniformly grey or brown with lines of granules like a corn-cob on the plates. This animal has a narrow, fleshy margin covered in delicate, overlapping rectangular scales; the shell width ratio is 4:5. Up to ¾ in. (20 mm) long.

Tonicella rubra
The shining red or brown-red shell plates have a central keel and the fleshy edge is granular in appearance. Shell width ratio 3:4, size about ½ in. (13 mm) long.

Tonicella marmorea
Less common than *T. rubra* but similar, though the fleshy edge does not appear granular, even through a hand lens. Up to 1 in. (25 mm) long.

The well-protected chitons

Small and inconspicuous animals, chitons are very common on and under rocks on the lower shore and below. The flattened, elongated body is protected by a shell made up of eight articulating plates whose ends are buried in a fleshy shell edge. This arrangement allows chitons to bend and cling tightly to rough and uneven rock surfaces. When removed or dislodged by waves, they curl up like woodlice. The armoured appearance – rather like the links in chain mail – is the reason for their alternative name, coat-of-mail shells.

Species can be distinguished only by careful examination of the shape, size, colour and the ratio of shell width to the total width of the animal. Chitons feed on vegetation such as algae which they scrape off rocks by means of a tongue bearing hard, rasping teeth called a radula.

Common acorn barnacle

Balanus balanoides

The low, greyish-white cone has six plates; one end-plate is much wider than the other. The diamond-shaped aperture is closed by four plates and the joint lines meet at an oblique angle. Up to ⅝ in. (15 mm) across.

Star acorn barnacle

Chthamalus stellatus and *C. montagui*

Similar in size and shape to the common barnacle, these almost identical species, which live on the upper shores of exposed west and south-west coasts, have a kite-shaped opening with joint lines that meet at right-angles. Both end-plates are narrow. Up to ⅝ in. (15 mm) across.

Australian barnacle

Elminius modestus

An introduced species common in the Channel near freshwater outlets; it is also found farther north. The flattened greyish-white shell has only four plates. Up to ⅜ in. (10 mm) across.

Goose barnacle

Lepas anatifera

The shell of five shiny white plates with a bluish-grey sheen is attached by a tough, grey-brown stalk to floating debris and is often found cast ashore. The shell may be 2 in. (50 mm) long, the stalk 8 in. (20 cm).

Asymmetrical barnacle

Verruca stroemia

The species lives singly under stones or on shells on the lower shore and below. The four indistinct grey, white or brown plates are all different sizes and have longitudinal ribs. The shell opening is also indistinct. Up to ⅕ in. (5 mm) across.

Balanus crenatus

The tall conical shell of six greyish-white, ridged plates clings to rocks, shells and debris on the lower shore. When removed from rocks, it leaves a circular white base. Up to ¾ in. (20 mm) across.

How to identify barnacles

Although they resemble molluscs, barnacles are crustaceans – the group that includes crabs and prawns – that have taken up an attached existence. They are among the most familiar animals on the seashore where they are seen covering large areas of rock. They appear lifeless until the tide is in, when they extend feathery legs through the opening in the top of the shell and catch suspended food particles. Different species are identified by the size, number and arrangement of the outer plates, and the shape of the top opening. Apart from the star and Australian barnacles, the species shown here are common all round Britain.

Marine life on piers and breakwaters

Man-made objects such as piers, jetties and breakwaters provide opportunities for many plants and animals that spend their lives attached to hard surfaces. Such creatures reproduce by means of floating or crawling spores and larvae, so what settles on a newly available surface depends partly on chance: a pier built over rocks covered in plant and animal life will attract a greater variety of life than a pier built over a sandy shore or one in a heavily polluted dock.

In the summer many rocky shore organisms produce large numbers of larvae, and the positions of various parts of a pier or breakwater affect what will settle and grow on them. Pilings exposed to air for long periods will support fewer species than those permanently or almost permanently covered. Some vertical pilings have different species living at different levels. The top parts are covered by green seaweeds such as *Enteromorpha* and *Cladophora* while below are barnacles and, near the low-water mark, mussels. At the levels rarely uncovered by the tide grow anemones, dead man's fingers, sponges and other encrusting animals. Heavy growths, or 'fouling', can cause damage to structures.

Green and brown seaweeds grow profusely on well-lit parts of the pier, such as the tops of pilings and on upper surfaces of horizontal spars. Deep beneath the pier, only red seaweeds adapted to low light levels can grow.

Anemones and dead man's fingers prefer to remain submerged and grow in profusion on pilings near the low-water mark, along with delicate pink and brown sea firs. At and below low water, many other colonial animals, such as sea squirts, sea mats and tube worms, find food and shelter.

Wooden parts of the pier near low-tide level often become riddled by the shallow tunnels of the gribble – a small crustacean. Old timbers may become infested by the shipworm *Teredo* whose mining activities may cause the structure to collapse. Mussels attach themselves and build up into large clusters. When the tide is in, the common jellyfish (*Aurelia aurita*) may be seen in the shelter of the pilings.

Mussels are one of the commonest animals found on piers. They cling firmly by tough threads called byssus threads, resisting the strongest waves.

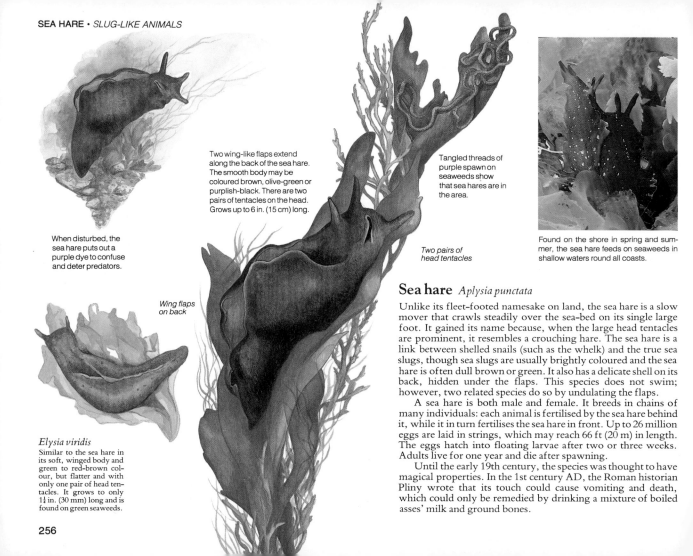

When disturbed, the sea hare puts out a purple dye to confuse and deter predators.

Two wing-like flaps extend along the back of the sea hare. The smooth body may be coloured brown, olive-green or purplish-black. There are two pairs of tentacles on the head. Grows up to 6 in. (15 cm) long.

Tangled threads of purple spawn on seaweeds show that sea hares are in the area.

Two pairs of head tentacles

Found on the shore in spring and summer, the sea hare feeds on seaweeds in shallow waters round all coasts.

Wing flaps on back

Elysia viridis
Similar to the sea hare in its soft, winged body and green to red-brown colour, but flatter and with only one pair of head tentacles. It grows to only 1¼ in. (30 mm) long and is found on green seaweeds.

Sea hare *Aplysia punctata*

Unlike its fleet-footed namesake on land, the sea hare is a slow mover that crawls steadily over the sea-bed on its single large foot. It gained its name because, when the large head tentacles are prominent, it resembles a crouching hare. The sea hare is a link between shelled snails (such as the whelk) and the true sea slugs, though sea slugs are usually brightly coloured and the sea hare is often dull brown or green. It also has a delicate shell on its back, hidden under the flaps. This species does not swim; however, two related species do so by undulating the flaps.

A sea hare is both male and female. It breeds in chains of many individuals: each animal is fertilised by the sea hare behind it, while it in turn fertilises the sea hare in front. Up to 26 million eggs are laid in strings, which may reach 66 ft (20 m) in length. The eggs hatch into floating larvae after two or three weeks. Adults live for one year and die after spawning.

Until the early 19th century, the species was thought to have magical properties. In the 1st century AD, the Roman historian Pliny wrote that its touch could cause vomiting and death, which could only be remedied by drinking a mixture of boiled asses' milk and ground bones.

Six common sea slugs

The soft, often intricately shaped bodies and bright colours of the many sea slugs are seen at their best in water, and the slugs are more easily identified there. The species illustrated are found on shores all round Britain.

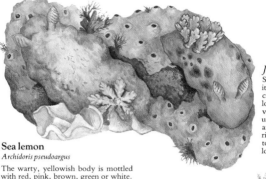

Jorunna tomentosa

Similar to the sea lemon, it also feeds on breadcrumb sponges. Its skin looks smooth but is covered with many small, uniform tubercles. There are up to 17 gills in the ring on the back. Grows to about 2½ in. (64 mm) long.

Common grey sea slug
Aeolidia papillosa

Long, fleshy projections called cerata cover the back in a dense mass. A pale crescent shows on the head, which has four tentacles. This slug is found among rocks, where it feeds on beadlet anemones, storing their stinging cells in its cerata for its own use. It may reach 4¾ in. (12 cm) in length.

Sea lemon
Archidoris pseudoargus

The warty, yellowish body is mottled with red, pink, brown, green or white. There are two head tentacles and a ring of eight or nine feathery, branched gills on the back. Up to 4¾ in. (12 cm) long.

Polycera quadrilineata

This small slug is spotted with yellow or orange and often with black as well. The head has four pointed projections and two tentacles. Commonly found on kelp, feeding on the encrusting sea mats, it grows up to 1¼ in. (30 mm) long.

Onchidoris bilamellata

Fawn-coloured, club-like tubercles cover the brown-blotched body. There are up to 29 gills in the ring on the back. This slug feeds on barnacles. Up to 1½ in. (40 mm) long.

Facelina bostoniensis

Eight clusters of long, soft fleshy projections called cerata project from each side of the body. There are two pairs of head tentacles. The slug feeds on sea firs. Grows up to 2 in. (50 mm) long.

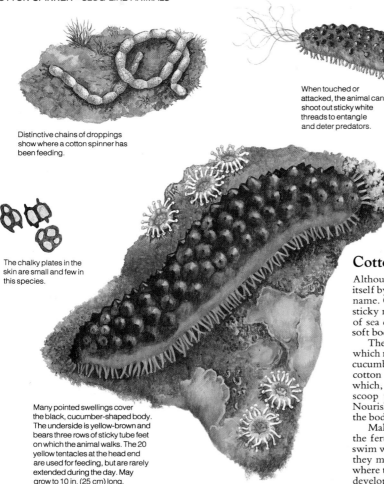

Distinctive chains of droppings show where a cotton spinner has been feeding.

When touched or attacked, the animal can shoot out sticky white threads to entangle and deter predators.

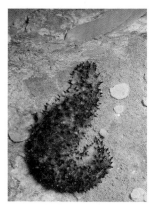

Cotton spinners live on parts of the Atlantic and Channel coasts among rocks, in sea grass or on sediments.

The chalky plates in the skin are small and few in this species.

Many pointed swellings cover the black, cucumber-shaped body. The underside is yellow-brown and bears three rows of sticky tube feet on which the animal walks. The 20 yellow tentacles at the head end are used for feeding, but are rarely extended during the day. May grow to 10 in. (25 cm) long.

Cotton spinner *Holothuria forskali*

Although it has few natural enemies, the cotton spinner defends itself by squirting out thin white threads from its rear – hence its name. Once emitted, the threads swell up and elongate to form a sticky mass that distracts or entangles the enemy. Other species of sea cucumber can eject the entire gut, and then renew it. Its soft body is protected by chalky plates embedded in the skin.

The starfish and sea urchin are related to the cotton spinner, which resembles a large black slug. It is also known as a black sea cucumber and is the largest of the sea cucumbers in Britain. The cotton spinner moves slowly forward on sucker-like tube feet which, at the mouth end, have also been adapted as tentacles to scoop up mud and sand from the sea-bed into the mouth. Nourishment is extracted and undigested remains pass through the body.

Males and females simultaneously spawn into the sea, and the fertilised eggs develop into minute larvae. The larvae can swim with the aid of rhythmically beating hairs, called cilia, but they mainly drift in the currents of the upper level of the sea, where they feed on plankton until they settle on the sea-bed to develop into adults.

Four common sea cucumbers

Many sea cucumbers hide in rock crevices or bury themselves in mud and sand. Their feeding tentacles may be the only visible part and these can quickly be retracted to safety. Some species can be identified only with the help of a microscope, which reveals the distinctive arrangement of chalky plates embedded in the skin. The species shown can be seen on the shore.

Labidoplax digitata

Similar in size and habitat to the worm cucumber, but a darker or brownish-pink and with only four branches on each tentacle tip. It is found on the west and south-west coasts.

Worm cucumber
Leptosynapta inhaerens

Minute hooks in the skin give a sticky feel to the pink body, which grows to 7 in. (18 cm) or more. There are 12 feeding tentacles with many finger-like branches. Worm cucumbers burrow in sand, mud and sea grass, on south-west, west and north-east coasts.

Sea gherkin
Cucumaria saxicola

The smooth, white body has ten darker, branched tentacles, bears five rows of tube feet which are zigzag on the upper surface, and grows to 6 in. (15 cm) long. It clings beneath rocks and crevices, and sometimes to seaweed. It inhabits the south-west coasts of Britain and Ireland, and Scotland's west coast.

Cucumaria normani

Finely branched black tentacles extend from the rough, wrinkled browny-white body. The tube feet are in double rows. The size, habitat and distribution are like the sea gherkin's.

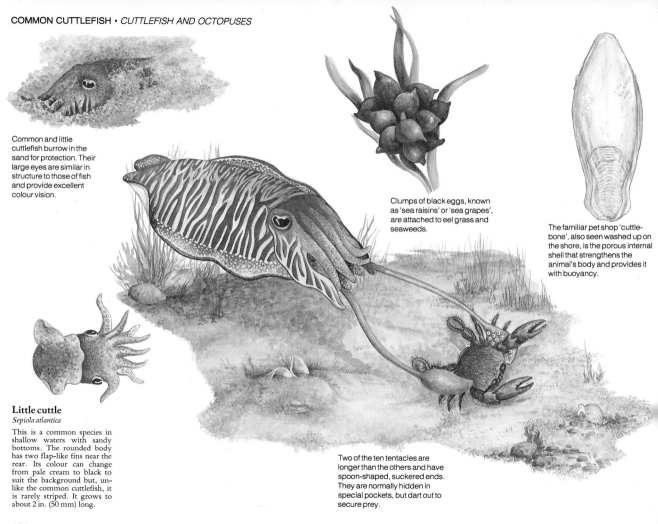

Common and little cuttlefish burrow in the sand for protection. Their large eyes are similar in structure to those of fish and provide excellent colour vision.

Clumps of black eggs, known as 'sea raisins' or 'sea grapes', are attached to eel grass and seaweeds.

The familiar pet shop 'cuttlebone', also seen washed up on the shore, is the porous internal shell that strengthens the animal's body and provides it with buoyancy.

Little cuttle
Sepiola atlantica

This is a common species in shallow waters with sandy bottoms. The rounded body has two flap-like fins near the rear. Its colour can change from pale cream to black to suit the background but, unlike the common cuttlefish, it is rarely striped. It grows to about 2 in. (50 mm) long.

Two of the ten tentacles are longer than the others and have spoon-shaped, suckered ends. They are normally hidden in special pockets, but dart out to secure prey.

A fringe of lateral fins, used for normal forward propulsion, runs round the broad, flattened body. For rapid backward movement, the cuttlefish expels a powerful jet of water from a funnel on the underside of the head. The animal can change colour quickly, but a zebra-stripe pattern is typical. Adults may grow up to 12 in. (30 cm) long.

The suckered tentacles are sometimes held together to form a protective shield as the animal swims forwards.

In summer, the common cuttlefish can be found in estuaries, sandy bays and eel grass beds all round the coast.

Common cuttlefish *Sepia officinalis*

The cuttlefish is a highly developed, carnivorous mollusc, with excellent eyesight, a good turn of speed, and full control over its buoyancy and colour. Like its relative the octopus, it is able to protect itself by producing a cloud of ink.

The internal shell, or cuttlebone, is composed of small chambers which act as buoyancy tanks. During the day the chambers are flooded with water, and the animal sinks to the sea-bed to lie hidden in the sand. At night the water is pumped out, and the cuttlefish emerges to hunt. Propelled by lateral fins it approaches its prey cautiously, hovering in the water; when within range long hunting arms shoot out to gather the prey into shorter tentacles which surround the beak-like jaws.

The skin is equipped with a range of colour cells, controlled by the nerves, which open and close to display or conceal the pigment. This allows the cuttlefish to change colour very rapidly. When it is hunting, waves of colour pass over the head and tentacles, and may confuse or distract the prey. Normally, when swimming, it adopts a zebra-stripe camouflage which is very effective in the broken light of eel grass beds, which are a favourite habitat.

Sand-dunes provide cover for rabbits, rats, mice and hedgehogs which will sometimes venture onto the shore to scavenge among the debris. Dead animals such as fish, seabirds and even seals attract rats and foxes. Sheep, cattle and deer sometimes come onto the beach to feed on seaweed, especially on isolated beaches in Scotland.

Large masses of wave-torn seaweed are thrown up on the strandline after a storm, and in autumn kelp plants shed their fronds which also get washed up, sometimes covering a whole beach. Mixed in with the seaweed are bottles, cans, rope and tar. The rotting material attracts flies and sandhoppers, small relatives of shrimps which jump erratically when disturbed.

The debris and litter of the strandline

Along the strandline there is much evidence of the animals that live below the tide level. Common finds are the cast shells and broken limbs of crabs and lobsters, sea urchin tests, worm tubes, and a wide variety of shells and cuttlebones. The latter are rarely found intact as seabirds love to peck at them. Many beaches are formed of shell sand, the minute particles of shells broken and crushed by the action of the sea over many years.

At each high tide the sea deposits on the shore an accumulation of debris ranging from man-made rubbish to marine life torn from offshore rocks and the remains of the sea's inhabitants. This straggling collection of seaweed, plastic bottles, shells and dead creatures is called the strandline. It shows up best on flat, sandy beaches, where there will often be more than one strandline since the tide reaches different levels each day.

Although sometimes unsightly and smelly, strandlines contain many interesting items. Winter is often the best time to search them, when storms deposit large quantities of debris. There may be timbers from shipwrecks and piers, and if these have been floating for some time they may be riddled with shipworm holes or the holes of the gribble, a woodlouse-like creature. Some of the material – such as tropical seeds – may have been transported thousands of miles.

Following spring tides, debris is thrown up near the top of the beach, and becomes home for a thriving community of sandhoppers, sea slaters, worms and small crabs – until the next spring tide two weeks later deposits a fresh crop of litter.

Tough protective egg cases of fish and molluscs survive after the young have hatched and are often attached to debris. The mermaid's purse of skates and rays has horn-like corners; the dogfish case has long tendrils. Sea wash balls are the sponge-like egg cases of the whelk.

Jellyfish are often driven ashore by strong winds. They should not be touched as some, like the Portuguese man-of-war, can still sting. Other visitors include goose barnacles attached to bottles and cans and tropical beans from distant shores.

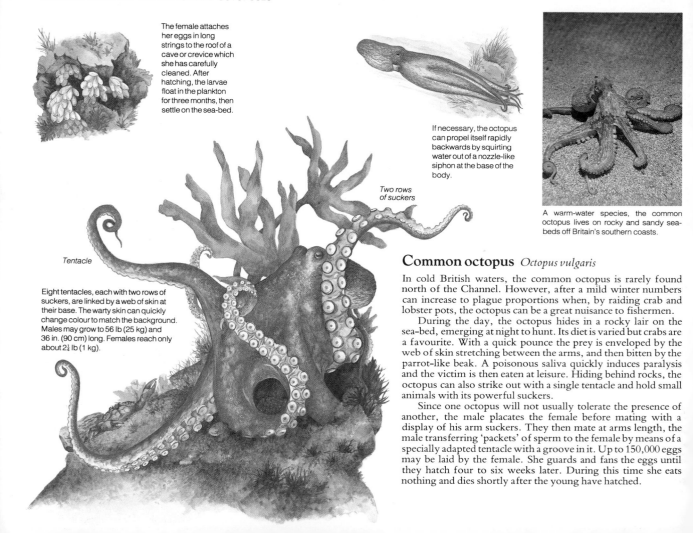

The female attaches her eggs in long strings to the roof of a cave or crevice which she has carefully cleaned. After hatching, the larvae float in the plankton for three months, then settle on the sea-bed.

If necessary, the octopus can propel itself rapidly backwards by squirting water out of a nozzle-like siphon at the base of the body.

Two rows of suckers

A warm-water species, the common octopus lives on rocky and sandy sea-beds off Britain's southern coasts.

Tentacle

Eight tentacles, each with two rows of suckers, are linked by a web of skin at their base. The warty skin can quickly change colour to match the background. Males may grow to 56 lb (25 kg) and 36 in. (90 cm) long. Females reach only about 2¼ lb (1 kg).

Common octopus *Octopus vulgaris*

In cold British waters, the common octopus is rarely found north of the Channel. However, after a mild winter numbers can increase to plague proportions when, by raiding crab and lobster pots, the octopus can be a great nuisance to fishermen.

During the day, the octopus hides in a rocky lair on the sea-bed, emerging at night to hunt. Its diet is varied but crabs are a favourite. With a quick pounce the prey is enveloped by the web of skin stretching between the arms, and then bitten by the parrot-like beak. A poisonous saliva quickly induces paralysis and the victim is then eaten at leisure. Hiding behind rocks, the octopus can also strike out with a single tentacle and hold small animals with its powerful suckers.

Since one octopus will not usually tolerate the presence of another, the male placates the female before mating with a display of his arm suckers. They then mate at arms length, the male transferring 'packets' of sperm to the female by means of a specially adapted tentacle with a groove in it. Up to 150,000 eggs may be laid by the female. She guards and fans the eggs until they hatch four to six weeks later. During this time she eats nothing and dies shortly after the young have hatched.

The powerful suckers enable the octopus to catch and hold prey, as well as helping it to grip and move quite nimbly over the sea-bed.

Discarded pieces of crab shell littering the sea-bed may indicate the entrance to an octopus's lair – usually a hole in the rocks.

Found all round Britain's rockier coasts, the lesser octopus is most common in northern waters.

Unlike the common octopus, the lesser has only a single line of suckers on the underside of each tentacle. It is also smaller. Although able to camouflage itself by changing colour quickly, it is usually reddish-brown. Grows up to 20 in. (50 cm) long.

Lesser octopus *Eledone cirrhosa*

Despite its name the lesser octopus is more widespread than the common octopus, and is often seen by divers and netted by trawlermen. The two species have similar habits and diets. Both are large-brained and capable of learning and carrying out complex activities that are beyond the scope of other invertebrates. They can also change colour rapidly to match their background. The size of the pigment cells of the skin alters after the large eyes – structured much like a vertebrate's – have assessed the colour to be imitated. If an octopus is jetting rapidly away from a predator, waves of colour may flow over its body in response to the changing background colours.

As a further defence against fish, conger eels, seals and dolphins, a cloud of black fluid can be discharged by the octopus to hide it and confuse the predator. Congers and other fish are attracted by the smell of the dispersing 'ink', and follow this instead of the octopus.

The female lays her eggs in bunches in a rock cranny or hole. If she cannot find one, she will use a strong water jet from her siphon to blow sand away from beneath a rock. She lays fewer eggs than the common octopus and does not guard them.

265

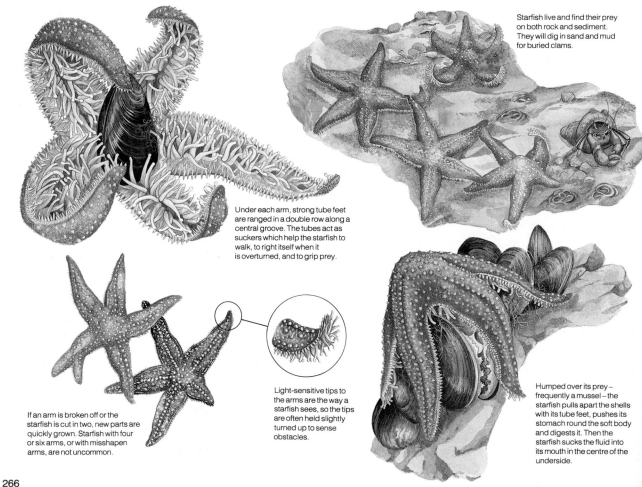

Starfish live and find their prey on both rock and sediment. They will dig in sand and mud for buried clams.

Under each arm, strong tube feet are ranged in a double row along a central groove. The tubes act as suckers which help the starfish to walk, to right itself when it is overturned, and to grip prey.

If an arm is broken off or the starfish is cut in two, new parts are quickly grown. Starfish with four or six arms, or with misshapen arms, are not uncommon.

Light-sensitive tips to the arms are the way a starfish sees, so the tips are often held slightly turned up to sense obstacles.

Humped over its prey – frequently a mussel – the starfish pulls apart the shells with its tube feet, pushes its stomach round the soft body and digests it. Then the starfish sucks the fluid into its mouth in the centre of the underside.

Five plump, tapering arms radiate symmetrically from a central disc. Yellow-brown is the typical colour, but some starfish are red-brown and a few are purple. Scattered, small white spines make the skin feel rough. The spines form a line down the middle of each arm. A shore specimen is usually 2–4 in. (5–10 cm) across, but some reach 20 in. (50 cm).

Common starfish live up to their name, being found all round Britain, both inshore and in deeper water. They are pests in mussel and oyster beds.

Common starfish *Asterias rubens*

The strangely shaped starfish belong to the echinoderms, a group of animals whose name means spiny-skinned. The spines arise from a skeleton just beneath the skin. In the starfish the skeleton is formed of separate bony plates.

The starfish has no head but a five-rayed symmetry like the petals on a flower. Any arm can take the lead when the starfish walks on its tube feet. The feet are filled with fluid connected to an internal water vascular system, and work by hydraulic pressure. They are very powerful and can pull apart the shells of mussels and other bivalves which starfish eat.

A single arm with part of the central disc attached will re-grow the other four arms. Before this regenerative power was recognised, fishermen would cut up any starfish they dredged from their mussel beds and throw the pieces back – unwittingly multiplying the pest that was eating their mussels.

Males and females shed sperm and eggs into the water in spring and summer, the presence of each sex stimulating the other. One female can release 2 million eggs in two hours. Fertilised eggs hatch into larvae that float in the plankton for about three weeks, and then settle on the sea-bed to develop.

267

Recognising other colourful starfish

These diverse, colourful and slow-moving predators of the shores and shallow seas are identified by the number and shape of their arms, their colour, relative size, number and type of spines, and where they live. Within the starfish class, groups with characteristics in common are: those with five distinct arms; pentagonal starfish with rounded points rather than arms; and starfish with more than five arms.

Starlet
Asterina gibbosa

A compact shore resident only about 1 in. (25 mm) wide. Above, it varies from olive-green to brown, with red, green and cream tints; below it is yellowish. It clings under stones and rock overhangs on the lower shores of the south and west coasts.

Spiny starfish
Marthasterias glacialis

Lines of bulbous, grey-green spines run along the brown-yellow arms, which are often tipped with red or purple. The underside is yellowy-white. This rigid starfish may reach 32 in. (80 cm) across. It lives in rocky areas on the lower shore and below, and is most common on the west and south-west coasts of Britain.

Each spine is ringed by pedicellariae, small pincer-like structures which the animal uses to clean itself.

Bloody Henry
Henricia sanguinolenta and *Henricia oculata*

Blood-red marks often found on these two virtually identical species give them their common name. Their main colour may be orange, purple or red, with a sandy underside. They are hard and stiff, have plump arms and grow to 4 in. (10 cm) across. They live on rock, muddy sand or pebbles on and off shore all round Britain.

Sand star
Astropecten irregularis

Distinct cream-coloured plates and a fringe of spines border the stiff, flattened arms. The sand star varies from pink to brick red, is white below, and may reach 4½ in. (12 cm) across. It burrows into sand at low-tide level all round Britain; some specimens can be found washed ashore from deeper water.

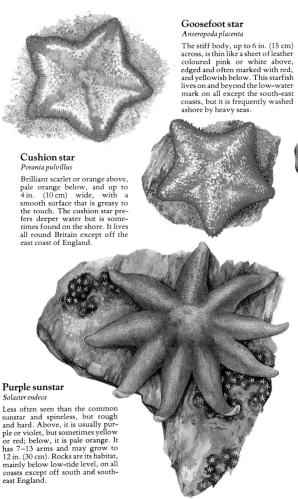

Goosefoot star
Anseropoda placenta

The stiff body, up to 6 in. (15 cm) across, is thin like a sheet of leather coloured pink or white above, edged and often marked with red, and yellowish below. This starfish lives on and beyond the low-water mark on all except the south-east coasts, but it is frequently washed ashore by heavy seas.

Cushion star
Porania pulvillus

Brilliant scarlet or orange above, pale orange below, and up to 4 in. (10 cm) wide, with a smooth surface that is greasy to the touch. The cushion star prefers deeper water but is sometimes found on the shore. It lives all round Britain except off the east coast of England.

Common sunstar
Crossaster papposus

Usually brownish-red marked with white, but sometimes purplish-red; the underside is pale yellow. The sunstar is covered with distinct short spines, has 8–13 short arms, and may grow to 10 in. (25 cm) across. It is common all round Britain on rocky shores and below low-tide level.

Purple sunstar
Solaster endeca

Less often seen than the common sunstar and spineless, but rough and hard. Above, it is usually purple or violet, but sometimes yellow or red; below, it is pale orange. It has 7–13 arms and may grow to 12 in. (30 cm). Rocks are its habitat, mainly below low-tide level, on all coasts except off south and south-east England.

Seven-armed starfish
Luidia ciliaris

A large orange-red starfish, white below, that always has seven arms fringed with white spines. It can grow up to 16 in. (40 cm) across. It lives beyond the low-tide mark, round all coasts except the southern east coast of England. Occasionally it is washed onto the beach.

269

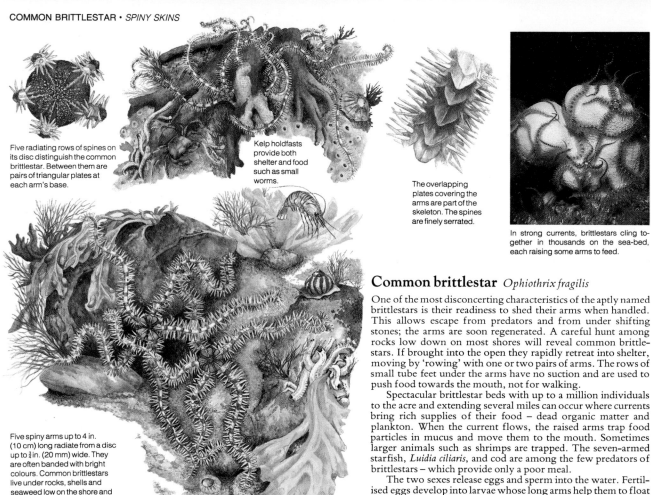

Five radiating rows of spines on its disc distinguish the common brittlestar. Between them are pairs of triangular plates at each arm's base.

Kelp holdfasts provide both shelter and food such as small worms.

The overlapping plates covering the arms are part of the skeleton. The spines are finely serrated.

In strong currents, brittlestars cling together in thousands on the sea-bed, each raising some arms to feed.

Five spiny arms up to 4 in. (10 cm) long radiate from a disc up to ¾ in. (20 mm) wide. They are often banded with bright colours. Common brittlestars live under rocks, shells and seaweed low on the shore and in deeper water.

Common brittlestar *Ophiothrix fragilis*

One of the most disconcerting characteristics of the aptly named brittlestars is their readiness to shed their arms when handled. This allows escape from predators and from under shifting stones; the arms are soon regenerated. A careful hunt among rocks low down on most shores will reveal common brittlestars. If brought into the open they rapidly retreat into shelter, moving by 'rowing' with one or two pairs of arms. The rows of small tube feet under the arms have no suction and are used to push food towards the mouth, not for walking.

Spectacular brittlestar beds with up to a million individuals to the acre and extending several miles can occur where currents bring rich supplies of their food – dead organic matter and plankton. When the current flows, the raised arms trap food particles in mucus and move them to the mouth. Sometimes larger animals such as shrimps are trapped. The seven-armed starfish, *Luidia ciliaris*, and cod are among the few predators of brittlestars – which provide only a poor meal.

The two sexes release eggs and sperm into the water. Fertilised eggs develop into larvae whose long arms help them to float in the plankton before becoming bottom-living adults.

Brittlestars and featherstars

Black brittlestar
Ophiocomina nigra

Usually dark – although albinos are seen – with a ¾ in. (20 mm) disc with no spines but finely granular. The black brittlestar lives all round Britain, except possibly off the east coast, among and under shoreline boulders and in dense beds on sediments offshore.

Long-armed brittlestar
Acrocnida brachiata

Its very long arms make this grey-brown brittlestar easy to identify. The arms are up to 6 in. (15 cm) long on a disc about ⅜ in. (10 mm) across. The tips are seen sticking up from the shallow covering of sand under which the species lives. It occurs on the middle and lower shore and in deeper water on most coasts.

Ophiopholis aculeata

Similar to the common brittlestar, but the disc, up to ¾ in. (20 mm) wide, is without spines and bluntly pointed between the arms. It is usually red, but sometimes bluish, with dark-banded arms which may be 3¼ in. (80 mm) long. It lives all round Britain in crevices and under stones on and offshore.

Amphipholis squamata

A greyish-white or bluish-grey brittlestar with a disc only ⅕ in. (5 mm) across and arms up to ¾ in. (20 mm) long. It is common in rock pools, and also lives in sand and gravel and under stones on all British coasts.

Featherstar
Antedon bifida

The ten mobile, feathery arms projecting from the cup-shaped body are 2¾–6 in. (7–15 cm) long, usually red-brown and white-banded, but individuals vary in colour. The featherstar can use its arms to swim. It is common on rocky shores and offshore along all coasts except the east coast's southern end.

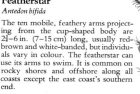

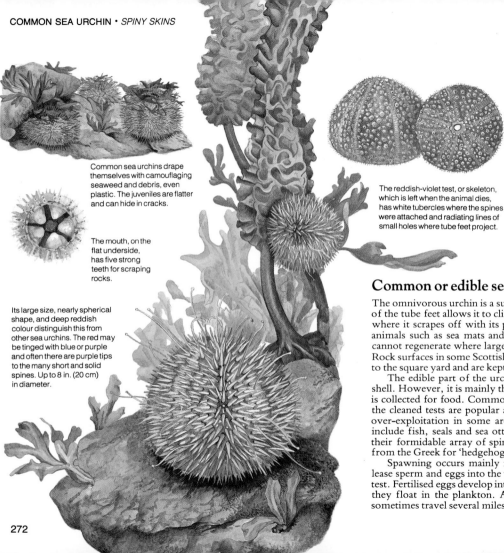

Common sea urchins drape themselves with camouflaging seaweed and debris, even plastic. The juveniles are flatter and can hide in cracks.

The mouth, on the flat underside, has five strong teeth for scraping rocks.

Its large size, nearly spherical shape, and deep reddish colour distinguish this from other sea urchins. The red may be tinged with blue or purple and often there are purple tips to the many short and solid spines. Up to 8 in. (20 cm) in diameter.

The reddish-violet test, or skeleton, which is left when the animal dies, has white tubercles where the spines were attached and radiating lines of small holes where tube feet project.

Common sea urchins live mostly below low-water mark on rocky coasts all round Britain, especially in Scotland.

Common or edible sea urchin *Echinus esculentus*

The omnivorous urchin is a supreme grazer. The strong suction of the tube feet allows it to climb vertical rocks and kelp stems, where it scrapes off with its powerful teeth seaweeds and tiny animals such as sea mats and sea firs. The plants and animals cannot regenerate where large numbers of urchins are feeding. Rock surfaces in some Scottish waters may have several urchins to the square yard and are kept bare.

The edible part of the urchin is the gonad or roe inside the shell. However, it is mainly the smaller *Paracentrotus lividus* that is collected for food. Common sea urchins face another threat; the cleaned tests are popular as souvenirs and there is a risk of over-exploitation in some areas. Predators of young urchins include fish, seals and sea otters. Older ones are protected by their formidable array of spines. The scientific name *Echinus*, from the Greek for 'hedgehog', is apt.

Spawning occurs mainly in spring. Males and females release sperm and eggs into the water through special holes in the test. Fertilised eggs develop into minute larvae which disperse as they float in the plankton. Adults are usually sedentary, but sometimes travel several miles. They live for about ten years.

Three species of sea urchin

The test varies in colour to match the spines. It is up to 2⅜ in. (60 mm) across.

Paracentrotus lividus

Sharp spines, 1¼ in. (30 mm) long, vary from dark green through purple to dark brown on different individuals. This gregarious, edible urchin lives in rocky shore pools and sea to 100 ft (30 m) deep. It bores into rocks to hide. It is found only on Ireland's west coast, rarely in Devon, Cornwall, and the Hebrides.

Strongylocentrotus drobachiensis

The flattened globular form is covered with ¾ in. (20 mm) spines, usually bright green but occasionally violet-red tipped with white. This urchin is found only on Britain's east coast and in the Shetlands and Orkneys. It lives among rocks and seaweed on the lower shore and in the sea down to about 3,930 ft (1,200 m).

The test is a pronounced green and up to 3¼ in. (80 mm) across.

The cleaned test is greenish, but not as bright as that of *Strongylocentrotus*.

Green sea urchin
Psammechinus miliaris

Strong, short, purple-tipped green spines cover the flattened globe shape, 2 in. (50 mm) across. This urchin is widespread round British coasts, clinging under stones and rock overhangs and to kelp holdfasts; it lives on the lower shore and in the sea to a depth of about 330 ft (100 m).

273

Urchins that come to the surface in the daytime are usually eaten by seabirds.

Only the fragile, off-white test, or skeleton, remains after the animal dies. Five petal-shaped rows of holes on the top show where tube feet extended. Broken tests are common on the shore.

The toothless mouth is on the underside. Tube feet pass food into it.

Heart urchins live buried in sand; only the tests are normally seen. They are found low on the shore and offshore.

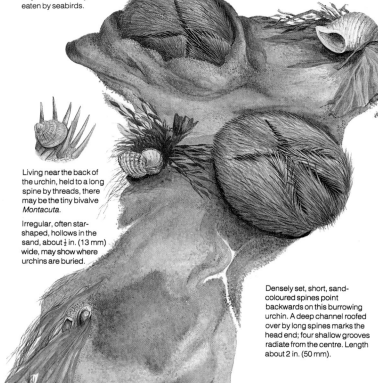

Living near the back of the urchin, held to a long spine by threads, there may be the tiny bivalve *Montacuta*.

Irregular, often star-shaped, hollows in the sand, about ½ in. (13 mm) wide, may show where urchins are buried.

Densely set, short, sand-coloured spines point backwards on this burrowing urchin. A deep channel roofed over by long spines marks the head end; four shallow grooves radiate from the centre. Length about 2 in. (50 mm).

Heart urchin or sea potato *Echinocardium cordatum*

The shape of the test gives this urchin its common name. The fragile test of a dead urchin is all that will ever be seen of the animal without digging near the low-water mark. Heart urchins have a definite head end. The spines serve not for defence but for digging, and those on the underside behind the mouth are broad and spade-like for this purpose. Once an urchin has burrowed down vertically, it ploughs slowly forwards, eating small particles of plant and animal matter which are passed to the simple mouth by specialised tube feet. The buried urchin keeps contact with the surface through a shaft lined with mucus and maintained by especially long tube feet. It draws in clean water and breathes through the shaft. It is a very clean animal and other long tube feet maintain a horizontal shaft into which waste products are passed. None of its tube feet are used for movement as they are in regular urchins.

Heart urchins live gregariously – which increases their reproductive success since like other urchins the two sexes shed their eggs and sperm freely into the water, about midsummer. The larvae float in the plankton before settling on the sea-bed and changing into young urchins.

Identifying other heart urchins

Large tubercles on the grey-white test show where the long spines were connected.

Purple heart urchin
Spatangus purpureus

A large urchin, up to 4¾ in. (12 cm) long, covered in distinctive violet short spines, with longer pale spines scattered over the back. It is common in coarse shell sand and gravel on coasts all round Britain.

Lyre urchin
Brissopsis lyrifera

Short, brownish-red spines give a dense, fur-like cover to the lyre urchin. It is not found on the south coast but elsewhere occurs in large numbers on muddy shores. It grows to 2¾ in. (70 mm) long.

On the underside of the yellow-grey test is a characteristic pattern of plates.

Pea urchin or cake urchin
Echinocyamus pusillus

An oval urchin only ⅗ in. (15 mm) long. Its short, close-set spines are greyish or bright green. It is found all round Britain on coarse sand or gravel shores and in deeper water.

Absence of notches or grooves distinguishes the grey-white test from those of young heart urchins.

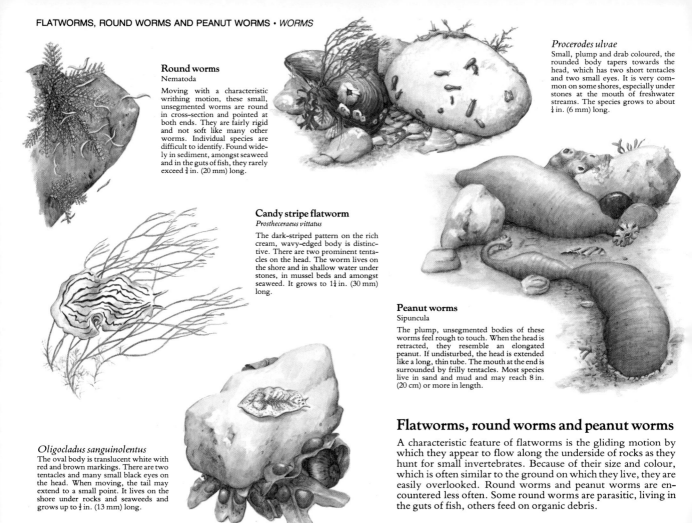

Round worms
Nematoda

Moving with a characteristic writhing motion, these small, unsegmented worms are round in cross-section and pointed at both ends. They are fairly rigid and not soft like many other worms. Individual species are difficult to identify. Found widely in sediment, amongst seaweed and in the guts of fish, they rarely exceed ¾ in. (20 mm) long.

Procerodes ulvae

Small, plump and drab coloured, the rounded body tapers towards the head, which has two short tentacles and two small eyes. It is very common on some shores, especially under stones at the mouth of freshwater streams. The species grows to about ¼ in. (6 mm) long.

Candy stripe flatworm
Prostheceraeus vittatus

The dark-striped pattern on the rich cream, wavy-edged body is distinctive. There are two prominent tentacles on the head. The worm lives on the shore and in shallow water under stones, in mussel beds and amongst seaweed. It grows to 1¼ in. (30 mm) long.

Peanut worms
Sipuncula

The plump, unsegmented bodies of these worms feel rough to touch. When the head is retracted, they resemble an elongated peanut. If undisturbed, the head is extended like a long, thin tube. The mouth at the end is surrounded by frilly tentacles. Most species live in sand and mud and may reach 8 in. (20 cm) or more in length.

Oligocladus sanguinolentus

The oval body is translucent white with red and brown markings. There are two tentacles and many small black eyes on the head. When moving, the tail may extend to a small point. It lives on the shore under rocks and seaweeds and grows up to ½ in. (13 mm) long.

Flatworms, round worms and peanut worms

A characteristic feature of flatworms is the gliding motion by which they appear to flow along the underside of rocks as they hunt for small invertebrates. Because of their size and colour, which is often similar to the ground on which they live, they are easily overlooked. Round worms and peanut worms are encountered less often. Some round worms are parasitic, living in the guts of fish, others feed on organic debris.

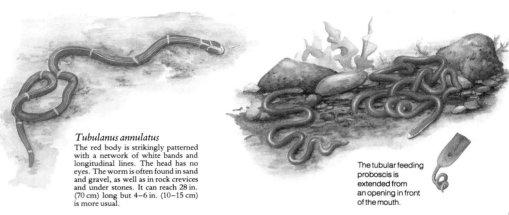

Tubulanus annulatus

The red body is strikingly patterned with a network of white bands and longitudinal lines. The head has no eyes. The worm is often found in sand and gravel, as well as in rock crevices and under stones. It can reach 28 in. (70 cm) long but 4–6 in. (10–15 cm) is more usual.

The tubular feeding proboscis is extended from an opening in front of the mouth.

Red ribbon worm
Lineus ruber

The commonest shore species, this small red-brown worm has two to eight eyes in a row on either side of the head. The species is found under stones and boulders on muddy sand. As it tolerates low salinities, it is common in estuaries. It reaches only 6 in. (15 cm) in length.

Lineus bilineatus

A close-set, double white line runs the length of the reddish-brown or chocolate body. The broad, eyeless head bears two deep slits along the edges. The worm is often found in kelp holdfasts and among coralline seaweeds in rock pools, as well as under stones. It normally grows between 3–6 in. (7·5–15 cm) long but large ones may be 18 in. (45 cm) in length.

Bootlace worm
Lineus longissimus

This aptly named worm can reach enormous lengths. Blackish-brown and slimy, it is usually found in a tangled mass under stones, especially on mud. The head has numerous eyes on either side but a hand lens may be needed to see them. Worms up to 100 ft (over 30 m) long have been recorded and 13 ft (4 m) is common.

Four common ribbon worms

As the name suggests, these worms are long, thin and unsegmented. Their smooth, shiny bodies extend and contract enormously. Examination of the head and eyes with a hand lens will help to identify different species. Ribbon worms are found under stones, among seaweed and in muds and sands. They are carnivores and scavengers, and feed on a variety of invertebrates and small fish.

277

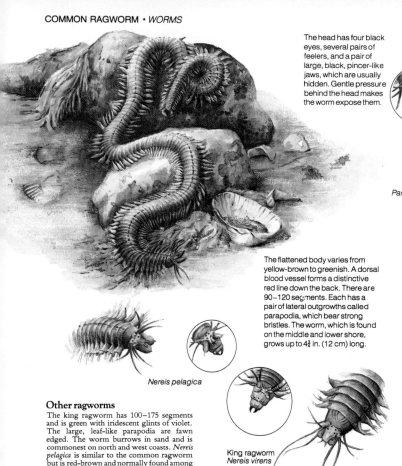

The head has four black eyes, several pairs of feelers, and a pair of large, black, pincer-like jaws, which are usually hidden. Gentle pressure behind the head makes the worm expose them.

Parapodium

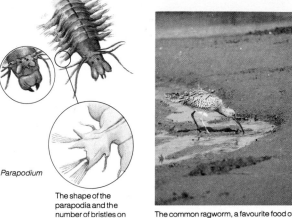

The shape of the parapodia and the number of bristles on each one help to identify the species.

The common ragworm, a favourite food of long-billed wading birds, is often found in the mud of estuaries.

The flattened body varies from yellow-brown to greenish. A dorsal blood vessel forms a distinctive red line down the back. There are 90–120 segments. Each has a pair of lateral outgrowths called parapodia, which bear strong bristles. The worm, which is found on the middle and lower shore, grows up to 4¾ in. (12 cm) long.

Nereis pelagica

King ragworm
Nereis virens

Other ragworms

The king ragworm has 100–175 segments and is green with iridescent glints of violet. The large, leaf-like parapodia are fawn edged. The worm burrows in sand and is commonest on north and west coasts. *Nereis pelagica* is similar to the common ragworm but is red-brown and normally found among rocks, shells and seaweed. The structures on the head are also arranged differently. King ragworm about 8 in. (20 cm) long but may reach 20 in. (50 cm). *N. pelagica* 4½ in. (12 cm) long.

Common ragworm *Nereis diversicolor*

Many of the marine worms of muddy and sandy shores stay in permanent burrows or tubes. But the ragworm is a free-living species, able to swim or crawl over and through the sediment with the aid of numerous leg-like outgrowths called parapodia. Expanded by fluid pressure from the body cavity, each parapodium is equipped with bristles which grip the sediment. Each one also acts as a gill, absorbing oxygen from the water.

An active hunter and scavenger, the ragworm locates small animals, such as crustaceans, with its eyes and antennae and seizes them with strong jaws which are shot out on the end of a muscular proboscis. Large specimens are powerful enough to bite humans.

Ragworms spawn in unison when the water temperature reaches at least 41°F (5°C). The parapodia and bristles of the tail end enlarge and flatten, and the worms emerge from the mud to swim freely in the water. In due course the tail end of each worm breaks open to release eggs and sperm which mix in the water. The fertilised eggs develop into minute larvae; these do not float in the plankton but live near the bottom, and in this way a large and separate population of adults builds up in one area.

Errant worms that hunt and scavenge

A large number of active, or errant, worms can be found beneath rocks and in sediment where they prey on small animals and scavenge on carrion. Most need a very careful examination using a hand lens before they can be accurately identified. Those on this page are common representatives of the major types.

Scale worm
Harmothoë impar

The back is covered by 15 pairs of overlapping, kidney-shaped scales which fall off easily if the animal is handled. All are fringed with hairs. Scale worms, which are difficult to identify, are found under stones and seaweeds on the lower shore. This species grows to about 1¼ in. (30 mm) long.

Catworm
Nephtys hombergi

The pinkish, flattened body has an iridescent pearly sheen. The tail ends in a single thread and the bristles on the parapodia are dense and short. The worm lives buried in sand or muddy gravel; when exposed, a large proboscis, or elongated mouth, armed with jaws is thrust repeatedly in and out. Grows up to 4 in. (10 cm) long.

Green leaf worm
Eulalia viridis

Living in rock crevices in shallow water, this bright green member of the paddle worm family can be found crawling over rocks when the tide is out. Its name derives from the large, leaf-like false legs (parapodia) which can be used for swimming. The small, rounded head has two eyes. In spring, green gelatinous egg sacs are attached to damp rocks and seaweed. Grows to about 4–6 in. (10–15 cm) long.

Sea mouse
Aphrodite aculeata

The back of the broad, bulky body is concealed by a dense felt of fine grey hairs so that the body segments can only be seen on the underside. Coarse iridescent green and golden hairs and brown bristles run along the sides. This species is found in and on sediment near low-tide mark and below, and is washed up after storms. Grows about 4 in. (10 cm) long.

Marphysa sanguinea

The fat, flattened body has about 300 segments. Bunches of red gill filaments arise from each segment after about the fourth, though those near the head are small. The tail ends in two threads. It lives in rock crevices on the lower shore and below. Up to 12 in. (30 cm) long.

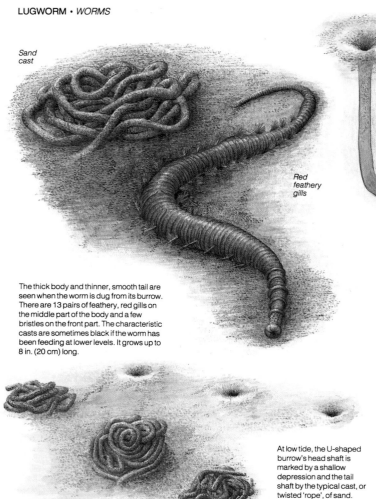

Sand cast

Red feathery gills

The thick body and thinner, smooth tail are seen when the worm is dug from its burrow. There are 13 pairs of feathery, red gills on the middle part of the body and a few bristles on the front part. The characteristic casts are sometimes black if the worm has been feeding at lower levels. It grows up to 8 in. (20 cm) long.

At low tide, the U-shaped burrow's head shaft is marked by a shallow depression and the tail shaft by the typical cast, or twisted 'rope', of sand.

The worm lines the walls of the U-shaped burrow with mucus to keep them intact.

The familiar casts are a common sight on most sandy beaches from the middle shore downwards to the sea.

Lugworm *Arenicola marina*

The distinctive casts of the lugworm are as familiar on sandy beaches as earthworm casts on the garden lawn. Both animals feed in the same way, by swallowing the sand or earth, digesting the food material which it contains and ejecting the rest onto the surface.

The higher the organic content of the sand, the more lugworms, and muddy beaches may support populations of over 13,000 per acre. Each animal lives in the bottom of a U-shaped burrow, eating the sand which is washed into the head shaft and periodically moving backwards to void the indigestible material at the top of the tail shaft. A current of water created by the lugworm keeps the sand in the head shaft loose, and ensures a supply of fresh water containing oxygen for respiration. Each new tide washes away the old casts and brings in fresh sand, food and water, and the lugworm can live for weeks in the same burrow, provided it is not dug up for use as bait by a fisherman.

Lugworm populations spawn simultaneously into the sea over a period of about two weeks – towards the end of October off Britain's south coast. The fertilised eggs hatch into larvae which remain near the sea-bed, where they develop into adults.

Burrowing and sediment-living tube worms

Several kinds of worms construct protective tubes in which they live buried or semi-buried in sands or muds. The tubes are made from mud or sand held together by mucus, from limy material or from slime secreted by their bodies.

Amphitrite johnstoni

The worm lives in a hole lined with slime in muddy sand with only the long, sticky, orange or pink tentacles visible, searching the surface for food. The fat, soft, buff or brown-buff body has dark red gills and 90–100 segments – 24 of them with bristles. It grows up to 10 in. (25 cm) long.

Parchment worm

Chaetopterus variopedatus

The worm, which is found in muddy sand on the lower shore and below, lives hidden within a tough, parchment-like tube. The whitish ends of the tube project above the surface and often have seaweeds attached. The tube is up to 16 in. (40 cm) long.

Myxicola infundibulum

Found in mud or muddy sand, the worm lies buried in a delicate tube of mucus. All that is usually visible is the characteristic crown of tentacles which unite to form a cup with dark points round the lip. The tube is up to 6 in. (15 cm) long.

Pectinaria koreni

The delicate, cornet-shaped tube is made from sand grains stuck together with mucus. The worm lives upside down with its head buried in the sand. The tubes, usually seen lying empty on the beach, may be up to 2 in. (50 mm) long.

Sand mason worm

Lanice conchilega

The tube is made of sand grains and shell debris. Its much-branched end supports the fine, sticky, feeding tentacles of the worm which has a fat, pink, red or brown body (right). Three pairs of red tufted gills lie below the tentacles. The tube may be up to 10 in. (25 cm) long but only a small part protrudes from the sand.

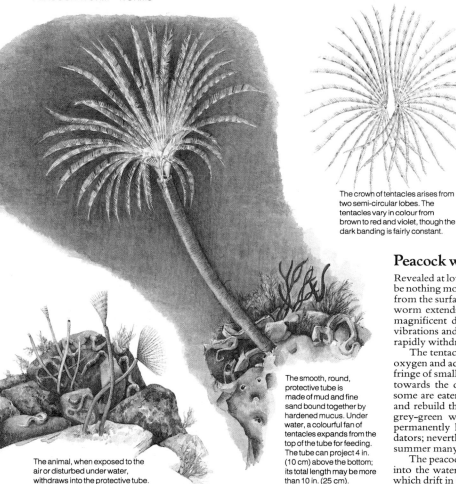

The crown of tentacles arises from two semi-circular lobes. The tentacles vary in colour from brown to red and violet, though the dark banding is fairly constant.

The colourful worm builds its tube in shallow water on fine sand, mud or, in deeper water, on silt-covered rocks.

The smooth, round, protective tube is made of mud and fine sand bound together by hardened mucus. Under water, a colourful fan of tentacles expands from the top of the tube for feeding. The tube can project 4 in. (10 cm) above the bottom; its total length may be more than 10 in. (25 cm).

The animal, when exposed to the air or disturbed under water, withdraws into the protective tube.

Peacock worm *Sabella pavonina*

Revealed at low tide, a community of peacock worms appears to be nothing more than a forest of mud-coloured tubes protruding from the surface of the beach. When submerged, however, each worm extends a fan of multi-coloured feathery tentacles in a magnificent display. The tentacles are extremely sensitive to vibrations and changes in light level; they are immediately and rapidly withdrawn if danger threatens.

The tentacles are in fact highly modified gills, which absorb oxygen and act as a filter to trap waterborne particles. Each has a fringe of small beating hairs called cilia, which drive the particles towards the centre of the fan. There the particles are sorted; some are eaten, some are rejected, and some are used to repair and rebuild the tube. The segmented body of the worm, pale grey-green with orange and violet tinting near the tail, is permanently hidden by the tube which protects it from predators; nevertheless, when the worms are exposed by the tide in summer many may be killed by high temperatures.

The peacock worm breeds by shedding both sperm and eggs into the water. After fertilisation the eggs develop into larvae which drift in the plankton before their sedentary stage.

Other tube worms attached to rocks

Many sedentary worms live in hard, limy, protective tubes firmly attached to rocks and seaweeds. Those shown below can all be found on the shore but the feeding tentacles emerge only when the worm is under water.

Coiled tube worm
Spirorbis borealis

Large numbers often occur on wracks and kelps, as well as on rocks. The spiral tube is twisted clockwise from the middle outwards. A hand lens is needed to see the circle of green feeding tentacles under water, as the tube is less than $\frac{1}{5}$ in. (5 mm) in diameter.

Honeycomb worm
Sabellaria alveolata

This worm builds its tube from sand grains. It lives in colonies which form large honeycomb-like structures, about 40 in. (100 cm) across. They are found attached to boulders and rocks on beaches where there is sand present. Each worm is about 1½ in. (40 mm) long.

Bispira volutacornis

The animal is normally seen as a crown of tentacles, arranged in two spirals, emerging from a rock crevice. The tube, made of mud and mucus, has a grey-coloured opening and projects for a short distance; it may only be seen when the animal is disturbed and withdraws rapidly. The tube may be up to 8 in. (20 cm) long.

Eupolymnia nebulosa

Found under stones – especially those on muddy gravel – the worm lives in a tube of sticky slime with bits of shell and sand attached. The fat, soft body tapers at the rear and is a characteristic bright orange with white dots. The tentacles are fine and sticky, and writhe around. It may grow up to 12 in. (30 cm) long.

Keel worm
Pomatoceros triqueter

The irregularly bent, tapering tubes occur in large numbers on rocks, shells and even the backs of crabs. Each tube, triangular in cross-section, has a keel along the back in live specimens, ending in a sharp point at the front end. The crown of feeding tentacles is banded and the tube 'stopper' is trumpet-shaped. It grows about 1½ in. (30 mm) long but may sometimes be larger.

Filograna implexa

Colonies usually grow in lumps of very fine, white, irregularly twisted, limy tubes. The mouth of each tube is sometimes slightly bell-shaped. It is a common 'fouling' animal, attaching itself to rocks, large seaweeds and marine structures. Colonies grow to at least 10 in. (25 cm) high.

283

The mud-flat – home of a vast community

The featureless and apparently desolate landscape of a mud-flat may seem to be a dull and barren habitat, yet beneath its glistening surface lies a wealth of animal life. Mud-flats are found in areas sheltered from the stirring actions of waves and tidal streams, where fine particles of sediment can settle as the tide retreats – areas such as estuaries, Scottish sea lochs, sheltered bays, creeks and harbours. Even in industrialised localities such as Southampton Water small creatures live and thrive in the mud.

What lives on and in the mud depends on how fine the sediment is, and whether there are pebbles and shells on the surface to provide a foothold for small animals and an attachment for plants. As the mud becomes finer towards the head of an estuary the variety of creatures living there decreases. Animals living in fine mud have specially adapted feeding and breathing mechanisms that would otherwise become clogged by the fine particles. But although there may be only a few species present, they exist in huge numbers. There may, for instance, be more than 1,000 ragworms to a square yard.

The daisy anemone lives in the mud, its column firmly attached to a buried stone or shellfish. The tentacles remain expanded even when the tide is out. Much finer, very extensive red or white tentacles belong to soft-bodied worms such as *Amphitrite* which occupy tunnels in the sediment and use the sticky threads to gather food from the mud surface.

Pebbles, cobbles and shell fragments provide a foothold for many organisms such as sea squirts, small sponges, brown wracks and other seaweeds which are not adapted to living on the mud itself. Green seaweeds such as *Ulva* (sea lettuce) and *Enteromorpha* grow directly on the mud surface or on tiny shell fragments and may form extensive mats in summer.

Muddy tubes protruding from the surface at low tide belong to suspension feeders such as the peacock worm, which sifts small organisms from the water using a pair of feathery tentacle fans.

Beneath the surface the mud is often densely colonised by ragworms and catworms which emerge to feed when the tide is in, and the cast-forming lugworm which lives by eating the sediment in which it has its U-shaped burrow. Two-shelled molluscs such as carpet shells and peppery furrow shells are common.

Edible winkles and shore crabs flourish in a wide variety of habitats and are frequently found crawling over the surface of mud-flats.

A few stones may provide the foundation for a colony of slipper limpets, which cling to each other's shells to form chains on the mud surface. The smaller shells at the end of the chain are males, which change into females as they grow.

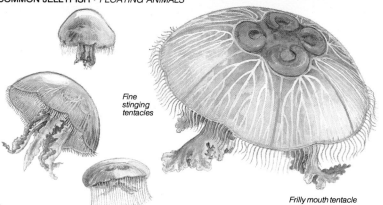

Fine stinging tentacles

Frilly mouth tentacle

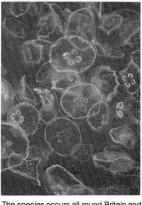

The four pale violet, circular or horseshoe-shaped reproductive organs grouped at the centre of the transparent bell are clearly visible from above.

The species occurs all round Britain and often enters docks, estuaries and bays in large swarms.

The transparent umbrella-shaped bell, tinged with blue and white, is fringed with short, fine stinging tentacles which are rarely painful to humans. Four much longer, frilly mouth tentacles hang down below the bell, which may be up to 12 in. (30 cm) in diameter.

The young jellyfish, called an ephyra, has eight divided arms and is almost transparent. It is about ⅛ in. (3 mm) in diameter.

The young develop from a fixed stage called a scyphistoma, about ⅜ in. (10 mm) high, which is attached to rocky overhangs or pier pilings. It grows into a stack of star-shaped larval jellyfish which break off one at a time and float away.

Common jellyfish *Aurelia aurita*

A jellyfish found cast up on the beach is more likely than not to be *Aurelia aurita*. It has an umbrella-shaped body, or bell, consisting mostly of jelly, sandwiched between two layers of cells; the jelly acts as a primitive skeleton. The bell can contract to expel water and this propels the animal along slowly. Normally, however, the jellyfish drifts with the current, feeding on small fish and swimming crustaceans which brush against the numerous stinging tentacles. These stun the prey, which is pushed into the central mouth by the mouth tentacles.

The sting is rarely powerful enough to be painful to a human, and some fish, such as the young of whiting, haddock and horse mackerel, appear to be quite immune to it and to the stings of other jellyfish. They shelter from predators among the tentacles, and may benefit their host by luring in other fish.

The common jellyfish has a complex life-cycle. The adult male releases sperm into the water to be drawn into the female. The fertilised egg then develops into a minute, hairy swimming larva, or planula. The larva fixes itself to a rock, where it changes into an anemone-like form called a scyphistoma. Immature jellyfish bud off from this and grow into adults.

Recognising other jellyfish

The soft, umbrella-shaped bodies of jellyfish have a central mouth on the underside, bordered by extended lips called mouth tentacles. Many species are equipped with stinging tentacles which trap prey and protect the jellyfish against predators. The jellyfish shown here are common all round the coasts of Britain.

Lion's mane jellyfish
Cyanea capillata

Numerous fine tentacles, which can inflict a painful sting, hang from the edge of the browny-red bell and may trail for several yards. There are four frilly mouth tentacles. The bell can reach 40 in. (100 cm) across.

Compass jellyfish
Chrysaora hysoscella

Dark lines radiate from the centre of the bell which has a border of 32 dark brown lobes and a fringe of 24 fine tentacles arranged in threes. Four frilly mouth tentacles hang from the centre. The species is harmless and may grow to 12 in. (30 cm) across.

Pelagia noctiluca

The mushroom-shaped, translucent bell is slightly warty, frilled at the edge, and speckled with purple and red-brown. There are eight slender tentacles hanging from the edge and four large mouth tentacles at the centre. This stinging species is strongly luminescent when disturbed at night. It reaches only about 4 in. (10 cm) across.

Rhizostoma pulmo

The tall, domed, yellowish-white bell has a purple-tinted, lobed edge. Eight heavy mouth tentacles, joined together for part of their length, hang from the centre. The bell of this harmless species may be up to 36 in. (90 cm) in diameter.

Cyanea lamarki

Similar to the lion's mane, this smaller, purple-blue species has a less virulent sting. It rarely exceeds 12 in. (30 cm) across.

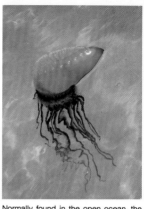

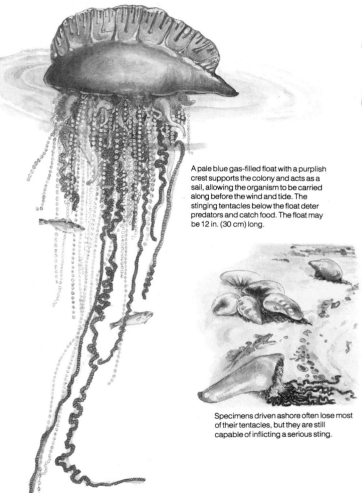

A pale blue gas-filled float with a purplish crest supports the colony and acts as a sail, allowing the organism to be carried along before the wind and tide. The stinging tentacles below the float deter predators and catch food. The float may be 12 in. (30 cm) long.

The mass of tentacles which may trail below the float for several yards consists of three individual types of polyp, differently shaped according to their function.

Normally found in the open ocean, the man-of-war is sometimes driven ashore by prolonged south-west winds.

Specimens driven ashore often lose most of their tentacles, but they are still capable of inflicting a serious sting.

Portuguese man-of-war *Physalia physalis*

Although the Portuguese man-of-war looks like a jellyfish, it belongs to a very different group called the siphonophores – complicated colonial animals made up of many individuals, or polyps, modified to carry out different functions. Unlike the true jellyfish it cannot swim but drifts with the wind and current, supported by a single polyp inflated by gas.

The colony is well protected by the stinging polyps, each of which has a single long tentacle which trails beneath the float. These tentacles are so long that a bather may be stung without ever seeing the animal, and the sting is powerful enough to cause severe injury if the victim becomes entangled. Prey such as small fish and shrimps are trapped and paralysed by the stinging polyps, gathered up and passed to the feeding polyps for digestion. Interconnections between the feeding polyps and other parts of the colony enable nourishment to reach every part of the organism.

Branched reproductive polyps produce very small jellyfish-like medusae which release eggs and sperm into the water. The fertilised eggs develop into larvae which split into the individual polyps of the adult colony.

Wind and current-borne drifters

The North Atlantic Drift – part of the warm Gulf Stream – carries many floating animals from warmer waters to Britain's shores. They are often washed up on south-west coasts and some reach the west coast of Scotland. A few of the commonest are shown here. All, except the violet sea snail, are colonial and are made up of many small individuals.

The sail, seen from above, runs obliquely across the oval body.

Violet sea snail

By-the-wind-sailor
Velella velella

A flat, bluish, oval float with an erect triangular sail, allows the animal to drift with the wind and tide. It traps food with a fringe of stinging tentacles and transfers it to the digestive bodies which hang from the centre of the float. These in turn pass nourishment to the other parts of the colony. Up to 4 in. (10 cm) across.

Stranded specimens resemble blue plastic discs.

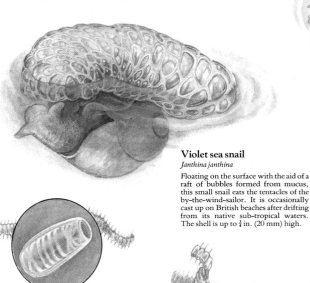

Violet sea snail
Janthina janthina

Floating on the surface with the aid of a raft of bubbles formed from mucus, this small snail eats the tentacles of the by-the-wind-sailor. It is occasionally cast up on British beaches after drifting from its native sub-tropical waters. The shell is up to ¾ in. (20 mm) high.

Salp
Salpa fusiformis

Closely related to sea squirts, the individuals which make up the salp colony have barrel-shaped, transparent bodies with clearly visible internal organs. Although capable of a solitary existence, they normally form floating chains up to 40 in. (100 cm) long.

Siphonophore
Apolemia uvaria

Comprising a long string of individuals, called polyps, trailing from a small float, *Apolemia* is an elongated version of the Portuguese man-of-war. The colony, which resembles a line of washing, may be 65 ft (20 m) long.

289

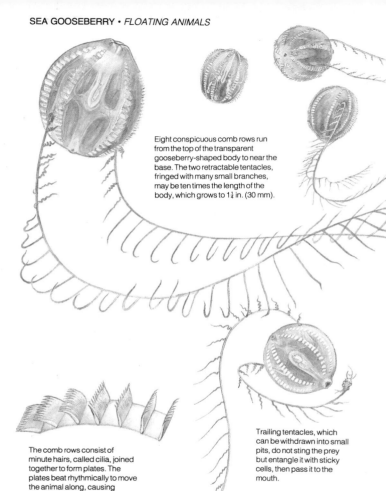

Eight conspicuous comb rows run from the top of the transparent gooseberry-shaped body to near the base. The two retractable tentacles, fringed with many small branches, may be ten times the length of the body, which grows to 1¼ in. (30 mm).

The sea gooseberry is common in open water all round Britain. Occasionally, it is found in rock pools.

The comb rows consist of minute hairs, called cilia, joined together to form plates. The plates beat rhythmically to move the animal along, causing ripples of iridescence along the body.

Trailing tentacles, which can be withdrawn into small pits, do not sting the prey but entangle it with sticky cells, then pass it to the mouth.

Sea gooseberry *Pleurobrachia pileus*

The delicate sea gooseberry resembles a miniature rounded jellyfish, but it belongs to a separate group called the ctenophores or comb jellies. The name refers to the comb rows which run the length of the body, each composed of plates resembling small combs with the teeth joined together. The transparent body consists of jelly sandwiched between two layers of cells.

A voracious predator, the sea gooseberry feeds on small floating animals such as copepods, shrimps and young fish. The animal is moved mouth first through the water by the beating of the comb rows, while the two long tentacles are trailed behind like fishing lines. When anything touches the wide net of side branches it is caught by special cells which secrete a sticky material, trapping the prey rather like a spider's web. The tentacles are then hauled back in and the prey is wiped off into the mouth.

The sea gooseberry is a hermaphrodite, and in late summer and autumn it sheds both eggs and sperm into the sea. The fertilised eggs hatch into floating larvae which develop gradually into the adult form.

Small animals in the plankton

Many different kinds of small animals live a permanent floating existence in the plankton. Their small size and transparent bodies make them difficult to see but some can be found in rock pools or caught in fine nets.

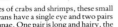

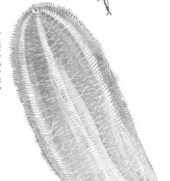

Comb jelly
Beroë cucumis

Eight comb-rows run the length of the pink-tinged body, through which shows a network of branched, inner canals. There are no trailing tentacles. Instead, a very wide mouth enables the comb jelly to engulf prey almost as big as itself. It may grow to 5 in. (12·5 cm) long.

Copepods
Copepoda

Relatives of crabs and shrimps, these small crustaceans have a single eye and two pairs of antennae. One pair is long and hairy, the other is short. The segmented, oval body has six pairs of limbs. Females often carry egg sacs. About ¹⁄₂₅ in. (1 mm) long.

Arrow worms
Sagitta species

When feeding, these transparent torpedo-shaped animals seize their prey using large, curved spines attached to the head. They live in the upper layers of the sea and grow to 1 in. (25 mm) long.

Comb jelly
Bolinopsis infundibulum

Almost transparent, this delicate, oval species has two rounded lobes on either side of the mouth; they extend at least half the length of the body. The two short, trailing tentacles have no sheaths. It reaches 6 in. (15 cm) in length.

Sea butterfly
Clione limacina

Two small flaps, or parapodia, enable this free-floating relative of the sea slug to swim. Two pairs of short tentacles on the head detect prey. Up to ¾ in. (20 mm) long.

When the sponge is living below low-tide mark – for example on kelp stems in areas of fast current, a common habitat – it is usually yellow and varies widely in shape and size.

Oscula are about ⅜ in. (10 mm) high. Through them the sponge discharges water from which food has been filtered.

Green breadcrumb sponge is found encrusting rocks on the middle and lower shore all round Britain.

Patches of this green sponge may be 8 in. (20 cm) across and ¾ in. (20 mm) thick. The smooth surface is broken fairly regularly by volcano-shaped openings called oscula. Patches shaded from the light are yellow, not green.

Breadcrumb sponge *Halichondria panicea*

Many people, familiar only with bath sponges, are astonished at the variety in colour, shape and size of British sponges. Although some grow quite large, they are useless commercially because the tiny spicules that make up the skeleton are of sharp, brittle silica or lime. Mediterranean bath sponges, however, have flexible horny skeletons that stay intact after the animals die. It was not until 1825 that sponges were known for certain to be animals. Spicules, which vary in shape from simple spears to complex stars, identify sponge species even when other features, such as overall shape, alter in different habitats. The breadcrumb sponge's spicules are of one main type called oxea – fat rods pointed at both ends.

The way it crumbles when handled has given the breadcrumb sponge its name. Its unusual colour normally makes it easy to recognise; it is due to algae living in the sponge. Patches in deep shade have no algae and are yellow. The sponge has a distinctive pungent smell. Small worms and crustaceans live in its oscula. Broken fragments can grow into new sponges but spawning also occurs. Fertilised eggs develop into larvae which swim in the plankton; this disperses the species.

Six common sponges

Many sponges vary their shape, size and colour under different conditions, and this makes them difficult to identify. A hand lens is essential for examining the smaller species. Some can be identified exactly only under greater magnification which reveals the shape of the spicules that make up the skeleton. The species illustrated are among the commonest.

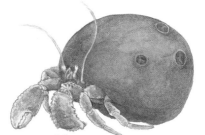

Leucosolenia species

Delicate branching and connecting tubes, off-white or grey and up to 1¼ in. (30 mm) high, grow in clumps and patches in sheltered places on seaweeds, shells and rocks.

Hymeniacidon perleve

This thin, encrusting sponge varies greatly in shape and size, sometimes reaching 20 in. (50 cm) across. Its blood-red to orange surface has many small openings at irregular intervals and is broken by furrows or small lumps. It usually grows in bare rock clefts, under overhangs or among kelp.

Sea orange or sulphur sponge
Suberites domuncula

Like a lumpy, misshapen orange, this sponge often grows on hermit crabs. It also grows on rocks, where it may reach 12 in. (30 cm) in diameter, but it is usually smaller. Oscula are few and large.

Elephant's ear sponge
Pachymatisma johnstonia

The hard, greyish, lumpy growths feel smooth to the touch. The oscula, or openings, are arranged in irregular rows. This sponge prefers to live on vertical rock surfaces. It is often found in inaccessible crevices and overhangs. It may reach 6 in. (15 cm) thick and 24 in. (60 cm) across.

Purse sponge
Scypha compressa

These white, stalked, vase-shaped sponges hang from shady, damp rocks. Out of water, they look like empty hot-water bottles. About ¾–2 in. (20–50 mm) long.

Purse sponge
Scypha ciliatum

Off-white or grey vase-shaped cylinders, up to 2 in. (50 mm) high, grow horizontally or upright on rocks or seaweeds. The surface is slightly hairy and long hairs ring the osculum, or opening, at the free end.

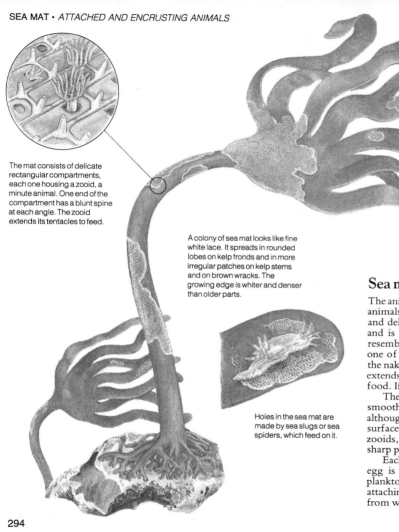

The mat consists of delicate rectangular compartments, each one housing a zooid, a minute animal. One end of the compartment has a blunt spine at each angle. The zooid extends its tentacles to feed.

A colony of sea mat looks like fine white lace. It spreads in rounded lobes on kelp fronds and in more irregular patches on kelp stems and on brown wracks. The growing edge is whiter and denser than older parts.

Holes in the sea mat are made by sea slugs or sea spiders, which feed on it.

Sea-mat colonies often encrust kelp fronds. When a frond is torn loose by winter gales, the sea mat dies.

Sea mat *Membranipora membranacea*

The animals of the large colonial group known as bryozoa (moss animals) come in many shapes, from flat sheets to hard lumps and delicate, plant-like tufts. The sea mat is part of this group and is easily distinguished from the sea firs and seaweeds it resembles because the rectangular compartments, each housing one of the zooids which make up the colony, can be seen with the naked eye. When the colony is covered with water, the zooid extends a ring of feeding tentacles to trap minute particles of food. If threatened, the zooid withdraws its tentacles quickly.

The sea mat is often found encrusting kelp fronds, the smooth surface of which is ideal for the flat, spreading colonies, although individual species of bryozoa prefer different firm surfaces. Sea slugs and sea spiders feed extensively on the zooids, the spiders puncturing the protective box with their sharp proboscis.

Each sea-mat colony contains both males and females. An egg is fertilised as it leaves the colony and develops into a planktonic larva. For several weeks it swims and feeds before attaching itself to a suitable surface. It then changes into a zooid, from which the colony develops through budding.

Six common moss animals

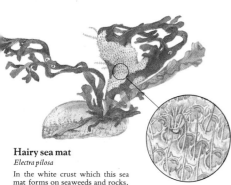

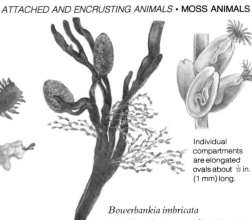

Individual compartments are elongated ovals about $\frac{1}{32}$ in. (1 mm) long.

Hairy sea mat
Electra pilosa

In the white crust which this sea mat forms on seaweeds and rocks, individual compartments are just visible. The hairy sea mat grows in more irregularly shaped patches than the common sea mat.

Each compartment housing a zooid is only $\frac{1}{64}$ in. (0·5 mm) long, is rounded and has long spines.

Alcyonidium species

Smooth, firmly gelatinous branches of variable shape, often with lobes and fingers, grow on rocks and kelp stems. Some colonies may grow to 12 in. (30 cm) but many reach only an inch or two long. Raised spots indicate individual compartments.

Bowerbankia imbricata

Branching colonies up to $2\frac{3}{4}$ in. (70 mm) long grow on seaweeds, especially brown wracks and *Corallina*. Individual compartments grow in tufts at intervals along the branches.

Hornwrack
Flustra foliacea

Colonies grow on rocks, forming bushy clumps with flexible, flat branches. The branches, which fork regularly, are at least $\frac{1}{3}$ in. (10 mm) wide and usually 4–6 in. (10–15 cm) high. Hornwrack lives in fast currents and is often seen cast up on the shore.

The compartments covering both sides of the branches are about $\frac{1}{64}$ in. (0·5 mm) long.

Bugula species

Branches are arranged spirally as on a Christmas tree. The colonies, up to 2 in. (50 mm) long, grow in small tufts, and often hang from rock shelves among seaweeds.

Ross coral or rose coral
Pentapora foliacea

The colonies form hard, brittle wavy-edged plates growing from rock faces. They break easily but those growing in sheltered deeper water can form rounded growths of 20 in. (50 cm) or more across. The oval individual compartments can just be seen with the naked eye.

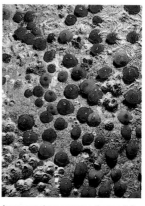

A common shore anemone, the beadlet lives in shallow waters on all coasts and is easily found at low tide.

Red beadlets are the most common and most aggressive, but there are green, brown and orange-brown varieties. The expanded column, up to 3 in. (75 mm) high, and the tentacles, up to 2¾ in. (70 mm) across, are of one colour. Blue warts encircle the tentacles, which number about 200, and a thin blue line sometimes rings the column base.

Strawberry anemone
Actinia fragacea

Small greenish-yellow dots pattern the red column of this anemone, which is so like the beadlet that it was formerly thought to be only a colour variety of it. It likes the same habitat as the beadlet, but is found only on Channel and south-western coasts of England.

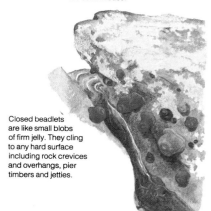

Closed beadlets are like small blobs of firm jelly. They cling to any hard surface including rock crevices and overhangs, pier timbers and jetties.

Beadlet anemone *Actinia equina*

The most numerous and widespread anemones on Britain's rocky shores are mere blobs of red or green jelly while the tide is out, but when covered by the returning sea they quickly expand their colourful tentacles to sting prey. Each sting is coiled in its cell until a hair-like trigger is touched, then a barbed thread is shot out into the victim – a shrimp or small fish. Food is pushed whole into the hollow sac-like body through the mouth, and waste is ejected through the mouth.

If two beadlets touch tentacles, one often attacks the other. The blue warts swell and the anemone stretches up, leans over, and stings its opponent's column with the warts, which contain powerful stinging cells. Larger anemones usually defeat smaller ones, and the red and brown varieties defeat the greens. The loser slowly retreats. This behaviour spaces out individuals, perhaps to avoid competition for living space and food.

Eggs are fertilised inside the female after the male has spawned into the surrounding water. Minute swimming larvae hatch from the eggs and develop into tiny anemones inside the female. Although they live where food is digested, they are rarely harmed and are finally released through the mouth.

Common anemones and cup coral

Colour, shape, and the number of tentacles differentiate anemone species. These features are seen best when the animal is covered by water, but the colour is very variable in many species. Five of the most readily identifiable are shown. They are common on the shore.

Plumose anemone
Metridium senile

A tall species, occasionally reaching 12 in. (30 cm), with lobed plumes of fine tentacles arising above a distinct collar on the column. Plain white, orange, brown and cream individuals live in rock crevices and overhangs, and on pier pilings on all coasts.

Devonshire cup coral
Caryophyllia smithii

The hard, brownish-white skeleton shows through the translucent white, orange, pink, red, brown or green body, only 1 in. (25 mm) across. A contrasting colour rings the slit mouth. The almost clear tentacles are knobbed. Cup corals live singly on shaded rocks round south-west and west coasts of Britain, and sometimes on the east coast of Scotland.

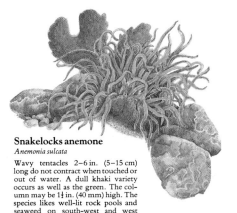

Snakelocks anemone
Anemonia sulcata

Wavy tentacles 2–6 in. (5–15 cm) long do not contract when touched or out of water. A dull khaki variety occurs as well as the green. The column may be 1½ in. (40 mm) high. The species likes well-lit rock pools and seaweed on south-west and west coasts.

Dahlia anemone
Urticina felina (syn. *Tealia felina*)

Exists in almost any colour combination. The many short, blunt tentacles are often banded and have red-lined bases. Shell and gravel adhere to the sticky warts on the column, which may be up to 5 in. (12·5 cm) wide and 2 in. (50 mm) high.

Gem anemone
Bunodactis verrucosa

Six rows of white warts show when the column, up to 1 in. (25 mm) across, is closed, but they are not sticky like the Dahlia anemone's. Stiff, banded tentacles surround the greenish mouth area. This species lives in pools and among rocks on west and south-west shores of Britain.

Jewel anemone
Corynactis viridis

Brightly coloured colonies live on vertical or overhanging rocks. The column, only ⅜ in. (10 mm) across, may be green, orange, scarlet, pink, grey or white. The short, knobbed tentacles are translucent. Jewel anemones hide from light on the lower shore and in deeper water on England's south-west and west coasts, and all round Ireland.

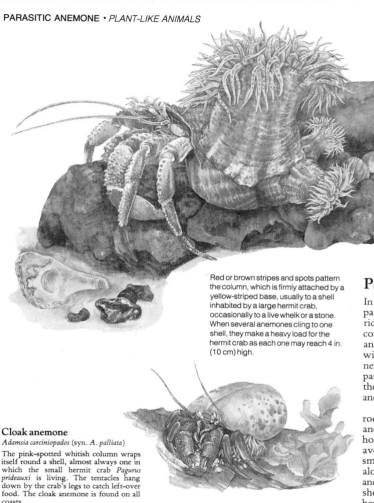

This species, restricted to the Channel and across to southern Ireland, is less widespread than its hermit crab host.

Red or brown stripes and spots pattern the column, which is firmly attached by a yellow-striped base, usually to a shell inhabited by a large hermit crab, occasionally to a live whelk or a stone. When several anemones cling to one shell, they make a heavy load for the hermit crab as each one may reach 4 in. (10 cm) high.

Cloak anemone
Adamsia carciniopados (syn. *A. palliata*)

The pink-spotted whitish column wraps itself round a shell, almost always one in which the small hermit crab *Pagurus prideauxi* is living. The tentacles hang down by the crab's legs to catch left-over food. The cloak anemone is found on all coasts.

Parasitic anemone *Calliactis parasitica*

In spite of its common name, the parasitic anemone is not a true parasite, for it does no harm to its hermit crab host. By taking a ride on the hermit's shell, the anemone assures itself of a constant food supply. Hermit crabs are messy eaters and the anemone needs only to bend over and sweep up food fragments with its tentacles, whereas other species of anemone, permanently attached to rocks, must wait for small animals to swim past or be carried within reach. The hermit crab benefits from the relationship – called a commensal association – since the anemone's stinging tentacles deter predators.

A parasitic anemone can live without a host, attached to a rock or stone, but if a suitable host comes within reach the anemone slowly transfers itself to the shell. When a hermit crab host moves to a larger shell, the anemone has to move quickly to avoid being left behind. The cloak anemone and its host, the small hermit crab, are much closer; neither can easily survive alone. The crab never has to change shells as it grows, because its anemone secretes a horny substance that effectively enlarges the shell. Thus the anemone gains a steady food supply and the hermit has a permanent, extending home.

Burrowing anemones

Anemones that live buried in sediment, often attached to a hidden stone or shell, usually have to be identified by the disc and tentacles – the only visible parts. The species shown are found on the shore as well as offshore.

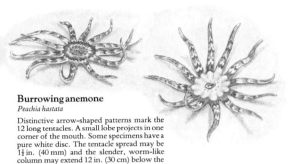

Burrowing anemone
Peachia hastata

Distinctive arrow-shaped patterns mark the 12 long tentacles. A small lobe projects in one corner of the mouth. Some specimens have a pure white disc. The tentacle spread may be 1½ in. (40 mm) and the slender, worm-like column may extend 12 in. (30 cm) below the sand or gravel. It is not attached to a stone and the species lives on all coasts.

Tube anemone
Cerianthus lloydii

Its two rings of different tentacles distinguish the species, which lives partially buried in muddy sand on all coasts. The inner tentacles are short and erect. The outer ones, about 60 in number, are slender and span about 2¾ in. (70 mm). Some specimens are white, though brown is the commonest colour. If touched, the anemone draws back into the slimy tube in which it lives.

Halcampa chrysanthellum

Horse-shoe marks in reddish-brown pattern the base of the 12 short, blunt tentacles, and there are other brown bands on them. The tentacle spread is only ⅜ in. (10 mm) and the buried column 2¾ in. (70 mm) long. The species lives on all coasts in mud, sand, gravel, and especially in beds of sea grass.

Sagartiogeton undatus

The pale, graceful tentacles are long and translucent. They number about 200 and span about 2¾ in. (70 mm). Light brown markings pattern the disc. The species lives on all British coasts and is seen with its column partly emerging from the mud, sand or gravel.

Tentacles are withdrawn only when the animal is disturbed.

Daisy anemone
Cereus pedunculatus

The numerous brown-grey tentacles are short and limp, and arranged only on the edge of the disc, which may reach a maximum diameter of 6 in. (15 cm). Various speckled patterns occur. The long column is inserted in a crevice in a rock pool, in mud or in sand. Daisy anemones are common on west and south coasts.

Obelia geniculata

Miniature forests of white threads up to 1¼ in. (30 mm) tall (left) rise from a lacy mesh of roots creeping over kelp fronds. The stems seem to zigzag because the heads are on alternate sides. The tentacled feeding heads (right) can withdraw into tiny cups.

Stalked jellyfish
Haliclystus and *Lucernariopsis*

The individuals in this group of almost identical, bell-shaped species of jellyfish never swim, but cling by a short sucker to seaweeds. The red or green bell may reach 2 in. (50 mm) long and has eight lobes, each bearing a bunch of tentacles.

Oaten pipes hydroid
Tubularia indivisa

Flask-shaped, reddish-pink heads (left) with white tentacles are held on tough, yellow-brown stems (right) up to 6 in. (15 cm) tall. The stems, rarely branched, rise from a mat of creeping roots. Large patches of this sea fir grow in areas where sea currents are strong.

Hydrachtinia echinata

The pale pink, fuzzy growth, ⅜ in. (10 mm) tall, is found only on shells in which a hermit crab is living (right). The individual polyps (left) are on white stems.

Club-headed hydroid
Clava multicornis

Pink clusters grow ⅜–¾ in. (10–20 mm) high on brown wrack seaweeds (left). The club-shaped polyps (above), rising on stems from a network of thin roots, have fine stinging tentacles to catch food.

Aglaeophenia pluma

Grey-brown, feather-shaped colonies grow on curving stems up to 1¼ in. (30 mm) tall and attached by an open root system to the pod weed, *Halidrys*. Feeding heads are on the side branches. Sometimes a cone-shaped reproductive head takes the place of a side branch.

Hydroids and stalked jellyfish

Because they bear a strong resemblance to miniature ferns and fir trees, hydroids are also known as sea firs. Anemone-like feeding heads or polyps line the interconnecting branches. Hydroids live in colonies; they are best seen in water through a magnifying glass. The species shown here, including the related stalked jellyfish, can be found on the shore attached to seaweeds and rocks.

Star sea squirt
Botryllus schlosseri

Vividly coloured star-shaped patterns of zooids each about $\frac{1}{16}$ in. (2 mm) long are embedded in a flat jelly-like layer. The patterns are often in contrasting colours. There is frequently a red spot at the tip of each zooid, especially in light varieties. This species encrusts rocks and stones, and sometimes forms thick, fleshy lobes attached to seaweeds.

Botrylloides leachi

Similar to the star sea squirt, but with petal-shaped zooids about $\frac{1}{16}$ in. (1·5 mm) long. They have no red spots and are grouped in long double lines and ovals. Colonies are usually yellow or grey, and form a fleshy layer over stones and rocks.

Morchellium argus

This species is difficult to distinguish from *Aplidium*, but the zooids are usually arranged in regular rows on the head of each club-shaped colony. Each zooid may have a red spot.

Sidnyum turbinatum

Rather like *Aplidium* except that the colonies form flat-topped egg-shaped heads $\frac{1}{3}-\frac{3}{4}$ in. (10–20 mm) tall, and are united at their bases with no true stalks. The individual zooids are roughly square and easily distinguishable.

Aplidium

A group of almost identical species that form fleshy, club-shaped colonies. They live attached to rock faces and overhangs by narrowing stalks. Individual zooids are small and indistinct within the colony, and are usually yellow or orange. Colonies grow up to $1\frac{1}{2}$ in. (40 mm) tall.

Colonial sea squirts

Large numbers of tiny individual sea squirts, or zooids, live together in colonies embedded in a firm jelly-like base. Some species, including those shown here, can be recognised by the shape of the colony or by the pattern in which the zooids are arranged. Others can be identified only by examining zooids under a microscope.

301

Mature colonies branch irregularly into sturdy, erect, blunt fingers at least 1¼ in. (30 mm) across. When the colony is feeding, translucent white polyps cover the surface. The colony may reach 10 in. (25 cm) high and spread widely.

Common on all coasts on kelp stems, rocks, piers and walls, this species thrives best in strongly moving currents.

Feeding polyps look like anemones but are only ⅛–⅜ in. (5–10 mm) long. Each has a slit mouth ringed by eight branched tentacles.

Orange colonies are common but the expanded polyps are always white. A resting colony with polyps withdrawn has a knobbly surface.

Young, developing colonies encrust stones, rocks and shells with small patches up to ⅜ in. (10 mm) thick.

Dead man's fingers *Alcyonium digitatum*

The macabre name of this soft coral is derived from the large pieces washed ashore after storms with all the feeding heads, or polyps, retracted so that they resemble bloated fingers. Soft corals are closely related to true corals and to anemones. However, soft corals do not have a hard skeleton; instead they have limy splinters, called spicules, scattered throughout the body of the colony to strengthen it.

Small colonies of dead man's fingers are sometimes found attached to rocks on the lower shore, although it more often grows totally submerged. Where strong currents flow or waves surge, more food is brought to the colonies – shipwrecks in such areas are often festooned by the orange or white fingers. The many tiny anemone–like polyps that constitute a colony are embedded in a tough, fleshy tissue and are interconnected by narrow canals. Food – planktonic animals – caught in the tentacles of each polyp can thus be shared by other polyps.

The eggs are fertilised inside each polyp by sperm drawn from the water through the mouth. The eggs develop into minute, hairy larvae that swim out of the parent polyp's mouth to find a new surface on which to settle and grow.

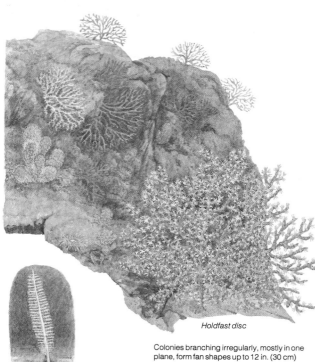

The horny skeleton left behind when the animal dies is quickly bleached from brown to white.

When the colony feeds, the branches are covered with anemone-like polyps about ⅛ in. (3 mm) high.

Sea fans grow below about 50 ft (15 m) on Britain's south and west coasts, often across currents to trap plankton.

Holdfast disc

Colonies branching irregularly, mostly in one plane, form fan shapes up to 12 in. (30 cm) high and 16 in. (40 cm) across. Most fans are pink but white ones occur. Branches are knobbly when the polyps are not feeding. A holdfast disc grips the rock.

Sea pen
Virgularia mirabilis

Shaped like a quill pen, the colony bears its feeding polyps on the side branches. Some sea pens may reach 24 in. (60 cm) high, but most are smaller. They grow on all coasts, and are found in sheltered muddy areas such as harbours.

Sea fan *Eunicella verrucosa*

Although common in tropical waters, sea fans are quite rare in Britain's cooler seas. However, off southern coasts (particularly Cornwall and Devon) and possibly west Scotland where the warm currents of the Gulf Stream flow, pink and – more rarely – white sea fans can be found. They grow very slowly – 12 in. (30 cm) in about 30 years. The intricate and beautiful skeleton – the animal is related to the soft corals – that supports the feeding heads, or polyps, remains after the colony dies.

As with dead man's fingers, each polyp has eight tentacles which are used to catch planktonic animals. When all the closely packed polyps extend their tentacles an efficient net is formed across food-bearing currents. The sea fan is therefore found where currents are strong enough to bring a constant supply of food. It cannot survive pounding waves and is not found on the shore. A small, pink sea slug, *Tritonia odhneri*, camouflaged to resemble a polyp, feeds on the sea fan.

The life-cycle of the sea fan is similar to that of dead man's fingers. Sperm from the surrounding water enters each polyp, inside which the eggs are fertilised. Later, the larvae swim out and settle on the sea-bed to start a new colony.

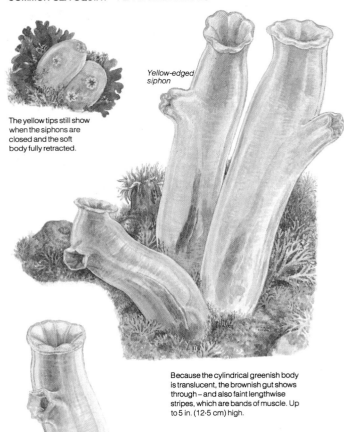

The yellow tips still show when the siphons are closed and the soft body fully retracted.

Yellow-edged siphon

Because the cylindrical greenish body is translucent, the brownish gut shows through – and also faint lengthwise stripes, which are bands of muscle. Up to 5 in. (12·5 cm) high.

There are two yellow-edged siphons – an eight-lobed intake siphon on the tip and a six-lobed outlet siphon on the side.

Some individuals have a rough, opaque covering and some are translucent orange.

Sea squirts are often numerous on sheltered rocks, harbour walls and pilings of piers and jetties all round the coast.

Common sea squirt *Ciona intestinalis*

The animal is very common and is widespread around Britain, as well as throughout the Mediterranean and off the coast of Norway. The adult is little more than a bag with a stout outer covering, or test, and with two openings, or siphons. It is just this simplicity that makes it one of the most successful of animals. It is aptly named because if it is squeezed or poked, it expels a jet of water through the siphons.

The sea squirt lives permanently attached to the sea-bed, filter feeding through its siphons. Water is drawn in through the top siphon, filtered through a fine-mesh net which occupies most of the body cavity, and is then passed out through the side siphon. Large species such as the common sea squirt can filter up to 44 gallons (200 litres) of water an hour. As it produces its own water currents it can live in very sheltered waters.

Adult squirts are hermaphrodite and release eggs and sperm, though at different times. Sperm is carried into the sea squirt on water currents and there fertilises the eggs. Minute larvae, shaped like tadpoles with primitive 'backbones', swim out through the side siphon of the parent to drift in the plankton until they settle on a new surface to develop into an adult.

Recognising sea squirts

Identifying sea squirts is often easier when they are under water. A blob of jelly found on the shore may expand into a delicate and beautiful creature once it is put in a rock pool. The species shown live attached to stones, shells, rocks and seaweeds on the lower shore and deeper, all round Britain.

Ascidia mentula

A thick, tough, dirty red to grey skin, or test, covers this species, which is one of Britain's largest. It is often attached not just at the base but at the side as well. The outlet siphon is halfway or more down the side. Both siphons have small lobes and they are often banded at the top with white marks. Grows to 6 in. (15 cm) high.

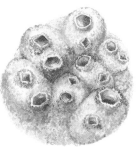

Molgula

Includes several virtually identical species, oval or globular. They live in sand and mud or on silt-covered rocks, often with only the siphons showing while sand grains or mud covers the body. They live separately but can be found in groups. About ⅛–¾ in. (5–20 mm) high.

Light-bulb sea squirt
Clavelina lepadiformis

Small individuals joined at the base in clusters. The soft, transparent body has white or yellow lines running down it and edging the siphons. Up to ¾ in. (20 mm) high.

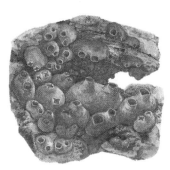

Gooseberry sea squirt
Dendrodoa grossularia

Bright or dull red individuals grow alone or in large groups. The siphons are squarish when open, and form a cross when closed. The rounded or flattened body is up to ¾ in. (20 mm) high.

Ascidiella scabra

Very similar to *Ascidiella aspersa* but with the two siphons close together and often bearing reddish marks. It grows mostly on other sea squirts and seaweeds. Up to 1¼ in. (30 mm) high.

Corella parallelogramma

Red-and-yellow flecks mark the transparent body through which the yellowy gut shows. The smooth, firm body is a flattened, rectangular shape. Up to 2 in. (50 mm) high.

Ascidiella aspersa

The firm, rough skin is usually a dirty grey. Other animals often attach themselves to it. The outlet siphon is about one-third of the way down the body, which grows to 3¼ in. (80 mm) high.

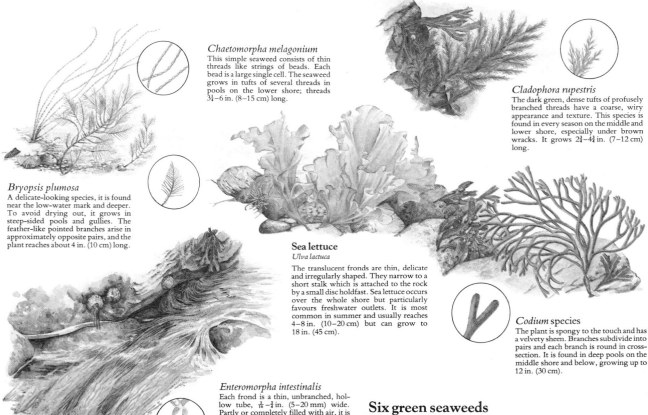

Chaetomorpha melagonium

This simple seaweed consists of thin threads like strings of beads. Each bead is a large single cell. The seaweed grows in tufts of several threads in pools on the lower shore; threads 3¼–6 in. (8–15 cm) long.

Cladophora rupestris

The dark green, dense tufts of profusely branched threads have a coarse, wiry appearance and texture. This species is found in every season on the middle and lower shore, especially under brown wracks. It grows 2¾–4¾ in. (7–12 cm) long.

Bryopsis plumosa

A delicate-looking species, it is found near the low-water mark and deeper. To avoid drying out, it grows in steep-sided pools and gullies. The feather-like pointed branches arise in approximately opposite pairs, and the plant reaches about 4 in. (10 cm) long.

Sea lettuce

Ulva lactuca

The translucent fronds are thin, delicate and irregularly shaped. They narrow to a short stalk which is attached to the rock by a small disc holdfast. Sea lettuce occurs over the whole shore but particularly favours freshwater outlets. It is most common in summer and usually reaches 4–8 in. (10–20 cm) but can grow to 18 in. (45 cm).

Codium species

The plant is spongy to the touch and has a velvety sheen. Branches subdivide into pairs and each branch is round in cross-section. It is found in deep pools on the middle shore and below, growing up to 12 in. (30 cm).

Enteromorpha intestinalis

Each frond is a thin, unbranched, hollow tube, ¹⁄₁₆–¾ in. (5–20 mm) wide. Partly or completely filled with air, it is attached in tufts to rocks or the back of limpets by a minute disc holdfast. It is found mainly on the upper shore where there is fresh water, but in summer may also cover the mud-flats of estuaries, attached to stones. It grows 2–12 in. (5–30 cm) long.

Six green seaweeds

These seaweeds are commonly found at most levels of Britain's rocky shores. Most species have thin and delicate fronds which can only be seen properly when under water. As there are many similar species it is often difficult to make anything other than a general identification.

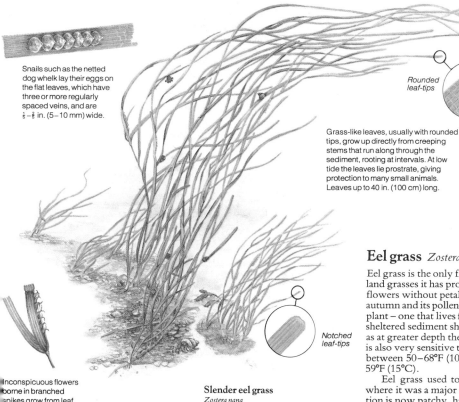

Snails such as the netted dog whelk lay their eggs on the flat leaves, which have three or more regularly spaced veins, and are $\frac{1}{5}-\frac{2}{5}$ in. (5–10 mm) wide.

Grass-like leaves, usually with rounded tips, grow up directly from creeping stems that run along through the sediment, rooting at intervals. At low tide the leaves lie prostrate, giving protection to many small animals. Leaves up to 40 in. (100 cm) long.

Rounded leaf-tips

Notched leaf-tips

Inconspicuous flowers borne in branched spikes grow from leaf bases during summer.

Extensive beds and small patches grow in mud or muddy sands in sheltered creeks, estuaries, and sea lochs.

Slender eel grass
Zostera nana

Similar in form to *Z. marina* but with very fine leaves with notched tips and irregularly spaced veins, *Z. nana* is common in several places on muddy shores in some estuaries and similar areas around Britain. The leaves are $\frac{1}{32}$ in. (1 mm) across and up to 6 in. (15 cm) long.

Eel grass *Zostera marina*

Eel grass is the only flowering plant that lives in the sea. As with land grasses it has proper roots, stems, leaves and inconspicuous flowers without petals. Seed spikes are produced in summer and autumn and its pollen is dispersed by water currents. A perennial plant – one that lives for an indefinite period – eel grass grows on sheltered sediment shores and in water less than 30 ft (9 m) deep, as at greater depth there is not enough light to sustain growth. It is also very sensitive to temperature and will actively grow only between 50–68°F (10–20°C) and will not fruit or flower above 59°F (15°C).

Eel grass used to cover vast areas, especially in estuaries, where it was a major food source for Brent geese. The distribution is now patchy, having been reduced by a disease which may have been caused by changes in the climate during the 1930s. As the beds are also susceptible to pollution, to storms and to damage from boat anchors, eel grass will not regain all its lost ground, even though some of the old beds are reviving.

Many small animals and species of seaweed live attached to eel grass leaves and among the roots. But apart from geese and sea urchins, few feed directly on the live plant.

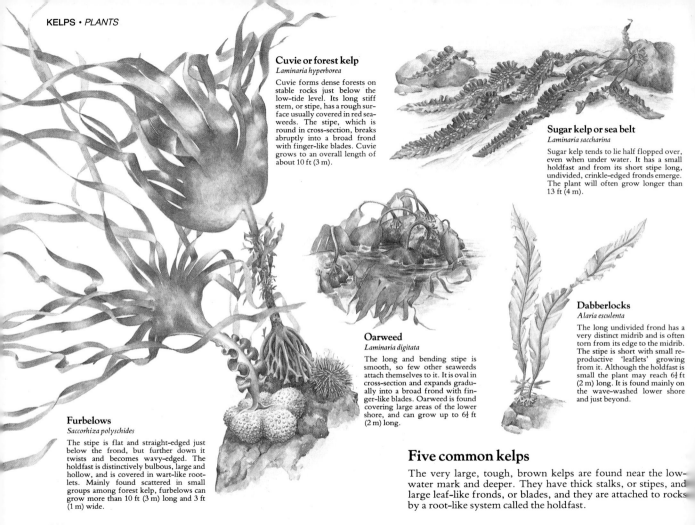

Cuvie or forest kelp
Laminaria hyperborea

Cuvie forms dense forests on stable rocks just below the low-tide level. Its long stiff stem, or stipe, has a rough surface usually covered in red seaweeds. The stipe, which is round in cross-section, breaks abruptly into a broad frond with finger-like blades. Cuvie grows to an overall length of about 10 ft (3 m).

Sugar kelp or sea belt
Laminaria saccharina

Sugar kelp tends to lie half flopped over, even when under water. It has a small holdfast and from its short stipe long, undivided, crinkle-edged fronds emerge. The plant will often grow longer than 13 ft (4 m).

Dabberlocks
Alaria esculenta

The long undivided frond has a very distinct midrib and is often torn from its edge to the midrib. The stipe is short with small reproductive 'leaflets' growing from it. Although the holdfast is small the plant may reach 6½ ft (2 m) long. It is found mainly on the wave-washed lower shore and just beyond.

Oarweed
Laminaria digitata

The long and bending stipe is smooth, so few other seaweeds attach themselves to it. It is oval in cross-section and expands gradually into a broad frond with finger-like blades. Oarweed is found covering large areas of the lower shore, and can grow up to 6½ ft (2 m) long.

Furbelows
Saccorhiza polyschides

The stipe is flat and straight-edged just below the frond, but further down it twists and becomes wavy-edged. The holdfast is distinctively bulbous, large and hollow, and is covered in wart-like rootlets. Mainly found scattered in small groups among forest kelp, furbelows can grow more than 10 ft (3 m) long and 3 ft (1 m) wide.

Five common kelps

The very large, tough, brown kelps are found near the low-water mark and deeper. They have thick stalks, or stipes, and large leaf-like fronds, or blades, and they are attached to rocks by a root-like system called the holdfast.

Channelled wrack
Pelvetia canaliculata

The wrack grows high on the shore and, in order to conserve water, the edges of the frond, which has no midrib or bladders, curl inwards to form a channel. Tufted bunches, looking dry and brittle, hang down from rocks with the channel side facing inwards. The reproductive ends to the fronds are often swollen and granular. Fronds are about 6 in. (15 cm) long.

Toothed wrack
Fucus serratus

Living in rock pools and on the lower shore, the fronds of this wrack have a distinct midrib, serrated edges and no air bladders. The slightly thickened patches at the ends of some branches are reproductive areas. May reach 24 in. (60 cm) long.

Knotted wrack
Ascophyllum nodosum

The largest of the wracks, it grows profusely on sheltered shores. The fronds have no midrib and are held up towards the light by oval bladders down the middle of each frond. Slender side branches bear yellow-green raisin-shaped fruiting bodies. The plant may grow to more than 10 ft (3 m) long.

Spiral wrack
Fucus spiralis

The broad frond has a smooth edge, no air bladders and a distinct midrib. It may grow with a slight twist. The tips often have clusters of heavy, swollen pod-like reproductive bodies. The plant grows mainly on the upper shore and, though usually smaller, may reach 16 in. (40 cm) in length.

Bladder wrack
Fucus vesiculosus

Found on the middle shore, the fronds of this wrack have a distinct midrib with pairs of air bladders on either side that float the fronds up towards the light. Side branches may have swollen tips with a granular appearance. These are reproductive pods. Bladder wrack may grow 6–40 in. (15–100 cm) long.

Brown seaweeds: wracks

Wracks are brown, leathery, slippery seaweeds found at different levels of rocky shores, according to species. Strap-like fronds arise from a disc holdfast and usually divide into a series of paired branches.

309

Thongweed
Himanthalia elongata

Long, strap–like fronds hang down in tangled masses from the rocks of the lower shore. The young plants are shaped like toadstools about 1½ in. (40 mm) high; the fronds arise from them and divide into paired branches. In spring, spots on the fronds indicate the plant is fertile. The fronds grow to 10 ft (3 m) long.

Pod weed or sea oak
Halidrys siliquosa

The slightly flattened branches grow on alternate sides of the main frond, giving it a zigzag appearance. The ends of the branches have distinctive air bladders similar to pointed seed-pods. Pod weed prefers shallow water and the deep rock pools of the lower shore. It grows to 3 ft (1 m) or more.

Leathesia difformis
Shiny lobe-like growths attach themselves to rocks or other sea-weeds. When small the capsules are solid but they become hollow and thick-walled as they grow. The oyster-thief seaweed (*Colpomenia peregrina*), although very similar, is not shiny and its spheres are always thin-walled hollows. *L. difformis* grows up to 2 in. (50 mm) in diameter.

Bifurcaria bifurcata
This tough, smooth plant is a native of the south and south-west coasts. It grows in bushy clumps in rock pools from mid-tide downwards. The frond is round in cross-section and unbranched near its base. The tips of some of its paired branches swell into elongated fruiting pods. Although usually 6–12 in. (15–30 cm) high it can grow up to 20 in. (50 cm).

Rainbow bladder weed
Cystoseira tamariscifolia

A prickly looking, shrubby perennial, it is only found in rock pools and shallow waters of the south and south-west coasts. Branches grow in an irregular mass and are covered with short spine-like branchlets. Single air bladders may be found near the branch tips, which end in oval, spiny fruiting bodies. Under water, the plant has a striking blue iridescence. Up to 18 in. (45 cm) long.

Common brown seaweeds

Apart from the kelps and wracks, a variety of other brown seaweeds are found on the shore. Those shown are mostly large, conspicuous plants but many other small, delicate species occur. The colours of red and brown seaweeds are due to special pigments which help them to survive where there is little light.

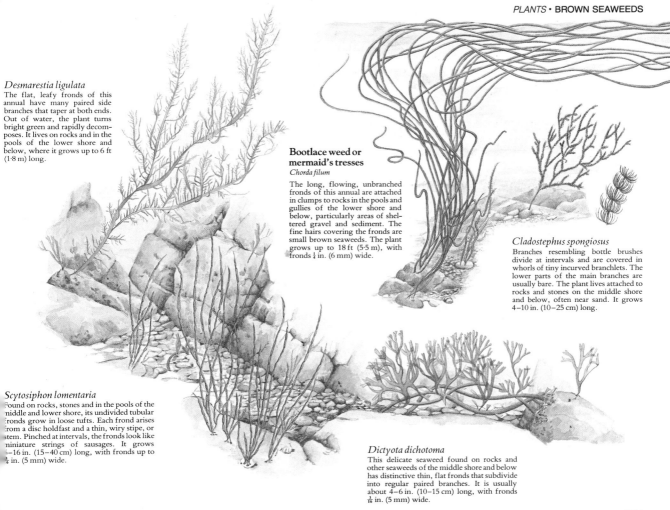

Desmarestia ligulata

The flat, leafy fronds of this annual have many paired side branches that taper at both ends. Out of water, the plant turns bright green and rapidly decomposes. It lives on rocks and in the pools of the lower shore and below, where it grows up to 6 ft (1·8 m) long.

Bootlace weed or mermaid's tresses
Chorda filum

The long, flowing, unbranched fronds of this annual are attached in clumps to rocks in the pools and gullies of the lower shore and below, particularly areas of sheltered gravel and sediment. The fine hairs covering the fronds are small brown seaweeds. The plant grows up to 18 ft (5·5 m), with fronds ¼ in. (6 mm) wide.

Cladostephus spongiosus

Branches resembling bottle brushes divide at intervals and are covered in whorls of tiny incurved branchlets. The lower parts of the main branches are usually bare. The plant lives attached to rocks and stones on the middle shore and below, often near sand. It grows 4–10 in. (10–25 cm) long.

Scytosiphon lomentaria

Found on rocks, stones and in the pools of the middle and lower shore, its undivided tubular fronds grow in loose tufts. Each frond arises from a disc holdfast and a thin, wiry stipe, or stem. Pinched at intervals, the fronds look like miniature strings of sausages. It grows 6–16 in. (15–40 cm) long, with fronds up to ¼ in. (5 mm) wide.

Dictyota dichotoma

This delicate seaweed found on rocks and other seaweeds of the middle shore and below has distinctive thin, flat fronds that subdivide into regular paired branches. It is usually about 4–6 in. (10–15 cm) long, with fronds ³⁄₁₆ in. (5 mm) wide.

311

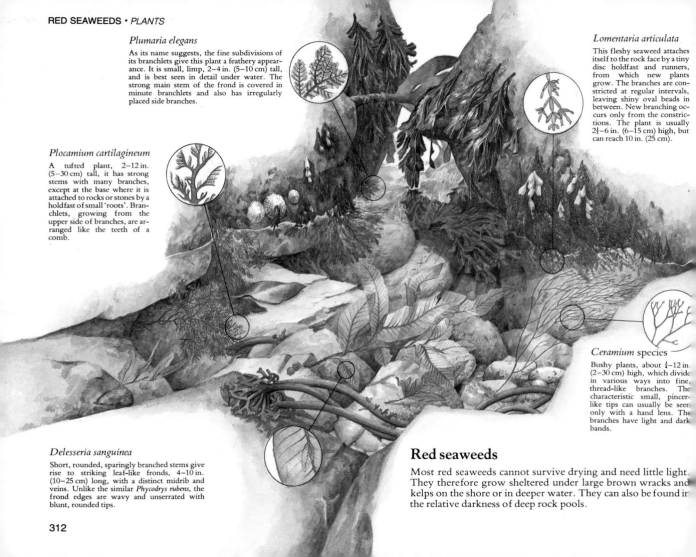

Plumaria elegans

As its name suggests, the fine subdivisions of its branchlets give this plant a feathery appearance. It is small, limp, 2–4 in. (5–10 cm) tall, and is best seen in detail under water. The strong main stem of the frond is covered in minute branchlets and also has irregularly placed side branches.

Lomentaria articulata

This fleshy seaweed attaches itself to the rock face by a tiny disc holdfast and runners, from which new plants grow. The branches are constricted at regular intervals, leaving shiny oval beads in between. New branching occurs only from the constrictions. The plant is usually 2¼–6 in. (6–15 cm) high, but can reach 10 in. (25 cm).

Plocamium cartilagineum

A tufted plant, 2–12 in. (5–30 cm) tall, it has strong stems with many branches, except at the base where it is attached to rocks or stones by a holdfast of small 'roots'. Branchlets, growing from the upper side of branches, are arranged like the teeth of a comb.

Ceramium species

Bushy plants, about ¾–12 in. (2–30 cm) high, which divide in various ways into fine, thread-like branches. The characteristic small, pincer-like tips can usually be seen only with a hand lens. The branches have light and dark bands.

Delesseria sanguinea

Short, rounded, sparingly branched stems give rise to striking leaf-like fronds, 4–10 in. (10–25 cm) long, with a distinct midrib and veins. Unlike the similar *Phycodrys rubens*, the frond edges are wavy and unserrated with blunt, rounded tips.

Red seaweeds

Most red seaweeds cannot survive drying and need little light. They therefore grow sheltered under large brown wracks and kelps on the shore or in deeper water. They can also be found in the relative darkness of deep rock pools.

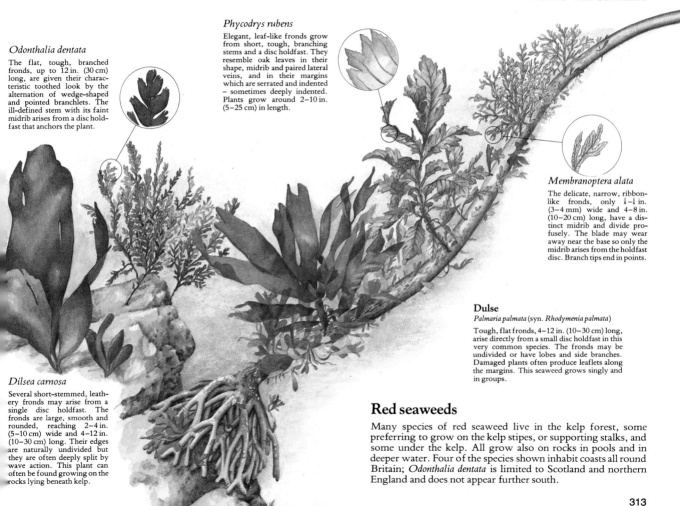

Odonthalia dentata

The flat, tough, branched fronds, up to 12 in. (30 cm) long, are given their characteristic toothed look by the alternation of wedge-shaped and pointed branchlets. The ill-defined stem with its faint midrib arises from a disc holdfast that anchors the plant.

Phycodrys rubens

Elegant, leaf-like fronds grow from short, tough, branching stems and a disc holdfast. They resemble oak leaves in their shape, midrib and paired lateral veins, and in their margins which are serrated and indented – sometimes deeply indented. Plants grow around 2–10 in. (5–25 cm) in length.

Membranoptera alata

The delicate, narrow, ribbon-like fronds, only ⅛–⅙ in. (3–4 mm) wide and 4–8 in. (10–20 cm) long, have a distinct midrib and divide profusely. The blade may wear away near the base so only the midrib arises from the holdfast disc. Branch tips end in points.

Dulse

Palmaria palmata (syn. *Rhodymenia palmata*)

Tough, flat fronds, 4–12 in. (10–30 cm) long, arise directly from a small disc holdfast in this very common species. The fronds may be undivided or have lobes and side branches. Damaged plants often produce leaflets along the margins. This seaweed grows singly and in groups.

Dilsea carnosa

Several short-stemmed, leathery fronds may arise from a single disc holdfast. The fronds are large, smooth and rounded, reaching 2–4 in. (5–10 cm) wide and 4–12 in. (10–30 cm) long. Their edges are naturally undivided but they are often deeply split by wave action. This plant can often be found growing on the rocks lying beneath kelp.

Red seaweeds

Many species of red seaweed live in the kelp forest, some preferring to grow on the kelp stipes, or supporting stalks, and some under the kelp. All grow also on rocks in pools and in deeper water. Four of the species shown inhabit coasts all round Britain; *Odonthalia dentata* is limited to Scotland and northern England and does not appear further south.

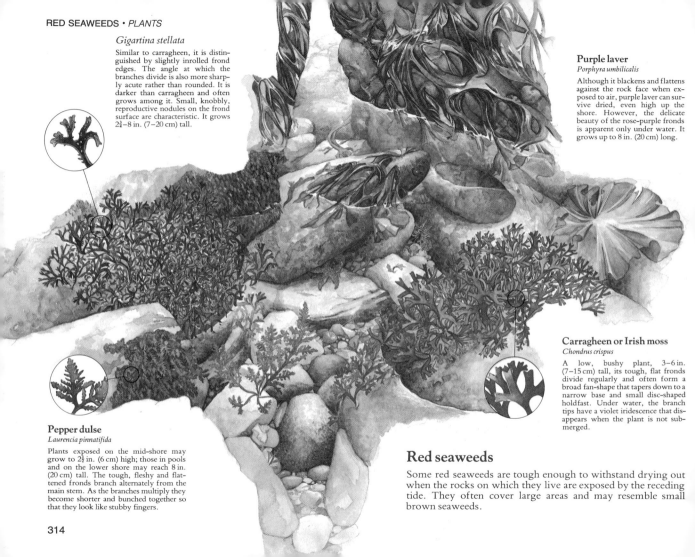

Gigartina stellata

Similar to carragheen, it is distinguished by slightly inrolled frond edges. The angle at which the branches divide is also more sharply acute rather than rounded. It is darker than carragheen and often grows among it. Small, knobbly, reproductive nodules on the frond surface are characteristic. It grows 2¾–8 in. (7–20 cm) tall.

Purple laver
Porphyra umbilicalis

Although it blackens and flattens against the rock face when exposed to air, purple laver can survive dried, even high up the shore. However, the delicate beauty of the rose-purple fronds is apparent only under water. It grows up to 8 in. (20 cm) long.

Carragheen or Irish moss
Chondrus crispus

A low, bushy plant, 3–6 in. (7–15 cm) tall, its tough, flat fronds divide regularly and often form a broad fan-shape that tapers down to a narrow base and small disc-shaped holdfast. Under water, the branch tips have a violet iridescence that disappears when the plant is not submerged.

Pepper dulse
Laurencia pinnatifida

Plants exposed on the mid-shore may grow to 2⅜ in. (6 cm) high; those in pools and on the lower shore may reach 8 in. (20 cm) tall. The tough, fleshy and flattened fronds branch alternately from the main stem. As the branches multiply they become shorter and bunched together so that they look like stubby fingers.

Red seaweeds

Some red seaweeds are tough enough to withstand drying out when the rocks on which they live are exposed by the receding tide. They often cover large areas and may resemble small brown seaweeds.

Gracilaria verrucosa

A long stringy plant growing up to 20 in. (50 cm), it is attached by a disc holdfast to stones or shells beneath the sand. The branches are very variable in length and number but always taper to a fine point. The wart-like reproductive pods on the branches make the plant easily recognisable in summer.

Lithothamnium/Lithophylum

This extremely widespread group of species is almost unrecognisable as seaweed. They form a hard, chalky, pink to purple crust over rocks and shells that is sometimes smooth or knobbly, or spreads as thin lobes with concentric growth rings.

Furcellaria fastigiata

An upright bushy plant, it grows 4–8 in. (10–20 cm) tall. Its shiny tubular branches subdivide into regular pairs of branches with an acute angle at the point of division. The paired offshoots are of equal length. The holdfast is a mass of small intertwined rootlets, and in summer the branch tips develop reproductive pods.

Polyides rotundus

Although similar to *Furcellaria fastigiata* this plant has a small disc holdfast. In addition, it has a wider angle of branching and the reproductive pods are oval lumps on the main branches rather than the branch tips. When picked it will appear red if held up to the light whereas *Furcellaria* appears blackish brown.

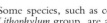

Coral weed

Corallina officinalis

Extensive tufts up to 4¾ in. (12 cm) high fill shallow rock pools. A chalky skeleton, which remains after death, gives the plant its characteristic stiff, brittle appearance. Each stem is made up of a chain of independently flexible bead-like segments from which branches and branchlets grow in opposite-facing pairs.

Red seaweeds

Some species, such as coral weed and the *Lithothamnium* and *Lithophylum* group, are able to live in rock pools high on the shore, while others, such as *Furcellaria*, *Polyides* and *Gracilaria*, attach themselves to small stones and shells in sand and gravel areas. Red seaweeds are usually small to moderate in size.

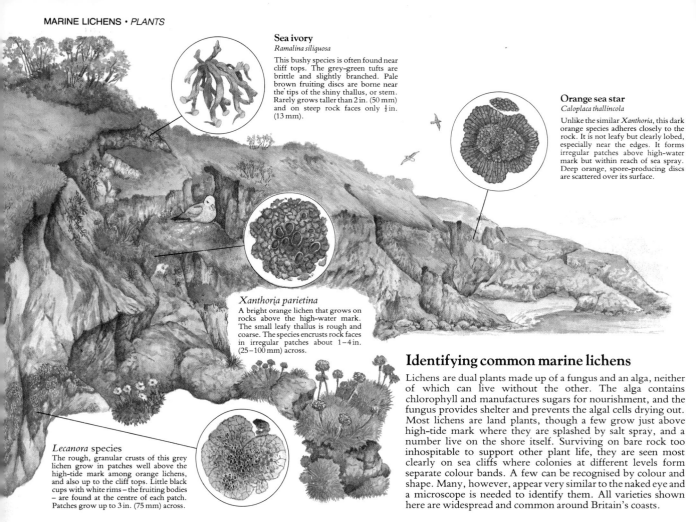

Sea ivory
Ramalina siliquosa

This bushy species is often found near cliff tops. The grey-green tufts are brittle and slightly branched. Pale brown fruiting discs are borne near the tips of the shiny thallus, or stem. Rarely grows taller than 2 in. (50 mm) and on steep rock faces only ½ in. (13 mm).

Orange sea star
Caloplaca thallincola

Unlike the similar *Xanthoria*, this dark orange species adheres closely to the rock. It is not leafy but clearly lobed, especially near the edges. It forms irregular patches above high-water mark but within reach of sea spray. Deep orange, spore-producing discs are scattered over its surface.

Xanthoria parietina

A bright orange lichen that grows on rocks above the high-water mark. The small leafy thallus is rough and coarse. The species encrusts rock faces in irregular patches about 1–4 in. (25–100 mm) across.

Lecanora species

The rough, granular crusts of this grey lichen grow in patches well above the high-tide mark among orange lichens, and also up to the cliff tops. Little black cups with white rims – the fruiting bodies – are found at the centre of each patch. Patches grow up to 3 in. (75 mm) across.

Identifying common marine lichens

Lichens are dual plants made up of a fungus and an alga, neither of which can live without the other. The alga contains chlorophyll and manufactures sugars for nourishment, and the fungus provides shelter and prevents the algal cells drying out. Most lichens are land plants, though a few grow just above high-tide mark where they are splashed by salt spray, and a number live on the shore itself. Surviving on bare rock too inhospitable to support other plant life, they are seen most clearly on sea cliffs where colonies at different levels form separate colour bands. A few can be recognised by colour and shape. Many, however, appear very similar to the naked eye and a microscope is needed to identify them. All varieties shown here are widespread and common around Britain's coasts.

Lichina pygmaea

Resembling a minute seaweed, this black, tufted species is found on rocks on the middle and upper shore, often beneath brown wrack. It also grows on steep, wave-exposed slopes where wrack cannot survive. Its flat, erect, branching lobes are about ⅓ in. (10 mm) high.

Orange sea lichen
Caloplaca marina

This species, though similar to orange sea star, is a darker orange and has small, irregular lobes which give the surface a granular appearance. The orange fruiting bodies have pale edges.

Tar lichen
Verrucaria maura

Just thick enough to be scraped off with a thumbnail, this closely encrusting species resembles oil or tar. It forms a black zone that covers large areas of rock at the top of the beach or base of cliffs.

Lichina confinis

Similar to, but smaller than, *Lichina pygmaea*, this species has fine, rounded branches which give it a furry appearance. It grows on the upper shore and on exposed cliffs even beyond the reach of spray. Branches are about ⅕ in. (5 mm) high.

Verrucaria mucosa

Although it resembles tar lichen, this green-black, closely encrusting species has a green tint when scraped off the rock. Also, it grows on the rocks of the middle shore in distinct smooth patches, each up to 12 in. (30 cm) across.

OBSERVING
AND CONSERVING
WATER LIFE

Migration patterns among water animals

Many water animals regularly move from one region to another to find food or ideal conditions for reproduction or to over-winter. These movements, or migrations, may occur seasonally or daily. Some involve travelling vast distances, or between fresh water and sea water and back again. Other migrations take place vertically, the animals moving from cool, dark depths to lighter, warmer surface waters, and then returning. Some animals may make only one or two migrations in their lives, but others repeat the journey time after time.

Migration to spawning areas

Most of the fish species important to man are strongly migratory. One example is the four separate stocks of plaice in the North Sea, each of which has a separate spawning area where the eggs are shed in winter. The larvae drift towards the coast, and after changing into the flattened adult form they go to the bottom, then move inshore on to intertidal 'nursery grounds'. In their first autumn, they move away from the shore but move back into shallow waters the following spring. In their second or third year, however, they reach sexual maturity and move to the deeper waters of the area where they were spawned.

The Atlantic cod also has several separate spawning populations, each with its own range. The Icelandic population spawns off south-west Iceland, and the young drift with the current towards southern Greenland. Several years later the mature fish migrate back to Iceland to spawn in their turn. Thereafter they migrate between Greenland and Iceland every year.

The amphibians – frogs, toads and newts – live on land for much of the year but migrate to particular spawning waters to breed. Common frogs, for example, normally return to the same pond year after year, guided by the smell of the water and of the vegetation. They may migrate half a mile overland to reach the breeding sites, normally travelling on mild, rainy nights when the vegetation is wet. After communal spawning, most of the frogs leave the pond and spend the summer and early autumn in damp places on land. Towards the end of October they return to hibernate in waters or damp places near to the original breeding grounds, emerging in late February to migrate back to the actual spawning ponds.

Some fish migrate between fresh water and the sea. This habit is quite rare, and restricted to about 120 of the 20,000 or so species of bony fish. The Atlantic salmon is the most famous example. The salmon spawns in gravel beds in rivers and streams, laying its eggs in depressions in the gravel called redds. The young fish, called parr, feed in the streams, then migrate to sea as smolts one to four years later. In the sea they probably follow well-defined routes to the feeding grounds where they spend one or more years. By this time they are sexually mature, and they move back to the coast to seek out freshwater streams.

There is strong evidence that each salmon returns to precisely the same tributary in which it was spawned, recognising the water by its distinctive smell. To reach the spawning grounds the adults will swim up rapids and even leap waterfalls, but they may be defeated by dams and other man-made obstacles. Today many such schemes incorporate by-pass fish ladders – flights of small waterfalls which provide the salmon with a route to the upper reaches.

Sea to river and back

The European eel also migrates between the sea and fresh waters – but in the opposite direction to the salmon. It spawns in the sea and moves into fresh water to grow. It is now agreed that adult eels, having migrated down rivers as silver eels, reach the sea and make a 2,000 mile oceanic migration to the Sargasso Sea in the central north Atlantic (named for the sargassum weed which is found floating there). They then spawn, somewhere near the sea-bed, and die.

The eggs float to the surface waters and hatch into leaf-like larvae, so unlike adult eels in appearance that they were once considered a separate species. These larvae drift passively in the currents across the north

Atlantic for two and a half years before they reach river estuaries on the European coast. They then change into slender, transparent elvers which, in their millions, make the upstream migration into tributaries and even overland into lakes and ponds. They grow for many years in fresh water before making the long journey back to the Sargasso Sea to spawn once, and then die.

The long migrations of whales

Many baleen whales breed in subtropical or temperate waters in winter, then migrate to polar or subpolar waters to enjoy the enormous production of plankton which takes place in these regions in the summer. There they feed intensively and store fat in preparation for the more frugal fare of warmer waters. The distances covered are huge. The humpback whale, for example, travels some 10,000 miles. Blue, fin and minke whales also make long annual migrations which may involve swimming day and night at speeds of over 15 knots (28 km/h).

Vertical migrations

Many types of small invertebrates and young fish in the oceans and in lakes migrate vertically through different water levels. These vertical movements occur every day, and are called diurnal or diel migrations. The normal pattern of movement involves a general rise to the levels near the surface at dusk, some dispersal of the population during the night and a concentration of the population at dawn, followed by rapid descent to the preferred depth during the hours of daylight. At dusk the population rises towards the surface again and the pattern is repeated every 24 hours. Some animals that live in the deep ocean are known to migrate diurnally through hundreds of feet. They are found nearer the surface at night and deeper down during the day.

The animals involved in these movements expend considerable energy in these ascents and descents and move quite quickly for their size – Calanus species (copepod crustaceans) are only a fraction of an inch long and yet swim upwards at 50 ft (15 m) an hour and downwards at 165 ft (50 m) an hour. In some species the depths and speeds of migration differ at different stages in the life history, and there are even variations in the behaviour of males and females of the same species.

The benefits of migration

Migration allows different stages in the life history of animals to exploit different food resources. Young salmon, for example, can use the food supply in freshwater streams while the adult fish live on different food in a different place – the sea. In this way competition between the young and adult members of the same species is avoided. The migration of baleen whales allows them to exploit the rich but highly seasonal food of the polar seas, and retreat to warmer waters when the food supply runs low.

The benefits of daily vertical migration are still not fully understood. It may give residents of the dark, non-productive deep waters the opportunity of feeding in well-lit water near the surface where algae are more plentiful. The animals may sink down in daylight to avoid being eaten, and come up to feed at night under cover of darkness. The problem with this theory is that the predators tend to migrate with the prey animals. Furthermore many species, on reaching surface waters in darkness, advertise their presence by being luminescent – a curious adaptation for animals trying to avoid attack. It is more probable that vertical movement through the layers of water, which are moving in different directions and at different speeds, redistributes the small organisms and minimises the risk of exhausting the food supply.

Finding the way

To arrive safely at their destinations, many migrating animals must be able to take up a direction by the instinctive use of clues in the environment. Despite intensive study little is known about these direction-finding abilities. Some whales are believed to find their way by recognising the shape of the coast. Fish may use the sun, the stars, the Earth's magnetic field, odours in the water, the direction of the current, or changes in water temperature and salt content. Some species probably use more than one type of clue, either at the same time or at different times. Salmon, for example, almost certainly use their sense of smell to find the tributary of their birth, though they probably use the sun or other clues during their oceanic migrations.

The problems of pollution

Every lake, river or sea supports a characteristic community of plants and animals which depends on particular conditions of chemical content, water temperature and light level. If any of these factors are altered by pollution, either by poisons or by introduced nutrients, the balance of the community is upset. Some organisms may prosper while others die, and the character of the water life changes. In severe cases all the original animals and plants are killed, and their place is taken by the few very specialised organisms which can survive in extreme conditions.

A major problem of disposal in Britain today is caused by the organic wastes produced by 55 million people, and by the rich organic effluents from food industries and intensive agriculture. If untreated sewage is pumped into a river, it partly settles on the bed in a blanket, eliminating the animals which live there and killing many plants. Smaller particles remain suspended in the water and cut down the light penetration, so weakening any plants which may survive.

At the same time the organic waste provides food for countless bacteria and fungi, for example sewage fungus – a mixture of bacteria, algae and fungi – which forms a coating like dirty yellowish-grey cotton wool over the bed of the river. Sludge worms and some other animals which feed on bacteria flourish in these conditions. The oxygen level falls rapidly as the oxygen-producing plants die and the oxygen users – bacteria and other animals – increase.

In the worst cases the water may become completely deoxygenated, so that only anaerobic bacteria can survive. These organisms obtain energy from chemical compounds in the water without the use of oxygen. Some types give off hydrogen sulphide gas, and others produce ammonia or other noxious gases.

Parts of the tidal Thames was in this state in the 1850s. While Britain celebrated the achievements of industry at the Great Exhibition of 1851, business was disrupted at the House of Commons because of the 'Great Stink' rising from the foul waters of the river. It was little more than an open sewer for the untreated waste of London, and a very overloaded one. With each incoming tide the sewage was washed back up through the city, instead of out to sea.

Today nearly all sewage is processed at the sewage works before disposal. Larger solid materials are allowed to settle out in tanks, and exposed to the digesting action of bacteria under controlled conditions. This treatment converts the complex organic substances into simple chemicals such as carbon dioxide, nitrogen and phosphates, which are dissolved in the water. This 'clean' effluent can be released into the river without polluting it, for the process of bacteriological decay has already taken place.

Unfortunately the effluent still upsets the chemical balance of the water. The chemical compounds produced by the sewage works are the same as those used by farmers and gardeners as fertiliser. In the enriched water aquatic plants can flourish, changing the character of the river and even choking the channel. A trout stream, for example, may be transformed into a richly vegetated river inhabited by coarse fish such as roach.

Agricultural fertilisers themselves are becoming a problem in many areas. They drain off the land into watercourses, enriching the water and increasing plant growth. Ponds and lakes are particularly vulnerable, and even mild enrichment of the water can have serious effects. The natural tendency of such waters to become overgrown and to silt up may be accelerated by the addition of enriching chemicals.

A very rich growth of plants can be disastrous, for when they die back the bacteriological decay of the dead organic material deoxygenates the water, and the effect is similar to that produced by the decay of raw sewage. This may even occur at sea, when treated sewage effluent becomes concentrated in sheltered bays and harbours and causes excessive growth of seaweeds, or turns the sea pink or green by encouraging a huge outburst 'bloom' of minute phytoplankton.

Blooms may occur on lakes, ponds, or on calm, warm seas, in such densities that they colour the water red, pink, brown, yellow or green. The blooms prevent light from reaching submerged plants, and many produce toxins which cause the death of fish in massive numbers. Lake water becomes poisonous to farm animals, and contaminated seafood can cause severe illness.

In the past many rivers, estuaries and coastal waters were severely polluted by toxic effluents from gasworks, paper mills, foundries and mine workings. Today there are strict controls on the material which may be dumped in rivers, but accidental and illegal pollution still occurs. Fresh waters are sometimes contaminated by herbicides, pesticides and heavy metals such as lead, mercury, copper and zinc. Chemicals such as DDT cannot be broken down by bacteria, and they build up in the bodies of animals, particularly predators which eat many contaminated animals every day. Fortunately DDT and some related compounds have been banned and replaced by less-persistent compounds. Road spillages of petrol find their way through the drainage network into streams, rivers and eventually the sea, causing the death of water life.

Large-scale spillages of petrol, oil or chemicals from industrial accidents or road crashes can cause spectacular destruction in rivers. As the pollutant moves downstream it may kill off all the plants and animals until it is dissipated by dilution, leaving behind a dead river. Within two or three years recolonisation restores the river to normal, but any rare species of water life may be permanently destroyed.

In the sea, massive pollution may be caused by spilt oil from shipping accidents or damage to drilling rigs, resulting in the deaths of many thousands of seabirds and seals. Since the first big disaster involving the *Torrey Canyon* in 1967, when 60,000 tons of oil escaped into the sea and on to the beaches of Cornwall, there have been over 25 major oil spills around Britain. Such events naturally receive a lot of publicity. Equally destructive, however, are the innumerable small spillages caused by carelessness and illegal dumping of waste oil from tankers, fishing boats and even pleasure boats. The oil kills water life and birds at sea, and drifts ashore to poison and smother animals and plants on the shore.

Unfortunately early attempts to deal with this problem caused more problems than they solved. Many beaches were treated with strong detergents, and it is now known that this caused far more damage to the wildlife than the oil itself. Apart from any direct toxic effects, detergents disperse and thin down the oil so that it penetrates under and into the shells, gills and feeding parts of animals. On sandy beaches, treated oil can form a layer beneath the surface which prevents recolonisation.

The pattern of water pollution varies from region to region, and changes constantly. Typically pollution is more marked in the lower reaches of rivers, in estuaries, and in the sea near river mouths. There are more industries in the lowland valleys and particularly on the estuaries which, in many cases, contain very dirty water indeed. This presents a barrier to migratory fish which must move between the sea and the upper reaches. This factor was responsible for the loss of salmon and sea trout runs in rivers such as the Thames, Tyne and Tees during the 19th century.

Before the 1950s the processes involved in pollution were poorly understood, and despite concern for public health little was done to improve the foul state of many major rivers. The tidal Thames remained virtually dead but for bacteria and sludge worms until the mid-1960s, when better control of effluents improved the oxygen content of the water and allowed tolerant fish such as bream and roach to re-colonise. Within ten years conditions had improved so much that migratory fish such as the smelt, flounder and even trout were to be found in the lower reaches. In 1982 salmon, last recorded in the Thames in 1833, were recorded migrating upstream to spawn in the upper reaches.

Despite this success there is still considerable cause for concern. Chemical effluents continue to pollute many rivers in industrial areas, and even warm water released from power stations has affected water life by killing some species and stimulating the growth of others. Fish stocks in many upland lakes are suffering from pollution of the water by acid rain – mild sulphuric acid formed by the contamination of rainwater with sulphurous smoke from industries and power stations. Nuclear waste is being discharged into coastal waters and dumped in the deep seas despite the misgivings of marine biologists, and radioactive pollution is becoming a problem on coasts near nuclear installations. The full consequences of deep-sea dumping are still unknown.

Conservation – observing the rules

Our coasts, seas, rivers and lakes are being increasingly used for recreation and natural history education. In Britain, where there are very few privately owned beaches, there is little restriction on the activities of marine naturalists, divers, fishermen and other sea users, and although some fresh waters are privately owned, there are still many rivers, streams and lakes with easy access (although for diving, permission may be required from the water authorities). If things are to remain this way it is very important to respect this right of access, and not endanger the very things people want to see or study by leaving litter lying around, disturbing the environment or wilful destruction.

As with all wildlife it is always best to LOOK and PHOTOGRAPH but DON'T TOUCH.

THE MENACE OF LITTER   Litter is not only an eyesore – it can and does kill animals. Fishing line and nets can trap and kill seabirds, seals, dolphins and other animals in fresh water as well as the sea. Collecting abandoned nylon fishing line will save the lives of many water birds. Discarded lead fishing weights poison swans and other birds that swallow them while feeding. Plastic bags, bottles, tin cans and other containers kill thousands of animals every year – they crawl in but cannot get out.

Boat users should remember that spilled boat fuel kills wildlife. Discarded motor oil kills millions of insects and fish every year by forming a thin film over the water and so stopping the passage of oxygen.

Never be tempted to empty or wash out weedkiller or pesticide containers in ponds or streams – even very low concentrations can kill wildlife.

If you suspect pollution of a stream or other watercourse because of the presence of dead fish, coloured water or odour, report it to the pollution inspector's office at your local water authority.

DISTURBANCE   Remember that damage can be caused by trampling and by car tyres, particularly on cliff tops, sand-dunes, salt-marshes and river banks. Walking too close to steep river banks may damage nesting sites if banks are eroded; in the sea, the flippers used by divers can knock off sea fans and other marine growths.

If you look under rocks on the shore or in rivers and lakes, always return them to their original positions. Animals living on the undersides of seashore rocks will die if left exposed to the sun, and seaweeds will have no light if they are hidden. If you expose an animal to photograph it, hide it again from its predators.

Fill in holes that you have dug for bait, and do not always dig in the same area.

Remember that the wash from a boat may flood burrows and destroy birds' nests in estuaries, rivers and lakes. River banks may be very badly damaged by fast moving boats with the loss of a variety of habitats.

It may be fun to build dams across rivers and streams, but changing the conditions may destroy breeding areas for fish.

DESTRUCTION   Living plants should not be picked, and animals should not be collected unless absolutely necessary. Some animals may take years to grow – a 12 in. (30 cm) high sea fan is 25 to 30 years old – and they should never be collected as souvenirs. It is always better to identify plants and animals by taking the book to them, and not taking them to the book.

If live animals are collected for examination in an aquarium, then they should later be returned to the place where they were collected or to a similar habitat – otherwise they will not survive.

Edible crustaceans such as crabs and lobsters should never be collected while they are carrying eggs.

LEGAL PROTECTION It is illegal to uproot any plant, however common, from the countryside without first getting the land-owner's permission. There are some rare plants which even the landowner may not uproot and these are listed in the Nature Conservancy Council booklet *Wildlife, the Law and You*. This law covers fresh-water plants but not seaweeds.

It is illegal to transfer fish from lake to lake, because it is easy to transfer parasites and disease to previously uninfected waters.

Fishing in fresh waters requires a water authority permit or permission from the owner, and is often only allowed in particular seasons. Spearfishing in fresh waters is illegal.

Commercial species such as crabs, lobsters and many fish have legal minimum size limits below which they must be returned to the water if caught.

Under the Wildlife and Countryside Act 1981 many animals are specifically pro-tected and this includes some water ani-mals. It is illegal to offer any British am-phibian for sale without a licence, and the great crested newt *Triturus cristatus* and the natterjack toad *Bufo calamita* may not even be handled without a licence. The bottle-nosed dolphin, common dolphin, com-mon porpoise and burbot are all specially protected animals. The booklet *Wildlife, the Law and You* explains in detail about all protected species.

Active conservation

There are several voluntary conservation bodies which are active in marine and freshwater matters, through which you can support the conservation movement. Their addresses are listed on p. 329.

These societies, as well as university and college extra-mural departments, run un-derwater biology, fish biology and natural history courses. Learning about water life in this way will help you understand more about how to protect your environment.

A great deal of information still needs to be collected before the water life of Britain can be adequately conserved and managed. Amateur naturalists can contribute by keep-ing careful records and taking part in record-ing projects. The Marine Conservation Soc-iety runs a number of national marine and freshwater recording projects in which any-one can participate.

Nature reserves and water parks

In Britain the official body concerned with conservation is the Nature Conservation Council, established by Act of Parliament in 1973. It has the power to set up nature reserves on land, which includes freshwater habitats such as rivers and lakes, but until recently this power did not extend beyond the low-water mark. There are over 180 national nature reserves in Britain and some of these include parts of the seashore and coast, sand-dunes, parts of estuaries, lakes and rivers. Local county trusts also own or lease a large number of sites as nature re-serves which contain good aquatic habitats.

Some councils and water authorities, in de-veloping water sites for other purposes, have integrated small nature reserves into their plans.

Recent years have seen a growth in the development of freshwater parks, such as the Nene Valley Water Park and the Cots-wold Water Park. Large man-made reser-voirs such as Graffham Water and Rutland Water have also been developed as water centres. Although most are used for fishing, sailing and other water sports, many also have nature reserve areas and are excellent for birdwatching.

In 1981 the Wildlife and Countryside Bill became law, and this allows the Nature Conservancy Council to set up statutory marine nature reserves on and off shore. Marine nature reserves matter because the wildlife of our seas is increasingly threatened by man's activities. Such threats include commercial exploitation for oil and gas, the construction of marinas and other coastal structures, dredging for minerals and gravel, effluent discharges and oil spills, modern fishing methods and the selective collection of some species by divers, bait diggers and, unfortunately, educational classes. These problems have been recog-nised for some years, and a number of vol-untary marine nature reserves have been set up through the agreement of all users includ-ing fishermen, divers, scientists and the pub-lic. Such areas, protected and managed by groups of enthusiasts, are in some cases equipped with study facilities and even marine wardens, and are excellent places to learn about marine life.

How to study water life

Ideally animals and plants should always be studied in their natural surroundings. It is far better to look, photograph and record than to collect wildlife for study elsewhere.

In the case of water life this is not always easy to do. Marine species which are active on the shore at low tide present no problem, but animals and plants which remain submerged can be very difficult to observe and identify. Glass-bottomed boxes are effective, particularly in running water where surface disturbance makes normal observation difficult. A diver's face mask held in the water is a good substitute. At sea, diving or snorkel equipment is invaluable for watching larger species, but in fresh waters permission may be required for their use. Furthermore, these techniques cannot be used for the close study of very small species, and it is useful to have some means of capturing live animals for brief examination before returning them to the water.

Before setting out to catch water life, check that no laws are violated. For example, it is illegal to use seine nets, drop nets, gill nets, fish traps or stunning devices in fresh water. Permission to catch fish and other water life should be obtained from the owner of the water rights.

Capturing water creatures

Animals such as shrimps, water beetles and small fish can be caught with a small net. The circular nets sold in seaside shops do not last long and it may be better to make your own using a strong wooden handle, thick, galvanised wire bent into an oval shape, and a piece of netting. The mesh size used will depend on the size of the organisms you want to catch, but for general use muslin of about 20 meshes per in. (25 mm) is suitable. It can be obtained from most large department stores or drapers' shops. Make the net a convenient size to carry. A long handle will make it easier to reach into less accessible places but can be unwieldy, so make it two sections which plug together. A piece of strong plastic tube that fits tightly over both sections makes a good connector.

Scavenging marine animals, such as crabs, prawns, whelks, and small fish can be caught with a drop net. This net can be baited and hung off a pier or boat.

Burrowing animals must be dug up. A fork is less likely to damage specimens than a spade. Sand, mud or stones can be sieved while washing through with water. An ordinary garden sieve will be adequate, lined with fine netting to catch small animals.

Many very small animals – the plankton – float freely in the sea and in ponds and lakes. Some can be caught with a simple plankton net. This is made from a long cone-shaped piece of muslin, or similar fine material, open at both ends. The wide mouth is held open by a circular wire frame, and a jam-jar is tied into the narrow end. The net is pulled through the water using a rope and the small animals trapped in the net are concentrated in the jar of water.

Having caught the animals, they must be kept in some sort of temporary aquarium, filled with sea, river or lake water as appropriate. Buckets, strong plastic bags, plastic boxes and washing-up bowls are suitable. Shallow white plastic trays, of the type used for developing photographic prints, are particularly good because small creatures show up well against the white background.

Studying your finds

Very small planktonic animals which need close examination with a lens are best kept in a small transparent container. The jam-jar of the plankton net is usually too big, and the image is often distorted by the curved sides. A simple solution is to use two small sheets of glass or clear plastic, about 2–4 in. (50–100 mm) square, separated by a length of rubber or a thick rubber band laid in a U-shape to contain the water. The sandwich is held together by two bulldog clips, making a watertight seal and allowing the container to be taken apart for cleaning.

A hand lens is useful for examining details of animals and plants, and is essential for looking at planktonic organisms, many of which cannot be seen with the naked eye. A lens with × 8 or × 10 magnification is the most useful type. A low-power microscope of × 50 or × 100 magnification will reveal even more, but it cannot be used in the field.

A camera is a useful tool for a naturalist. Many species can be identified from photographs, making it unnecessary to collect an animal or plant which cannot be identified on the spot. A waterproof camera bag is essential. Sea water is particularly corrosive, and sand in the camera mechanism will ruin it. When using a camera on the shore or near any water it is a good idea to protect the lens with a filter. The skylight (1A) type is most suitable when using colour film.

Many water animals are very small and the camera will have to be brought close to them to get any detail. Most cameras do not focus close enough, and a close-up lens or a close-focusing device may be needed. An ordinary reflex camera is the best type to use for such work, for it enables the photographer to see exactly what is in the picture, and whether it is in focus.

Photographs can be taken under water using an amphibious camera or a waterproof camera housing. High-quality equipment of this type is expensive, but small waterproof 'snapshot' cameras are available and provide an introduction to this specialised activity.

How to dry water plants

Many water plants are difficult to identify accurately, and it is very useful to make a dried collection to use as a reference. As with all wild plants, water plants should not be picked unless absolutely necessary, so look for loose specimens which have been dislodged by waves or by the current. With many small species, however, it is important to have the whole plant including the holdfast, stipe and frond of a seaweed, or the roots, stem and leaves of freshwater plants. The majority of freshwater plants also have flowers which are very useful for identification. Some freshwater plants are legally protected, so check before picking. (See p. 325.)

Spread the plant out in a shallow tray of sea water or fresh water, depending on its origin. Cut a piece of stiff white cartridge paper to use as a backing sheet, large enough for the plant but small enough to fit the plastic storage wallet (see below). Using pencil or Indian ink, write on the details of the plant: its name, if known, and where and when it was collected. Make sure that these details will not be obscured by the plant.

With delicate specimens, slide the paper underneath the plant in the tray and spread the branches out with a paint brush or a mounted needle. Lift the paper carefully at one end or tilt the tray to drain off the water, leaving the plant in the arranged position. Sturdy plants can be simply lifted out of the tray and arranged on the paper as desired.

Place a nappy liner, a piece of muslin or an old nylon stocking over the specimen and put it in a plant press, sandwiched between layers of drying paper such as newspaper or blotting paper. Change the drying paper after the first day, and again after two or three days. The plants should be ready after about a week and the protective muslin can be carefully peeled away leaving the plant stuck firmly to the paper. Large specimens will not stick and will have to be glued.

The dried, mounted plants can be stored in plastic wallets designed to fit in ring binders. It is best to get an expert, perhaps in a museum, to check the names of the plants, after which the collection will make a very useful 'book' of reference.

Collecting shells

Making a collection of stranded, dead or empty shells is a very good way of learning to recognise different species.

Empty freshwater shells need little treatment, but sea shells should be rinsed in fresh water before drying. Shells which still have dead animal matter in them can be cleaned by immersing in cold water, bringing to the boil and allowing to cool before picking out the remains. Soaking in a weak solution of household bleach will help to remove stains, and a thin coat of varnish will often bring out the colours of the shell.

Shells should be stored away from dust and strong sunlight, which will fade the colours. They are best kept in boxes or tubes, preferably of clear plastic. Boxes with small plastic drawers designed for nuts and bolts or fishing tackle make useful storage units. When the shells have been identified (see Shell chart on pp. 130–43) they should be labelled with the English and Latin names, and where and when found. Labels are easily lost, and ideally both shell and label should be stuck on to card or put together in a closed compartment. Other specimens such as sea urchin tests, or skeletons, worm tubes, caddis fly cases, egg cases and moulted dragonfly skins and crab shells can be mounted and stored in the same way.

Places to visit to see water life

The best way to study marine life in its natural habitat is to dive or snorkel. Alternatively, a visit to the shore at low tide can be just as rewarding. Do this at the low ebb of one of the twice-monthly spring tides, when the greatest area of shore is exposed.

Rocky shores sheltered from violent wave action support the widest variety of species. Dorset, Devon, Cornwall, Wales, southern Ireland and the Isle of Man all have extensive rocky shores with many rock pools. Searching damp places beneath overhangs and in cracks and crevices is usually rewarding, and many encrusting animals, small fish and crabs can be found under stones and boulders. (Remember to return rocks to their original positions.)

Shores composed of muddy sediments are excellent places to search for shells and strandline material. Empty shells can also be found in rock pools and gullies, swept there by waves and the incoming tide.

The animals which lie hidden beneath the surface of sandy and muddy shores may be hard to find, but they leave many clues to their presence, such as worm casts, tubes and depressions in the sand; and some emit occasional spurts of water.

The smallest fresh waters contain an amazing variety of animals and plants, and artificial garden ponds, water butts and even puddles can be rewarding places to look for water life. Village and farm ponds which have escaped major pollution are rich in species, and streams, dykes and drainage ditches contain many small animals.

Tracing the course of a river provides an opportunity to study a variety of running water habitats – from the fast-flowing streams in the hills, through the gentler middle reaches and finally to the broad, slow, mature lowland river. Each section supports its own characteristic plants and animals.

The deep lakes of upland Britain contain few species compared with lowland lakes. The water in such areas usually contains few plant nutrients. The thin vegetation that results supports a restricted range of animals. In contrast, lowland waters usually contain high concentrations of dissolved nutrients. They are often shallow, with gently sloping shores with plenty of space for plants, which in turn provide food and shelter for numerous animals.

Many old artificial lakes in the lowlands have developed into rich habitats for water life. The Norfolk Broads, for example, were formed by centuries of peat cutting. More recent excavations such as gravel pits often contain relatively few species in their early stages but the variety increases with age.

Dams and reservoirs

At some of the larger reservoirs and dams, such as Rutland Water and the power station at Trawsfynydd in North Wales, there are displays which describe the natural history of the water and the surrounding area. At Pitlochry on the River Tummel in Scotland there is a fish ladder for migrating salmon, with an underwater viewing chamber open to the public.

Fish farms

Many fish farms are open to the public and encourage people to look around. You can see the fish being fed and handled, and at some farms there are displays and information about the fish and other water life. Aquarists' shops often have excellent displays of freshwater, brackish water and marine fish and other water life.

Parks and gardens

Parks and botanic gardens often have ponds, streams and wet areas where freshwater plants can be seen, together with associated animal life. Many country parks have nature trails featuring streams and lakes.

Aquariums

Zoos and wildlife parks sometimes contain marine and freshwater aquariums, dolphinariums and vivariums (which house reptiles and amphibians). There are marine aquariums in some seaside towns, and the aquariums in marine biological stations are often open to the public.

Useful addresses

BRITISH HERPETOLOGICAL SOCIETY, *c/o Zoological Society of London, Regent's Park, London* NW1 4RY. Devoted to the scientific study of reptiles and amphibians.

BRITISH NATURALISTS' ASSOCIATION, *6 Chancery Place, The Green, Writtle, Essex* CM1 3DY. Parent body of local naturalists' organisations, which organise meetings, lectures and field trips.

BRITISH RECORDS CENTRE, *Monk's Wood Experimental Station, Abbots Ripton, Huntingdon, Cambridgeshire.* Runs recording card schemes for some groups of water life.

BRITISH TRUST FOR CONSERVATION VOLUNTEERS, *36 St Mary's Street, Wallingford, Oxfordshire* OX10 0EU. Involves volunteers of all ages in practical conservation work throughout the United Kingdom.

BRITISH WATERWAYS BOARD, *Melbury House, Melbury Terrace, London* NW1 6JX.

CETACEAN GROUP, *c/o Dr P. Evans, Mammal Society, Edward Grey Institute, Department of Zoology, South Parks Road, Oxford* OX1 3PS. Supplies information on whale distribution. All whale sightings in British waters should be reported to them.

CONCHOLOGICAL SOCIETY OF GREAT BRITAIN AND IRELAND, *c/o Mrs E. B. Rands, 51 Wychwood Avenue, Luton, Bedfordshire* LU2 7HT. Devoted to the study of molluscs.

COUNCIL FOR ENVIRONMENTAL CONSERVATION, *Zoological Gardens, Regent's Park, London* NW1 4RY. Acts as spokesman for voluntary environmental groups.

ESTUARINE AND BRACKISH WATER SCIENCES ASSOCIATION, *c/o R. S. K. Barnes, Department of Zoology, University of Cambridge, Downing Street, Cambridge* CB2 3EJ.

FAUNA AND FLORA PRESERVATION SOCIETY, *c/o Zoological Society of London, Regent's Park, London* NW1 4RY.

THE FIELD STUDIES COUNCIL, *Information Office, Preston Montford, Montford Bridge, Shrewsbury* SY4 1HW. Runs field courses on a variety of natural history topics including sea, seashore and fresh water.

FRESHWATER BIOLOGICAL ASSOCIATION, *The Ferry House, Far Sawry, Ambleside, Cumbria* LA22 0LP. Investigates the biology of plants and animals found in fresh and brackish water.

FRIENDS OF THE EARTH, *9 Poland Street, London* W1V 3DG. Promotes the rational use of natural resources and campaigns against environmental abuse.

GREENPEACE, *36 Graham Street, London* N1 8LL. Whale and seal conservation projects.

MARINE BIOLOGICAL ASSOCIATION OF THE UNITED KINGDOM, *The Laboratory, Citadel Hall, Plymouth, Devon* PL1 2PB. Investigates the biology of marine plants and animals.

MARINE CONSERVATION SOCIETY, *Candle Cottage, Kempley, Dymock, Gloucestershire.* Concerned with the study and conservation of water life. Runs courses and projects.

NATURE CONSERVANCY COUNCIL, *19–20 Belgrave Square, London* SW1X 8PY. Promotes nature conservation and manages national nature reserves. Produces a range of useful leaflets, posters and other publications.

PORCUPINE SOCIETY, *c/o Dr S. Smith, 17 Sydney Terrace, Edinburgh* EH7 6SR. Promotes interest in marine life of the NE Atlantic.

ROYAL SOCIETY FOR NATURE CONSERVATION, *The Green, Nettleham, Lincoln* LN2 2NR. The parent body of 46 local nature conservation trusts which own and manage nature reserves. Also a joint sponsor of the young people's conservation club WATCH.

SCOTTISH MARINE BIOLOGICAL ASSOCIATION, *Dunstaffnage Marine Research Laboratory, PO Box 3, Oban, Argyll* PA34 4AD.

UNDERWATER ASSOCIATION, *c/o J. Shand, Department of Zoology, University of Bristol, Woodland Road, Bristol* BS8 1UG. Promotes underwater science.

WORLD WILDLIFE FUND, *Panda House, 11–13 Ockford Road, Godalming, Surrey* GU7 1QU. Concerned with projects to conserve British species.

YOUNG ZOOLOGISTS' CLUB (XYZ CLUB), *The London Zoo, Regent's Park, London* NW1 4RY.

INDEX

Acknowledgments

Artwork in *Water Life of Britain* was supplied by the following artists:

3, 8–17 Sue Stitt · 18–27 Bob Bampton · 28–29 Jim Russell · 30–33 Denys Ovendon · 34–63 Mick Loates · 64–71 Stuart Lafford · 72–73 Dick Bonson · 74–75 Colin Newman · 76–81 Phil Weare · 82–83 Kevin Dean · 84–87 Andrew Robinson · 88–89 Mick Loates · 90–91 Norman Lacey · 92–95 Dick Bonson · 96–101 Andrew Robinson · 102–3 Kevin Dean · 104–5 Andrew Robinson · 106–7 Phil Weare · 108–9 Kevin Dean · 110–19 Tricia Newell · 126–7 Wendy Bramall · 128–9 Kevin Dean · 130–43 Jim Channell · 144–9 Colin Newman · 150–3 Peter Barrett · 154–9 Mick Loates · 160–7 Colin Newman · 168–9 Stuart Lafford · 170–1 Mick Loates · 172–3 Stuart Lafford · 174 Robin Armstrong · 175–7 Stuart Lafford · 178–9 Sue Stitt · 180–1 Stuart Lafford · 182–3 Mick Loates · 184–7 Stuart Lafford · 188–209 Colin Newman · 210–23 Dick Bonson · 224–5 Wendy Bramall · 226–7 Sue Wickison · 228–9 Ann Winterbotham · 230–53 Sue Wickison · 254–5 Sue Stitt · 256–75 Wendy Bramall · 276–83 Sue Stitt · 284–305 Ann Winterbotham · 306–15 Sue Stitt · 316–17 Wendy Bramall · 318–19 Jim Russell

Photographs in *Water Life of Britain* were supplied by the following photographers and agencies. Names of agencies are in CAPITAL LETTERS. Those in *italics* were commissioned by Reader's Digest. The following abbreviations are used:
KRD – King, Read and Doré
NHPA – Natural History Photographic Agency
NSP – Natural Science Photos
OSF – Oxford Scientific Films

30 Heather Angel · 31 J. Paling/OSF · 33 Heather Angel · 35 AIRVIEWS LTD · 37 Heather Angel · 41 BRUCE COLEMAN/J. E. Burton · 43 G. Doré/KRD · 44 Heather Angel · 45 NORTHERN IRELAND TOURIST BOARD · 46 Heather Angel · 47 S. & O. Mathews · 48 SEAPHOT/ G. & J. Lythgoe · 49 NATURE PHOTOGRAPHERS/A. A. Butcher · 50 S. & O. Mathews · 51 Derek Forss · 52 NATURE PHOTOGRAPHERS/J. Harrison · 53 *Neil Holmes* · 54 G. I. Bernard/OSF · 56–57 G. Doré/KRD · 58 R. Gibbons/ARDEA, LONDON · 59 NATURE PHOTOGRAPHERS/K. Handford · 61 BORD FAILTE/IRISH TOURIST BOARD · 62 Heather Angel · 63 Anne Powell · 64 Heather Angel · 65 P. Morris/ARDEA, LONDON · 66 R. Gibbons/ARDEA, LONDON · 67 Heather Angel · 68 NATURE PHOTOGRAPHERS/D. L. Sewell · 69 Heather Angel · 70–77 Heather Angel · 79 NSP/P. H. Ward. 80–81 Heather Angel · 84 Anne Powell · 86 BRUCE COLEMAN/E. Crichton · 90 NATURE PHOTOGRAPHERS/P. R. Sterry · 92 G. I. Bernard/OSF · 94 M. Read/KRD · 96–99 Heather Angel · 100 NHPA/S. Dalton · 104 Heather Angel · 105 J. Gooders/ARDEA, LONDON · 106 Heather Angel · 107 NATURE PHOTOGRAPHERS/F. V. Blackburn · 110 D. Thompson/OSF · 112 P. O'Toole/OSF · 113 G. I. Bernard/OSF · 114 BRUCE COLEMAN/J. Burton · 116 J. A. L. Cooke/OSF · 118 Pat Morris · 124 G. I. Bernard · 144 BRUCE COLEMAN/G. Williamson · 145 SEAPHOT/C. Bishop · 146 Z. Leszczynski/ANIMALS ANIMALS/ · 147 Pat Morris · 151 AQUILA/A. J. Bond · 153 BRUCE COLEMAN/J. Foott · 155 Peter Tatton · 160 Bernard Picton · 161 NHPA/J. Goodman · 162 SEAPHOT/G. & J. Lythgoe · 163 JACANA/J. L. S. Dubois · 167 Heather Angel · 168 NHPA/J. Goodman · 170 SEAPHOT/P. Scoones · 171 Heather Angel · 173 G. Thurston/OSF · 174 V. Taylor/ARDEA, LONDON · 175 E. & D. HOSKING/D. P. Wilson · 177 SEAPHOT/M. L. Buehr · 182 Heather Angel · 183 AQUILA/J. V. & G. R. Harrison · 184 Heather Angel · 185 SEAPHOT/A. Joyce · 186 NHPA/L. Campbell · 188–9 Heather Angel · 191 E. & D. Hosking · 192 E. & D. HOSKING/D. P. Wilson · 193 SEAPHOT/P. Scoones · 194 Heather Angel · 195 SEAPHOT/C. C. Hemmings · 197 John Cleare/MOUNTAIN CAMERA · 201 G. I. Bernard/OSF · 202 AQUILA/J. V. & G. R. Harrison · 203 E. & D. HOSKING/D. P. Wilson · 204 V. Taylor/ARDEA, LONDON · 205 E. & D. HOSKING/D. P. Wilson · 207 SEAPHOT/P. Clark · 208 *Patrick Thurston* · 209 JACANA/P. Laboute · 210 G. I. Bernard/OSF · 211 John Taylor · 212 SEAPHOT/D. George · 213–14 Frances Dipper · 215 Jo Jamieson · 216 John Taylor · 218–19 Bernard Picton · 220 SEAPHOT/R. Waller · 221 John Taylor · 222 Bernard Picton · 223 Heather Angel · 226 BRUCE COLEMAN/J. Burton · 230 Heather Angel · 232 E. & D. HOSKING/D. P. Wilson · 234 John Taylor · 236 Heather Angel · 238 BRUCE COLEMAN/G. Langsbury · 240–1 Heather Angel · 242 SEAPHOT/D. George · 244 Heather Angel · 245 P. Morris/ARDEA, LONDON · 246 John Taylor · 247 Heather Angel · 248 SEAPHOT/G. & J. Lythgoe · 250 NATURE PHOTOGRAPHERS/A. J. Cleave · 256 G. I. Bernard/OSF · 258 E. & D. HOSKING/D. P. Wilson · 261 SEAPHOT/P. Scoones · 264 SEAPHOT/A. Petron · 265 E. & D. HOSKING/D. P. Wilson · 267 SEAPHOT/A. Fvoboda. 270 SEAPHOT/J. Greenfield · 272 Bernard Picton · 274 BRUCE COLEMAN/J. Burton · 278 J. B. & S. Bottomley/ARDEA, LONDON · 280 NHPA/G. J. Cambridge · 282 E. & D. HOSKING/D. P. Wilson 286 SEAPHOT/G. & J. Lythgoe · 288 P. Parks/OSF · 290 SEAPHOT/J. King · 292 Heather Angel · 294 SEAPHOT/M. Laverack · 296 Heather Angel · 298 G. I. Bernard/OSF · 302–3 Bernard Picton · 304 E. & D. HOSKING/D. P. Wilson · 307 Frances Dipper

The publishers acknowledge their indebtedness to the following books which were consulted for reference:

An Angler's Entomology by J. R. Harris (Collins) · *British Bivalve Seashells* by N. Tebble (British Museum) · *British Shells* by N. F. McMillan (Warne) · *Coasts and Estuaries* by Richard Barnes (Hodder & Stoughton) · *Guide to the Freshwater Fishes of Britain and Europe* and *Guide to the Sea Fishes of Britain and North-western Europe* both by Bent J. Muus and Preben Dahlstrøm (Collins) · *Pocket Guide to the Sea Shore* by John Barrett and C. M. Yonge (Collins) · *A Field Guide to the Insects of Britain and Northern Europe* by Michael Chinery (Collins) · *The Fishes of the British Isles & N. W. Europe* by Alwyne Wheeler (MacMillan) · *Fishes of the Sea* by J. and G. Lythgoe (Blandford) · *Freshwater Life* by John Clegg (Warne) · *The Guinness Book of Mammals* by John A. Burton (Guinness Superlatives) · *The Guinness Book of Seashore Life* by Heather Angel (Guinness Superlatives) · *Guide to the Seashore and Shallow Seas of Britain and Europe* by A. C. Campbell (Hamlyn) · *The Handbook of British Mammals* by G. B. Corbet and H. N. Southern (Blackwell) · *A History of British Fishes* by William Yarrell (John Van Voorst, 1836) · *Jarrold Nature Series: Fresh and Saltwater Life* (Jarrold) · *Key to the Fishes of Northern Europe* by Alwyne Wheeler (Warne) · *Life in Lakes and Rivers* by T. T. Macan and E. B. Worthington (Collins) · *The Living Countryside* (Eaglemoss) · *Living Marine Molluscs* by C. M. Yonge and T. E. Thompson (Collins) · *Living Seashells* edited by B. E. Picton (Blandford) · *The Natural History of Britain and Ireland* (Michael Joseph) · *The Observer's Book of Insects* by E. F. Linssen (Warne) · *The Observer's Book of Pond Life* by John Clegg (Warne) · *The Open Sea: Fish and Fisheries* by Sir Alister Hardy (Collins) · *The Oxford Book of Flowerless Plants* by F. H. Brightman (Oxford University Press) · *The Oxford Book of Vertebrates* by M. Nixon and D. Whiteley (Oxford University Press) · *Oysters* by C. M. Yonge (Collins) · *Rivers and Lakes* by Eckart Pott (Chatto Nature Guides) · *Rivers, Lakes and Marshes* by Brian Whitton (Hodder & Stoughton) · *The Salmon* by J. W. Jones (Collins) · *The Sea Shore* by C. M. Yonge (Collins) · *The Young Specialist Looks at Marine Life* by W. de Haas and F. Knorr (Burke) · *The Young Specialist Looks at Molluscs* by H. Janus (Burke) · *The Young Specialist Looks at Pond Life* by W. Engelhardt (Burke) · *The Young Specialist Looks at the Seashore* by A. Kosch, H. Frieling and H. Janus (Burke)

Typesetting: VANTAGE PHOTOSETTING CO. LTD, EASTLEIGH · Separations: MULLIS MORGAN LTD, LONDON
Paper: KONINKLIJKE NEDERLANDSE PAPIERFABRIEKEN NV, MAASTRICHT · Printer/Binder: HAZELL WATSON & VINEY LTD, AYLESBURY
Cloth: REDBRIDGE BOOK CLOTH CO. LTD, BOLTON